# Advances in Self-Organising Maps

Springer
*London*
*Berlin*
*Heidelberg*
*New York*
*Barcelona*
*Hong Kong*
*Milan*
*Paris*
*Singapore*
*Tokyo*

Nigel Allinson, Hujun Yin, Lesley Allinson
and Jon Slack

# Advances in Self-Organising Maps

[illegible]

NOVEMBER 2003

Springer

Nigel Allinson
Hujun Yin
Department of Electrical Engineering and Electronics, UMIST, P O Box 88, Manchester M60 1QD

Lesley Allinson
Jon Slack
Department of Psychology, University of Lincolnshire and Humberside, Brayford Pool, Lincoln LN6 7TS

British Library Cataloguing in Publication Data
Advances in self-organising maps
1.Self-organizing maps - Congresses
I,Allinson, Nigel
006.4
ISBN 1852335114

Library of Congress Cataloging-in-Publication Data
A catalog record for this book is available from the Library of Congress.



ISBN 1-85233-511-4 Springer-Verlag London Berlin Heidelberg
a member of BertelsmannSpringer Science+Business Media GmbH
http://www.springer.co.uk


Printed in Great Britain



Typesetting: camera-ready by author
Printed and bound at the Athenæum Press Ltd., Gateshead, Tyne and Wear
34/3830-543210 Printed on acid-free paper SPIN 10838887

# PREFACE

This is the third Workshop on Self-Organising Maps (WSOM) and its related techniques. The previous two were held in Helsinki (1997 and 1999) and confirmed the vitality of the SOM as one of the most popular and powerful concepts for unsupervised pattern recognition and data visualisation. These meetings not only acted as a showcase for the latest advances in SOM theory and for illustrating its vast range of applicability, but also as venues where much informal and fruitful interaction could take place. It is interesting to observe the development of the original SOM, and this remarkable progress confirms the originality and insight of Teuvo Kohonen's pioneering work. With the range and quality of the papers in this volume, the stage is set for another very successful meeting.

This volume is a permanent record of all the contributions presented during WSOM'01 held at the University of Lincolnshire and Humberside, 13 – 15 June, 2001. The University is the newest of England's universities but it is situated in the heart of one of our oldest cities – founded by the Romans and overlooked by the towering mass of its medieval cathedral. Primarily Lincoln has always been a centre for the rich agricultural heartland of England; however, it was the birthplace, 186 years ago, of George Boole. So WSOM'01 is continuing Lincoln's long and honourable tradition of advancing scientific understanding.

We, the local organisers, would like to thank the Programme Committee and additional reviewers for so efficiently reviewing the contributed papers and providing helpful comments to the authors. Their work has helped to ensure this Workshop and its accompanying record will be of lasting value. We also wish to acknowledge the assistance of our secretarial support both in Manchester and Lincoln; together with the generous co-operation of the Conference Office at ULH. There are numerous individuals who assist in the successful organisation of scientific meetings–they are certainly not self-organising–we would wish to take this opportunity to thank them all.

Nigel Allinson
Hujun Yin
UMIST, Manchester

Lesley Allinson
Jon Slack
ULH, Lincoln

# WSOM 2001 PROGRAMME COMMITTEE

Nigel M. Allinson, UMIST, UK
Marie Cottrell, Universite Paris 1, France
Guido Deboeck, World Bank, USA
Bernd Fritzke, Dresden University of Technology, Germany
Colin Fyfe, University of Paisley, UK
Sami Kaski, Helsinki University of Technology, Finland
Stephen Luttrell, DERA Malvern, UK
Risto Miikkulainen, University of Texas at Austin, USA
Klaus Obermayer,Technische Universität Berlin, Germany
Erkki Oja, Helsinki University of Technology, Finland
José Principe, University of Florida, USA
Helge Ritter, Universität Bielefeld, Germany
Olli Simula, Helsinki University of Technology, Finland
Heizo Tokutaka, Tottori University, Japan
Lei Xu, Chinese University of Hong Kong
Hujun Yin, UMIST, UK

# ADDITIONAL REVIEWERS

Ralf Der, University of Leipzig, Germany
Dieter Merkl, Universitat Wien, Austria
Udo Seiffert, University of Magdeburg, Germany
Alfred Ultsch, University of Marburg, Germany
Marc M. Van Hulle, Katholieke Universiteit Leuven, Belgium

# TABLE OF CONTENTS

# Towards an information-theoretic approach to kernel-based topographic map formation

Marc M. Van Hulle

K.U.Leuven, Laboratorium voor Neuro- en Psychofysiologie, Campus Gasthuisberg, Herestraat, B-3000 Leuven, Belgium

**Abstract.** A new information-theoretic learning scheme is introduced for kernel-based topographic map formation. The kernel parameters are adjusted so as to maximize the differential entropies of the kernel outputs and, at the same time, to minimize the mutual information between these outputs. The learning scheme is based on *infomax* learning supplemented with a cooperative/competitive stage to achieve topographically-organized maps. As a potential application, we consider density estimation.

## 1 Introduction

In an attempt to improve the density estimation properties, the noise tolerance, or even the biological relevance of the Self-Organizing Map (SOM) [1,2], algorithms have been developed that can accommodate neurons with overlapping activation regions. Rather than having disjoint and uniform regions (Voronoi tessellation), the lattice neurons are equipped with local kernel functions, such as Gaussians. We further call such lattices *kernel-based* topographic maps (Fig. 1A). An early example is the elastic net of Durbin and Willshaw [3], which can be viewed as a Gaussian mixture density model, fitted to the data points by a penalized maximum likelihood term. In recent years, algorithms that adopt a probabilistic approach have been introduced by Graepel *et al.* [4] (fuzzy membership in clusters), Sum *et al.* [5] (maximizing local correlations), Bishop *et al.* [6] (Generative Topographic Map, based on constrained Gaussian mixture density modeling), and Van Hulle [7,8] (equiprobabilistic maps). András [9] proposed a still different approach by putting the development of the (Gaussian) kernels in the context of (kernel-based) Support Vector Machine (SVM) learning.

The question is now: how to develop kernel-based topographic maps that optimize an *information-theoretic* criterion, such as Linsker's principle of maximum information preservation — *infomax* for short. The obvious answer is to express the information integral in terms of the kernel output densities (or probabilities when they are discretized), and adjust the kernel parameters so that the integral is maximized. However, this rapidly becomes infeasible in practice (*e.g.*, see [10]).

We will introduce a new learning scheme that adjusts the kernel parameters in such a manner that the (differential) entropies of the neural outputs

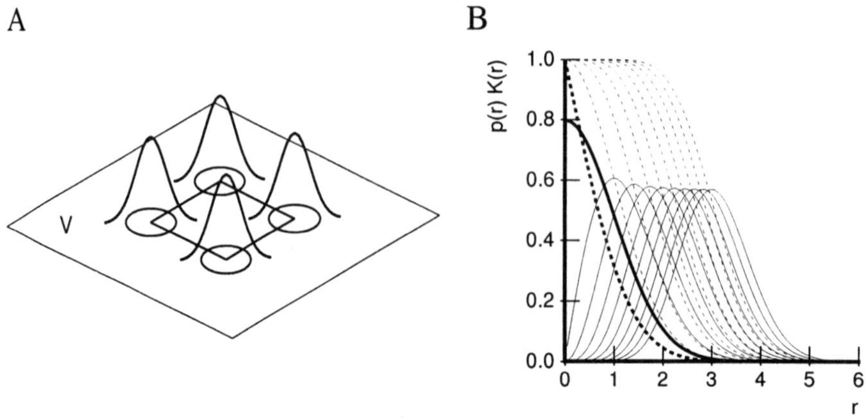


**Fig. 1.** (A) Kernel-based topographic maps. Example of a $2 \times 2$ lattice. Each neuron has a bell-shaped kernel as output function, represented in $V$-space, $V \subseteq \Re^d$, by a circle with center the neuron weight vector, and radius the kernel range.
(B) Distribution functions of the Euclidean distance from the mean, $r$, of a unit-variance, radially-symmetrical Gaussian, parameterized with respect to the dimensionality $d$ (thick and thin continuous lines), and the corresponding complements of the cumulative distribution functions (thick and thin dashed lines). The functions are plotted for $d = 1, \ldots, 10$ (from left to right); the thick continuous and dashed lines correspond to the $d = 1$ case.

are maximized and that, at the same time, the mutual information between the neural outputs is minimized. The former will be achieved by applying *infomax* learning and the latter by having a competitive stage in the learning process, the presence of which is, together with the neighborhood function, essential in topographic map formation.

## 2 Information-theoretic approach

As motivated by Bell and Sejnowski [11], the mutual information that the output $y_i \in \Re$ of neuron $i$ contains about its input $\mathbf{v} \in \Re^d$, $I(y_i, \mathbf{v})$, is maximized by maximizing the differential entropy of $y_i$, $H(y_i)$. When assuming that the output function has bounded support, $H(y_i)$ will be maximized when the output distribution is uniform. This is achieved when the output function corresponds to the cumulative distribution function of the input density.

Assume that the input density is a $d$-dimensional, radially-symmetrical Gaussian with mean $[\mu_j]$, $j = 1, \ldots, d$, and standard deviation $\sigma$. The distribution of the Euclidean distances to the mean of the Gaussian can be obtained as follows. For a unit-variance Gaussian, $x \triangleq \sum_{j=1}^{d} (v_j - \mu_j)^2$, $\mathbf{v} = [v_j]$, obeys the chi-squared distribution with $\theta = 2$ and $\alpha = \frac{d}{2}$ degrees of freedom [12]:

$$p_{\chi^2}(x) = \frac{x^{\frac{d}{2}-1} \exp(-\frac{x}{2})}{2^{\frac{d}{2}} \Gamma(\frac{d}{2})},$$

with $\Gamma(.)$ the gamma distribution. Let $r = \sqrt{x}$, then $p(r) = 2\ r\ p_{\chi^2}(r^2)$, following the fundamental law of probabilities. After some algebraic manipulations, and by generalizing to a Gaussian with standard deviation $\sigma$, we can write the distribution of the Euclidean distances as follows:

$$p(r) = \frac{2(\frac{r}{\sigma})^{d-1} \exp(-\frac{(\frac{r}{\sigma})^2}{2})}{2^{\frac{d}{2}} \Gamma(\frac{d}{2})}.$$

The distribution is exemplified in Fig. 1B (thick and thin continuous lines), and parameterized with respect to $d$. The cumulative distribution of $p(r)$ is the (complement of the) incomplete gamma distribution: $P(\frac{d}{2}, \frac{r^2}{2\sigma_i^2}) \equiv \frac{\Gamma(\frac{d}{2}, \frac{R^2}{2\sigma_i^2})}{\Gamma(\frac{d}{2})}$. Our kernel function is now defined in accordance with the latter, which becomes, in $V$-space:

$$y_i = K(\mathbf{v}, \mathbf{w}_i, \sigma_i) = P(\frac{d}{2}, \frac{\|\mathbf{w}_i - \mathbf{v}\|^2}{2\sigma_i^2}) \equiv \frac{\Gamma(\frac{d}{2}, \frac{\|\mathbf{w}_i - \mathbf{v}\|^2}{2\sigma_i^2})}{\Gamma(\frac{d}{2})},$$

with $\mathbf{w}_i$ the kernel center of neuron $i$ in $V$-space (*i.e.*, the weight vector of neuron $i$), and $\sigma_i$ the kernel radius.

## 3 Self-organizing learning algorithm

We will now derive an "on-line" learning rule that adapts the kernel parameters in such a way that the information transfer through the kernel is maximized. (We only present an overview; full details will be given elsewhere.) Let $A$ be a lattice of $N$ formal neurons. The entropy of the kernel output of neuron $i$ can be written as:

$$H(y_i) = -\int_0^{\infty} p_{y_i}(x) \ln p_{y_i}(x)\, dx,$$

with $p_{y_i}(.)$ the *p.d.f.* of the kernel output which, in turn, can be written as a function of the distribution of the Euclidean distance to the kernel center, $r$, $p_{y_i}(y_i) = \frac{p_r(r)}{|\frac{\partial y_i}{\partial r}|}$. By performing gradient ascent on the resulting $H(.)$, $\Delta \mathbf{w}_i \propto \frac{\partial H}{\partial r} \frac{\partial r}{\partial \mathbf{w}_i}$ and $\Delta \sigma_i \propto \frac{\partial H}{\partial \sigma_i}$, we obtain our learning rules, which we supplement with the neighborhood function:

$$\Delta \mathbf{w}_i = \eta_w\ \Lambda(i, i^*, \sigma_\Lambda) \frac{\mathbf{v} - \mathbf{w}_i}{\sigma_i^2},$$

$$\Delta \sigma_i = \eta_\sigma\ \Lambda(i, i^*, \sigma_\Lambda) \frac{1}{\sigma_i} \left( \frac{\|\mathbf{v} - \mathbf{w}_i\|^2}{\sigma_i^2} - 1 \right), \quad \forall i \in A,$$

with $\eta_w$ and $\eta_\sigma$ the learning rates, and with $i^* = \arg\max_{\forall i \in A}(y_i)$, *i.e.*, an activity-based "winner-takes-all" rule, rather than a minimum Euclidean distance based rule ($i^* = \arg\min_i \|\mathbf{w}_i - \mathbf{v}\|$). The term $\Lambda(.)$ stands for the neighborhood function, a monotonous decreasing function of the lattice distance from the winner, *e.g.*, a Gaussian: $\Lambda(i, i^*, \sigma_\Lambda) = \exp\left(-\frac{\|\mathbf{r}_i - \mathbf{r}_{i^*}\|^2}{2\sigma_\Lambda^2}\right)$, with $\sigma_\Lambda$ the neighborhood function range, and $\mathbf{r}_i$ neuron $i$'s lattice coordinate (we assume discrete lattices with rectangular topologies). We adopt the following neighborhood cooling scheme: $\sigma_\Lambda(t) = \sigma_{\Lambda 0} \exp\left(-2\sigma_{\Lambda 0} \frac{t}{t_{max}}\right)$, with $t$ the present and $t_{max}$ the maximum number of time steps, and $\sigma_{\Lambda 0}$ the range spanned at $t = 0$.

In order to show the lattice disentangling dynamics of the learning algorithm, we develop a 24×24 lattice in two-dimensional $V$-space by continuously drawing samples from the uniform *p.d.f.* in the unit square $(0, 1]^2$. The initial weights are initialized by sampling from the same *p.d.f.*; the radii are initialized by sampling the uniform distribution $(0, 0.1]$. We take $t_{max} = 2,000,000$, $\sigma_{\Lambda 0} = 12$, $\eta_w = 0.01$ and $\eta_\sigma = 10^{-4}\,\eta_w$. The results are shown in Fig. 2.

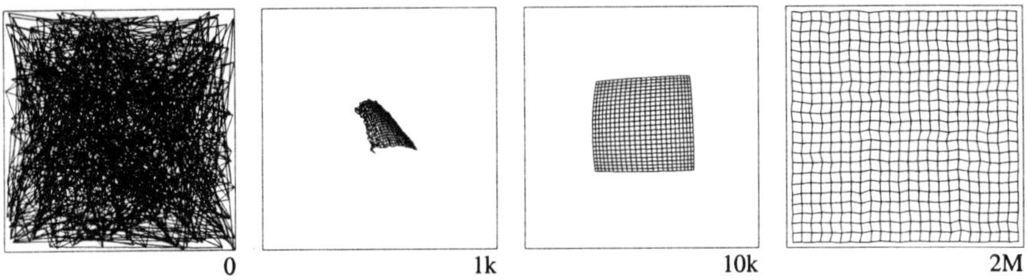


**Fig. 2.** Evolution of a 24 × 24 lattice as a function of time. The boxes outline the uniform *p.d.f.* The values given below them represent time.

## 4 Density estimation

Since each kernel is adapted under the assumption that the local density is Gaussian, with mean the corresponding neuron weight and standard deviation the kernel radius, we can estimate the input density $p(\mathbf{v})$ in a way similar to the classic *Variable Kernel* (VK) density estimation method [13]:

$$\widehat{p}(\mathbf{v}) = \sum_{i=1}^{N} \frac{\exp\left(-\frac{\|\mathbf{v}-\mathbf{w}_i\|^2}{2(\sigma_i)^2}\right)}{N(2\pi)^{\frac{d}{2}}\sigma_i^d}.$$

In order to systematically examine the density estimation performance, we consider an example for which the expected performance can be determined analytically. We will determine the joint entropy and the mutual information in terms of probabilities: we quantize the kernel outputs $y_i$, $\forall i \in A$, uniformly into $k$ equally-sized quantization intervals. Consider a Gaussian *p.d.f.* centered at the origin and with unit variance. The joint entropy JE and the mutual information MI become analytically, for the $k = 2$ case:

$$\mathrm{MI} = \log_2 2N + 2\left(\frac{N-1}{N}\log_2\frac{N}{N-1} + \frac{1}{N}\right) + \log_2\frac{1}{2N}$$

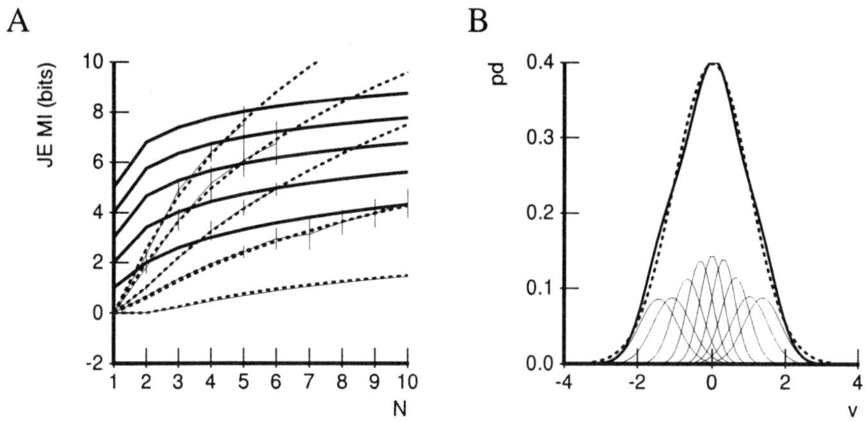


**Fig. 3.** (A) Theoretically-expected joint entropy (JE) (thick continuous lines) and mutual information (MI) (thick dashed lines) plotted as a function of the number of neurons $N$. The JE and MI are plotted for $k = 2, 4, \ldots, 32$ quantization levels, starting with $k = 2$ at the bottom and moving upwards. The thin continuous lines and their error bars correspond to the mean and standard deviations of the mutual information obtained in the simulations. The mean JE curves coincide with the theoretically-expected ones; the error bars are not drawn since they are smaller than $2.2\ 10^{-2}$ everywhere.
(B) One-dimensional, unit-variance Gaussian (thick dashed line) and the obtained estimate (thick continuous line) shown together with the local density estimates corresponding to the 9 neurons (thin continuous lines).

$$+(N-2)\left(\frac{3}{2N}\log_2\frac{2N}{3}+\frac{2N-3}{2N}\log_2\frac{2N}{2N-3}\right)\quad\text{(bits)},$$

$$\mathrm{JE} = \log_2 2N \quad \text{(bits)},$$

with the second term in MI being present for $N > 2$ only. Figure 3A shows JE and MI as a function $N$ and parameterized with respect to $k$; the plots for the $k > 2$ cases were determined by optimizing the kernel parameters numerically. We initialized the weights and radii by sampling the $(0, 1]$ interval uniformly and run our learning scheme 20 times for several $(N, k)$-combinations using our learning scheme and plotted the mean and standard deviations for the JE and MI obtained. We observe that the difference between the expected and the obtained JE is small (the curves almost completely coincide), and that MI continues to increase when $N$ increases, and quickly becomes larger than JE. The density estimate $\widehat{p}(\mathbf{v})$ and the original Gaussian distribution $p(\mathbf{v})$ are shown for $N = 9$ in Fig. 3B. We observe a good correspondence between $\widehat{p}(\mathbf{v})$ and $p(\mathbf{v})$. The mean squared error is $9.35\ 10^{-5}$ when calculated for 100 equidistant positions in the $[-4, 4]$ range; the joint entropy is 4.14 bits and the mutual information 1.32 bits (both for $k = 2$).

## 5 Conclusion

We have introduced a new learning scheme for kernel-based topographic map formation. The kernel parameters are adjusted so as to maximize the differential entropies of the kernel outputs and, at the same time, to minimize the mutual information between them. The former is achieved by applying *infomax* learning and the latter, in a heuristic sense, by having a competitive stage in the learning process. The kernel output functions approximate the cumulative distributions of the assumed Gaussian local input densities.

## Acknowledgments

M.M.V.H. is supported by research grants received from the Fund for Scientific Research (Belgium) (G.0185.96N), the National Lottery (Belgium) (9.0185.96), the Flemish Regional Ministry of Education (Belgium) (GOA 95/99-06; 2000/11), the Flemish Ministry for Science and Technology (VIS/98 /012), and the European Commission (QLG3-CT-2000-30161).

## References

1. Kohonen, T. (1982) Self-organized formation of topologically correct feature maps. Biol. Cybern., **43**, 59–69
2. Kohonen, T. (1995) Self-organizing maps. Springer, Berlin Heidelberg
3. Durbin, R., Willshaw, D. (1987) An analogue approach to the travelling salesman problem using an elastic net method. Nature, **326**, 689–691
4. Graepel, T., Burger, M., Obermayer, K. (1997) Phase transitions in stochastic self-organizing maps. Physical Rev. E, **56**(4), 3876–3890
5. Sum, J., Leung, C.-S., Chan, L.-W., Xu, L. (1997) Yet another algorithm which can generate topography map. IEEE TNNS, **8**(5), 1204–1207
6. Bishop, C.M., Svensén, M., Williams, C.K.I. (1998) GTM: The generative topographic mapping. Neural Computat., **10**, 215–234
7. Van Hulle, M.M. (1998) Kernel-based equiprobabilistic topographic map formation. Neural Computat., **10**(7), 1847–1871
8. Van Hulle, M.M. (2000) Faithful representations and topographic maps: From distortion- to information-based self-organization, Wiley, New York
9. András, P. (2001) Kernel-Kohonen networks. Int. J. Neural Systems, submitted
10. Linsker, R. (1989) How to generate ordered maps by maximizing the mutual information between input and output signals. Neural Computat., **1**, 402–411
11. Bell A.J., Sejnowski, T.J. (1995) An information-maximization approach to blind separation and blind deconvolution. Neural Computat., **7**, 1129–1159
12. Weisstein, E.W. (1999) CRC Concise Encyclopedia of Mathematics. Chapman and Hall, London
13. Silverman, B.W. (1992) Density estimation for statistics and data analysis. Chapman and Hall, London

# A Statistical Tool to Assess the Reliability of Self-Organizing Maps

M. Cottrell[1], E. de Bodt[2], M. Verleysen[3]

[1] Université Paris I, SAMOS-MATISSE, UMR CNRS 8595
90 rue de Tolbiac, F-75634 Paris Cedex 13, France
[2] Université catholique de Louvain, IAG-FIN, 1 pl. des Doyens,
B-1348 Louvain-la-Neuve, Belgium and
Université Lille 2, ESA, Place Deliot, BP 381, F-59020 Lille, France
[3] Université catholique de Louvain, DICE, 3, place du Levant,
B-1348 Louvain-la-Neuve, Belgium

Abstract

Making results reliable is one of the major concerns in artificial neural networks research. It is often argued that Self-Organizing Maps are less sensitive than other neural paradigms to problems related to convergence, local minima, etc. This paper introduces objective statistical measures that can be used to assess the stability of the results of SOM, both on the distortion and on the topology preservation points of views.

## 1 Introduction

Neural networks are powerful data analysis tools. Part of their interesting properties comes from their inherent non-linearities, in contrast to classical, linear tools. Nevertheless, the non-linear character of the methods has also its drawbacks: most neural network algorithms rely on the non-linear optimization of a criterion, leading to well-known problems or limitations concerning local minima, speed of convergence, etc.

It is commonly argued that vector quantization methods, and in particular self-organizing maps, are less sensitive to these limitations than other classical neural networks, like multi-layer perceptrons and radial-basis function networks. For this reason, self-organizing maps (SOM) [1] are often used in real applications, but rarely studied on the point of view of their reliability: one usually admits that, with some "proper" choice of convergence parameters (adaptation step and neighborhood), the SOM algorithm converges to an "adequate", or "useful", state.

This paper aims at defining objective criteria that may be used to measure the "reliability" of a SOM in a particular situation. The bootstrap methodology is used to measure the variability of both the quantization error and the organization of the map. Section 2 summarizes the main idea of the bootstrap, section 3 defines a measure of the variability of the quantization error, section 4 a measure of the variability of the organization of the map, and section 5 applies the concepts to a few simple distributions.

# 2 Bootstrap

The main idea of the bootstrap [2] is to use the so-called "plug-in principle". Let $F$ be a probability distribution depending on an unknown parameter vector $\boldsymbol{\theta}$. Let $\boldsymbol{x}=x_1, x_2,\ldots, x_n$ be the observed sample of data and $\hat{\boldsymbol{\theta}} = T(\boldsymbol{x})$ an estimate of $\boldsymbol{\theta}$. The bootstrap consists in using artificial samples (called *bootstrapped samples*) with the same empirical distribution as the initial data set in order to guess the distribution of $\hat{\boldsymbol{\theta}}$. Each *bootstrapped sample* consists in $n$ uniform drawings with repetitions from the initial sample. If $\boldsymbol{x}^*$ is a bootstrapped sample, $T(\boldsymbol{x}^*)$ will be a bootstrap replicate of $\hat{\boldsymbol{\theta}}$.

This main idea of the bootstrap may be declined in several ways. In particular, when the evaluation of $T(\boldsymbol{x})$ requires non-linear optimization, the well-known problems, or limitations, related to local minima and convergence are encountered. It may thus happen that different local minima are reached when $T(\boldsymbol{x}^*)$ is evaluated for different bootstrapped samples. This is clearly not what we are looking for: our purpose is to examine the variability (or the sampling distribution) of some parameters when they are evaluated through different (bootstrapped) samples, but keeping all other conditions unchanged. In order to overcome this problem, local bootstrap methods may be used, where the initial conditions for each evaluation of $\hat{\boldsymbol{\theta}}$ are kept fixed. In the following, we will speak about:

- Common Bootstrap (**CB**) when each evaluation of $\hat{\boldsymbol{\theta}}$ is initialized at random;
- Local Bootstrap (**LB**) when the initial values of each evaluation are kept fixed;
- Local Perturbed Bootstrap (**LPB**) when a small perturbation is applied to the initial conditions obtained as with the Local Bootstrap.

If we want to evaluate the influence of the convergence (only) during the evaluation of $\hat{\boldsymbol{\theta}}$, we will not bootstrap samples, but reiterate the evaluation of $T(\boldsymbol{x})$ with the same sample $\boldsymbol{x}$ and different initial conditions. In this case, we will speak about Monte-Carlo simulations instead of bootstrap, and we will use the same three variants as above: Common Monte-Carlo (**CMC**), Local Monte-Carlo (**LMC**) and Local Perturbed Monte-Carlo (**LPMC**).

# 3 Stability of the Quantization in the SOM

One of the two main goals of Self-Organizing Maps is to quantize the data space into a finite number of so-called *centroids* (or *code vectors*). Vector quantization is used in many areas to compress data over transmission links, to remove noise, etc. The distance between an observed data $x_i$ and its corresponding (nearest) centroid is te *quantization error*. Averaging this quantization error over all data leads to the *distortion* or *intra-class sum of squares* (which are different names for the same error, used respectively in the information theory domain and by statisticians):

$$SSIntra = \sum_{i=1}^{U} \sum_{x_j \in V_i} d^2(x_j, G_i) \tag{1}$$

where $U$ is the number of units in the SOM, $G_i$ is the $i$-th centroid, $d$ is the classical Euclidean distance, and $V_i$ is the Voronoi region associated to $G_i$, i.e. the region of the space nearer to $G_i$ than to any other centroid. Note that the objective function associated with the SOM algorithm for a constant size of neighborhood and finite data set is *the sum of squares intra-classes extended to the neighbor classes,* (see [3]). But actually, one usually ends with no neighbor for the last iterations of the SOM algorithm; at the end of its convergence, the SOM algorithm thus exactly minimizes the *SSIntra* function.

The Monte-Carlo and/or bootstrap methods will allow us to estimate the variability of *SSIntra*, in other words to assess if one may be confident in the stability of the quantization obtained by the SOM. Note that we do not speak about the value (location) of the centroids themselves, but on how they quantify the space in average. If the SOM is computed several times according to the Monte-Carlo or the Boostrap principle detailed in the previous section, one can calculate the mean $\mu_{SSIntra}$ and the standard deviation $\sigma_{SSIntra}$ of the distortion. The variability of *SSIntra* is the evaluated by its coefficient of variation $CV$ defined as follows:

$$CV(\theta) = 100 \frac{\sigma_\theta}{\mu_\theta} \tag{2}$$

where $\theta$ is the parameter to examine, here *SSIntra*.

# 4 Stability of the Neighborhood Relations in the SOM

The second main goal of the SOM is the so-called *topology preservation*, which means that close data in the input space will be quantized by either the same centroid, either two centroids that are close one from another on a predefined string or grid. Often, for example when the SOM is used as a visualization tool, it is desirable to have an objective measure of this *neighborhood* property. We then define for any pair of data $x_i$ and $x_j$,

$$STAB_{i,j}(r) = \frac{\sum_{b=1}^{B} NEIGH_{i,j}^{b}(r)}{B} \tag{3}$$

where $NEIGH^{b}{}_{i,j}(r)$ is an indicator function that returns 1 if the observations $x_i$ and $x_j$ are neighbor at the radius $r$ for the bootstrap sample $b$, and $B$ is the total number of bootstrapped samples. If the radius $r$ is 0, it means that we evaluate if the two data are projected on the same centroid; if $r = 1$, it means that we evaluate if the two data are projected on the same centroid *or* on the immediate neighboring centroids on the string or grid (2 on the string, 8 on the grid), etc.

A perfect stability would lead $STAB_{i,j}$ to always be 0 (never neighbor) or 1 (always neighbor). We can study the significance of the statistics $STAB_{ij}(r)$, by comparing it to the value it would have if the observations fell in the same class (or in two classes distant of less than $r$) in a completely random way. Let $U$ be the total number of classes and $v$ the size of the considered neighborhood. The size $v$ of the neighborhood can be computed from the radius $r$ by $v = (2r + 1)$ for a one-dimensional SOM map (a string); and $v = (2r + 1)^2$ for a two-dimensional SOM map (a grid), if edge effects are not taken into account. For a fixed pair of observations $x_i$ and $x_j$, with random drawings, the probability of neighboring would be $v/U$. If we define a Bernoulli random variable with probability of success $v/U$, (where success means: "$x_i$ and $x_j$ are neighbor"), the number $Y$ of successes on $B$ trials is distributed as a Binomial distribution, with parameters $B$ and $v/U$. Therefore, it is possible to build a test of the hypothesis $H_0$ "$x_i$ and $x_j$ are only randomly neighbors" against the hypothesis $H_1$ "the fact that whether $x_i$ and $x_j$ are neighbors or not is meaningful".

If $B$ is large enough (i.e. greater than 50), the binomial random variable can be approximated by a Gaussian variable, making the hypothesis test easier. For example, with a test level of 5%, we conclude to $H_1$ if $Y$ is less than $B\frac{v}{U} - 1.96\sqrt{B\frac{v}{U}\left(1 - \frac{v}{U}\right)}$ or greater than $B\frac{v}{U} + 1.96\sqrt{B\frac{v}{U}\left(1 - \frac{v}{U}\right)}$. Note that in the case of the bootstrap, B depends on the pair $(x_i, x_j)$, since the bootstrapped samples have to contain both data: we follow the same approach as in [4] which consists in evaluating $STAB_{i,j}(r)$ only on the samples that contain observations $x_i$ and $x_j$.

# 5 Experiments

The above described indicators have been evaluated on artificial and real databases; a selection of results follow.

## 5.1 Databases

Three artificial databases have been used: Gauss_1, Gauss_2 and Gauss_3. All three are two-dimensional data sets, obtained by random drawings on uncorrelated Gaussian distributions. They are respectively represented in figures 1, 2, and 3. Gauss_1 contains only one cluster of observations. Gauss_2 contains three clusters of equal variance and some overlap. Gauss_3 is also composed of three clusters, but of different variances and without overlap. Each data set has 500 observations. For data sets Gauss_2 and Gauss_ 3, observations 1-166, 167-333 and 334-500 are in the same cluster.

A real database, POP_84, was also used. It contains six ratios measured in 1984 on the macroeconomic situation of countries: annual population growth, mortality rate, analphabetism rate, population proportion in high school, GDP per head and GDP growth rate. This dataset has been already used in [5], and is available through [6].

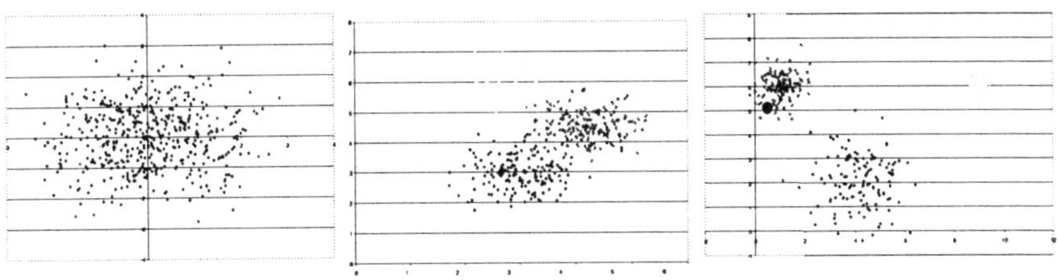

Figure 1: Gauss_1, Gauss_2 and Gauss_3 databases

## 5.2 Stability of the Distortion Error

Table 1 summarizes the results on the coefficient of variation (2) of the distortion (1), measured on the three artificial databases, and obtained by the CMC, LMC and LPMC methods on 5000 independent samples (note that such a large number of samples is not necessary in practice to obtain reliable results; 100 samples is already a good choice). The Kohonen map used in these simulations is a 3- or 6-units 1-dimensional string.

| method | CMC | | LMC | | LPMC | |
|---|---|---|---|---|---|---|
| # units | 3 | 6 | 3 | 6 | 3 | 6 |
| Gauss_1 | 5.2 | 4.5 | 5.3 | 4.4 | 5.2 | 4.5 |
| Gauss_2 | 5.1 | 4.6 | 4.9 | 4.5 | 5.1 | 4.6 |
| Gauss_3 | 7.6 | 10.1 | 6.4 | 10.3 | 6.7 | 10.1 |

Table 1: Coefficients of variation of *SSIntra*, obtained with 1-dimensional 3- and 6-units SOM, and with CMC, LMC and LPMC Monte-Carlo simulations.

An expected result is that the CV is low in each case, and does not seem to be influenced by the initialization method used (CMC, LMC or LPMC). This enforces the idea that the SOM is a reliable method, not falling too much into the traps of local minima encountered with other neural network models (see for example [7] for opposite results obtained with MLP).

A more surprising result at the first sight is the increase of the CV when switching from a 3- to a 6-units map on the Gauss_3 distribution. This can however be explained by the fact that the Gauss_3 distribution contains 3 well separated clusters. When using a 3-units map on this distribution, one centroid will obviously converge around the center of each cluster. Adding units will however lead to situation where the supplementary ones may be "captured" by one cluster or by another, increasing the variability of the situation after convergence. This is of course not (or less) true with the Gauss-1 and Gauss_2 distributions where there is overlap between clusters.

## 5.3 Assessing the Right Number of Units in the Map

This last comment makes think that it would be possible to estimate the number of units to include in a SOM on a particular database, by examining the evolution of the CV when increasing the number of units. This is an important question when using SOM, and may be considered as a by-product of the proposed measures. Table 2 shows the coefficient of variation of *SSIntra* obtained after Local Boostrap, on the three Gauss_x databases. Table 2 confirms the step between 3 and 4 units on the Gauss_3 databases, while the results on the other databases are less conclusive, as expected.

| # units | Gauss_1 | Gauss_2 | Gauss_3 | # units | Gauss_1 | Gauss_2 | Gauss_3 |
|---|---|---|---|---|---|---|---|
| 1 | 5.2 | 4.3 | 5.5 | 6 | 5.1 | 4.7 | 12.0 |
| 2 | 4.5 | 6.0 | 8.9 | 9 | 5.4 | 4.7 | 10.9 |
| 3 | 5.9 | 5.4 | 6.5 | 12 | 3.7 | 4.9 | 9.2 |
| 4 | 5.5 | 4.9 | 14.4 | 15 | 4.0 | 4.0 | 8.0 |
| 5 | 4.4 | 6.6 | 15.2 | | | | |

Table 2: Coefficients of variations of *SSIntra* obtained after Local Bootstrap.

## 5.4 Stability of the Neighborhood Relations

Table 3 shows a few results obtained on the POP_84 data set, on selected pairs of countries. The values in the table are those of the *STAB* indicator defined in equation (3). The two first columns refer to a 1-dimensional SOM with 6 units, and show respectively $STAB_{i,j}(0)$ and $STAB_{i,j}(1)$, where $i,j$ identifies the countries, while the third column refers to $STAB_{i,j}(2)$ on a 7x7 2-dimensional SOM. The results marked with * are those for which the hypothesis test (with 1% significance) defined in section 4 lead to conclude hypothesis H1, i.e. the neighborhood relation (presence or absence) between two countries is meaningful. As expected, countries with similar situations give high values of the *STAB* indicator. Countries with very opposite situations give very low values.

| Countries | $STAB_{i,j}(0)$ 6 units | $STAB_{i,j}(1)$ 6 units | Countries | $STAB_{i,j}(2)$ 7x7 units |
|---|---|---|---|---|
| Turkey/Upper Volta | 0.04* | 0.65* | Greece/France | 0.18* |
| Turkey/Cuba | 0* | 0.22* | Australia-France | 0.82* |
| Turkey/Sweden | 0* | 0.05* | Greece/Belgium | 0.21* |
| Turkey/France | 0* | 0* | Turkey/France | 0.02* |
| Turkey/Greece | 0* | 0.25* | Singapore/USA | 0.49 |
| Upper Volta/Cuba | 0* | 0* | Sweden/Japan | 0.73* |
| Upper Volta / Sweden | 0* | 0* | Greece/Malta | 1* |
| Upper Volta / France | 0* | 0* | Canada/France | 0.84* |
| Sweden/France | 1* | 1* | Sweden/France | 0.97* |
| Cuba / Sweden | 0.02* | 0.81* | USA/Zimbabwe | 0* |
| Cuba / France | 0.02* | 0.78* | USA/Finland | 0.85* |
| Cuba / Greece | 0.69* | 0.97* | USA/Australia | 0.68* |

Table 3: *STAB* indicator on the POP_84 dataset, for selected pairs of countries..

# 6 Discussion

This paper introduces measures to assess the stability of the SOM convergence, under the distortion and topology preservation point of views. Having objective measures of the stability may enforce the idea that SOM are less sensitive to problems related to convergence and local minima than other neural network paradigms. A by-product to the proposed measure is a way to estimate the number of units in a Kohonen map.

The measure of the stability of the neighborhood relations concerns a particular pair of data. It would be interesting to obtain a global measure of the topology preservation over the whole map. As "reliable" neighborhood relations mean values of the *STAB* indicator near to 0 or 1, a possibility is to draw an histogram of the *STAB* values, and to measure how this histogram is close or not to a "U"-shape (peaks at 0 and 1, and smooth curve with low values in between).

# References


1. Kohonen T., Self-Organizing Maps, Springer, Berlin, 1995
2. Efron B. & Tibshirani R., Bootstrap Methods for Standard Errors, Confidence Intervals, and Other Measures of Statistical Accuracy, Statistical Science, 1986, 1:54-77
3. Ritter H., Martinetz T. and Schulten K., Neural Computation and Self-Organizing Maps, Addison-Wesley, Reading, 1992
4. Efron B. & Tibshirani R., An Introduction to the Bootstrap, Chapman and Hall, 1993
5. Blayo F., Demartines P., Data Analysis: How to Compare Kohonen Neural Networks to Other Techniques? In A. Prieto ed., Artificial Neural Networks, Lecture Notes in Computer Science, 1991, 469-476
6. Available from http://panoramix.univ-paris1.fr/SAMOS/
7. Zapranis A. & Refenes A.P., Principles of Neural Model Identification, Selection and Adequacy, Springer, 1999

# A SOM Association Network

T. Yamakawa[†][*], K. Horio[†] and R. Kubota[‡]
[†]Department of Brain Science and Engineering
Kyushu Institute of Technology
[‡]Department of Control Engineering and Science
Kyushu Institute of Technology
680-4 Kawazu, Iizuka, Fukuoka 820-8502 Japan
yamakawa@ces.kyutech.ac.jp

Abstract

In this paper, we propose a self-organizing map association (SOMA) network which associates a perfect information from a part of the information. The processes of the SOMA network are divided into a learning mode and an association mode. In the learning mode, the similar perfect informations are represented by a few units on the competitive layer. In the association mode, when the information, whose parts are lost, is applied to the SOMA network, the lost part is associated. The performance of the SOMA network is evaluated by applying to the association of the data used in the orthodontics.

## 1 Introduction

In the real world, it is very difficult to obtain perfect information. Thus, it is important to associate the perfect information from some parts of this information. Almost all the traditional methods of association are based on the detection of mapping from the parts of the information (input) to the perfect information (output) by using the linear network, neural networks and so on. Although these methods are effective in case which fixed parts of the information are given, they can not associate the perfect information in case which the parts of the information as the inputs are changed.

The self-organizing map (SOM)[1][2] has a function that many kinds of input vectors are mapped to small number of units on the competitive layer. This feature facilitates that the similar input vectors are represented by typical vector.

In this paper, we propose a self-organizing map association (SOMA) network which associates the information from some parts of the information. The SOMA network is based on the feature of SOM, that many kinds of input vectors are represented by small number of typical vectors. The processes

*The author is also with the Fuzzy Logic Systems Institute (FLSI), Kawazu, Iizuka 820-0067 Japan.

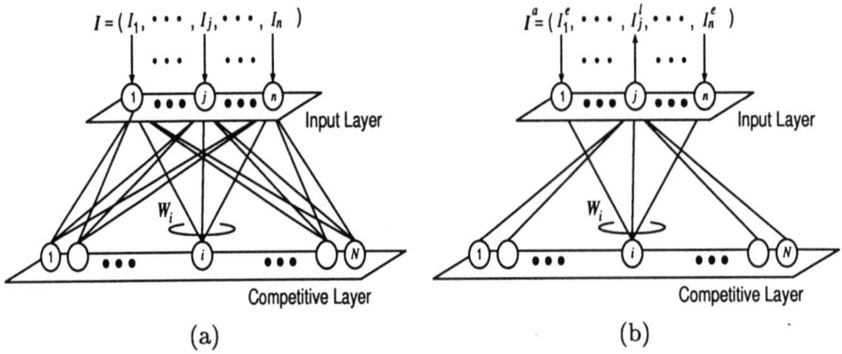


Figure 1: Structure of the self-organizing map association network. (a)Learning mode. (b)Association mode.

of the SOMA network are divided into a learning mode and an association mode. In the learning mode, the similar informations are represented by some units on the competitive layer. Thus, many kinds of informations are represented by small number of units. In the association mode, when the information whose parts are lost is applied to the SOMA network, the lost parts are associated by employing similarity between the input vector and weight vectors.

The effectiveness and the validity of the proposed method are verified by association of the data whose parts are lost.

## 2 Self-Organizing Map Association

The SOMA network is the network which can associate the perfect information from some parts of the information. The structure of the SOMA network is shown in Figure 1(a). The SOMA network consists of a input layer and a competitive layer that have $n$ and $N$ units, respectively. The $i$-th unit in the competitive layer is connected to all the units in the input layer by the weight vector $\mathbf{w}_i = (w_{i1}, w_{i2}, \cdots, w_{in})$. The processes of the SOMA network are divided into a learning mode and an association mode. The learning of the SOMA network is achieved as follows (learning mode).

step 1: The input vector $\mathbf{I} = (I_1, I_2, \cdots, I_n)$ which is a perfect information is applied to the input layer.

step 2: The Euclidean distance $d_{\mathbf{I}i}$ between the input vector $\mathbf{I}$ and the weight vector $\mathbf{w}_i$ is calculated by:

$$d_{\mathbf{I}i} = ||\mathbf{I} - \mathbf{w}_i|| = \sqrt{\sum_{j=1}^{n} (I_j - w_{ij})^2}. \tag{1}$$

step 3: The unit which has the minimum $d_{\mathbf{I}i}$ is defined as a winner unit $c$ by:

$$c = \arg\min_i \{d_{\mathbf{I}i}\}. \tag{2}$$

step 4: The weight vectors of the winner unit and its neighboring units are updated by:

$$\mathbf{w}_i^{new} = \mathbf{w}_i^{old} + \alpha(t)(\mathbf{I} - \mathbf{w}_i^{old}), \tag{3}$$

where $\mathbf{w}_i^{new}$ and $\mathbf{w}_i^{old}$ are the weight vectors after and before updating, respectively. $\alpha(t)$ is defined as a learning ratio, and monotonously decreases. For example, $\alpha(t)$ is defined by:

$$\alpha(t) = \alpha(0)(1 - \frac{t}{T}), \tag{4}$$

where $\alpha(0)$ is a initial value of the learning ratio. $t$ is a present time in the learning, and $T$ is the total number of learning iterations. The region of the neiborhood of the winner unit gradually decreases.

In the learning mode, many kinds of input vectors are represented by small number of typical weight vectors.

The association of the SOMA network illustrated in Figure 1(b) is achieved as follows (association mode).

step 1: The input vector $\mathbf{I}^a = (I_1^e, I_2^e, \cdots, I_j^l \cdots, I_n^e)$ whose some elements are lost is applied to the input layer. Super-script $e$ and $l$ mean that the elements with $e$ or $l$ exists or are lost information, respectively.

step 2: The Euclidean distance $d_{\mathbf{I}^a i}$ between input vector $\mathbf{I}^a$ and $\mathbf{w}_i$ is calculated by:

$$d_{\mathbf{I}^a i} = ||\mathbf{I}^a - \mathbf{w}_i|| = \sqrt{\sum_{j=1}^{n} (I_j^e - w_{ij})^2}. \tag{5}$$

step 3: $d_{\mathbf{I}^a i}$ is translated to the similarity $g_i$ by:

$$g_i = \exp\left(-||d_{\mathbf{I}^a i}||\right). \tag{6}$$

step 4: The lost element of the input vector $I_j^l$ is associated by:

$$I_j^l = \frac{\sum_{i=1}^{N} g_i w_{ji}}{\sum_{i=1}^{N} g_i}. \tag{7}$$

When another elements in the input vectors are lost, the association can be achieved with same algorithm. It is no need to adjust the weight vectors even if number of the lost elements is changed.

Table 1: The meaning of each elements in the data.

| Number of Element | Subject |
|---|---|
| 1 | Facial Convexity Angle: FCA |
| 2 | Maxillary Prognathism: Mxp |
| 3 | Mandibular Prognathism |
| 4 | Vertical Height Ratio: VHR |
| 5 | Naso-Labial Angle: NLA |
| 6 | Upper Lip Protrusion: ULP |
| 7 | Lower Lip Protrusion: LLP |
| 8 | Labio-Mental Sulcus: LMS |
| 9 | Vertical Lip-Chin Ratio: VCR |
| 10 | Maxillary Incisor Exposure |
| 11 | Interlabial Gap |
| 12 | Esthetic Plane (Ls) |
| 13 | Esthetic Plane (Li) |
| 14 | Merrifield's Z Angle |
| 15 | Sn Vertical Upper Lip |
| 16 | Sn Vertical Lower Lip |
| 17 | Sn Vertical Soft Pogonion |
| 18 | Naso-Facial Angle: NFA |
| 19 | Naso-Mental Angle: NMA |
| 20 | E Line to FH pl. |

# 3 Simulation

In order to verify the validity of the proposed SOMA, we indicate some experimental results in which data used in the orthodontics are associated. The data of a patient includes 20 elements that are detected from his/her roentgen photograph. The meaning of each element is shown in Table 1, and there are 44 patients' data. For example, the value of each element of three patients is shown in Figure 2. In the Figure 2, horizontal and vertical axes represent the element number and the value of the element, respectively. Consider that the some elements of the data can be obtained and the others have to be associated.

In the learning of the SOMA, the data of 43 patients except patient 1 are used as the input vectors. The number of the units on the input layer is equal to that of elements of the data (20). The number of the units in the competitive layer, the initial value of the learning rate and the number of learning iterations are 16 (4×4), 0.8 and 5000, respectively.

After the leaning, association of some elements of the patient 1 is achieved in the association mode. It is assumed that the elements from 1 to 15 are exist and those from 16 to 20 are lost. Figure 3(a) shows that the associated

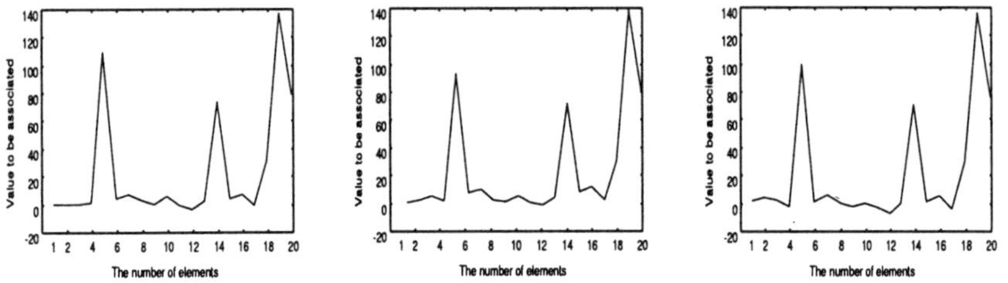


(a) (b) (c)

Figure 2: The three examples of patients' data used in the orthodontics. (a)Patient 1. (b)Patient 2. (c)Patient 3.

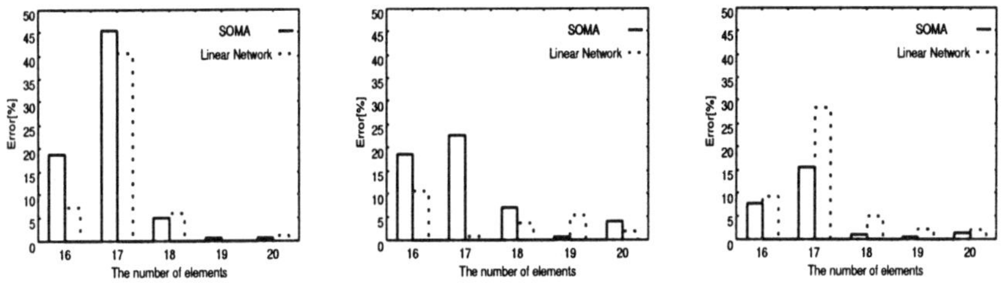


(a) (b) (c)

Figure 3: The associated results of each patient. (a)Patient 1. (b)Patient 2. (c)Patient 3.

results. Figure 3(a) also includes the results associated by the linear network which had 15 inputs (elements from 1 to 15) and 5 outputs (those from 16 to 20). The linear network is shown by:

$$\begin{bmatrix} I_{16}^l \\ I_{17}^l \\ \vdots \\ I_{20}^l \end{bmatrix} = \begin{bmatrix} a_{1,1} & a_{1,2} & \cdots & a_{1,15} \\ a_{2,1} & a_{2,2} & \cdots & a_{2,15} \\ \vdots & \vdots & \ddots & \vdots \\ a_{5,1} & a_{5,2} & \cdots & a_{5,15} \end{bmatrix} \begin{bmatrix} I_1^e \\ I_2^e \\ \vdots \\ I_{15}^e \end{bmatrix} + \begin{bmatrix} b_1 \\ b_2 \\ \vdots \\ b_{15} \end{bmatrix}, \tag{8}$$

where all of the $a$ and $b$ are the model parameters. The each parameter is updated by the gradient descent method. Figure 4 shows another simulation results in which the elements of the input vector (from 1 to 5) are lost and they are associated by the SOMA.

The performance of each network is evaluated by employing the root mean square error. The experimental result of the SOMA network was 4.4, and that of the linear network was 2.9. The experimental result of the SOMA network is almost equal to that of the linear network. Besides, the SOMA network can associate the lost elements, when the part of lost is changed. However, the linear network need to adjust the input/output relationships, when the parts of lost are changed.

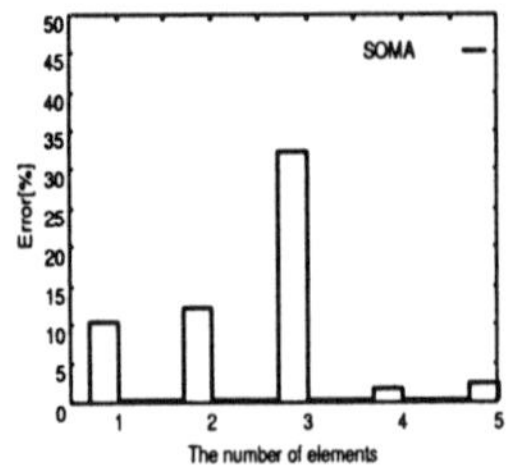


Figure 4: An associated result of other elements.

# 4 Conclusions

In this paper, we proposed the SOMA network which associates the perfect information from some parts of the information. The processes of the SOMA network are divided into the learning mode and the association mode. In the learning mode, many kinds of input vectors are represented by the small number of typical vectors. In the association mode, the lost elements of the input vector are associated. The association of the SOMA network can be achieved with same algorithm when another elements in the input vector are lost. So that it is no need to adjust the weight vectors even if the number of the lost elements is changed. The validity of the SOMA is verified by applying it to the association of the data used in the orthodontics.

# References

[1] T. Kohonen, "Self-organized formation of topologically correct feature maps", *Biol. Cybern.*, Vol.43, pp.59-69, 1982.

[2] T.Kohonen, *Self-organization and associative memory*, Second edition, Berlin: Springer-Verlag, 1988.

[3] Legan HL, Burstone CJ, "Soft tissue cephalometric analysis for orthognathic surgery", *J Oral Sung*, Vol. 38, pp.744-751, 1980.

# A Supervised Self-Organizing Map for Structured Data*

Markus Hagenbuchner[1], Ah Chung Tsoi[1], and Alessandro Sperduti[2]

[1] Faculty of Informatics, University of Wollongong, Australia
[2] Dip. di Informatica, Università di Pisa, Corso Italia, 40, 56125 Pisa, Italy

**Abstract.** A self organizing map (SOM) for processing of structured data, using an unsupervised learning approach, called SOM-SD, has recently been proposed. Here, we suggest a new version of SOM, using the supervised learning approach. We compare the supervised version and the unsupervised version of SOM-SD on a benchmark problem involving visual patterns. As may be expected, the supervised version is able to solve the classification problem using very compact networks.

## 1 Introduction

The processing of structured data is usually confined to the domain of symbolic systems [2]. Recently, however, supervised neural networks for adaptive processing of structured data have been proposed [3,7]. These neural networks, called Recursive Neural Networks [7], are capable of learning a mapping from a domain of ordered directed acyclic graphs (DAGs), with labels attached to each node, to the set of real numbers. This unique ability has successfully been exploited in Chemistry (e.g., QSAR studies), as well as in other application domains (for example [1,4]). Recursive neural networks are based on the idea of adaptive encoding of data structures. For example, in the special case of data structures being the set of labeled trees, the processing of the set of training samples begins from the leaf nodes, and progressively follows the graph structure until it reaches a root node. Each node could be represented by a multilayer feedforward neural network [7].

More recently, using a related approach, Self-Organizing Maps (SOMs) [6] have been extended to allow the mapping of structured objects into a topological map using the unsupervised learning approach[1]. It has been shown that this model, i.e., Self-Organizing Maps for Structured Data (SOM-SD), is able to arrange a set of labeled structures according both to structural similarities and similarities among the labels attached to nodes.

In this paper, we show how SOM-SD can also be trained using a supervised approach. This is a new direction in supervised learning to approach adaptive processing of structured data. So far, we are only aware of supervised learning methods using the Recursive Neural Network models. However, in this paper,

* Partially supported by MURST grant 9903244848 and MM09308497.
[1] *Self-Organizing Maps for Structured Data* manuscript in preparation.

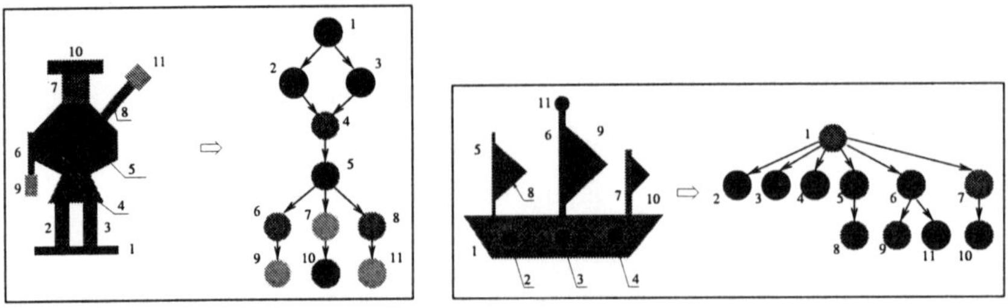

**Fig. 1.** Examples of DOAGs representing visual patterns. Labels are not shown.

we will show that it is possible to train the SOM-SD model using a supervised learning approach. We shall call this method sSOM-SD as opposed to the unsupervised version which we refer to as uSOM-SD.

There are several motivations for considering sSOM-SD instead of Recursive Neural Networks. First, the kind of supervision we are going to adopt is directly derived from the unsupervised algorithm, and, for this reason, it works with training sets where some of the examples may have unknown targets. In other words, sometimes the training examples may contain a mixture of target values, and missing target values. Moreover, sSOM-SD are able to map the training data structures to a low dimensional space and thus it is easier to understand how the network implements the solution to the computational task. Finally, with sSOM-SDs it is much easier to deal with many target classes, and it is easier to deal with the addition of new target classes.

## 2 Self-Organizing Maps for Structured Data

We shall use lowercase bold letters to denote vectors, uppercase bold letters to denote matrices, and calligraphic letters for representing graphs. A data structure $\mathcal{Y}$ is a directed acyclic graph (DAG) whose vertices are labeled by vectors of real-valued numbers which either represent numerical or categorical variables. Subscript notation will be used when referencing the labels attached to vertices in a data structure. Hence $\boldsymbol{y}_v$ denotes the vector labeling vertex $v \in \text{vert}(\mathcal{Y})$, where $\text{vert}(\cdot)$ denotes the set of vertices of the DAG $\mathcal{Y}$.

In this paper, it is assumed that instances in the learning domain are DOAGs (directed ordered acyclic graphs) or DPAGs (directed positional acyclic graphs). The *indegree* of node $v$ is the number of incoming edges to $v$, whereas the *outdegree* of $v$ is the number of outgoing edges from $v$.

We shall require the DAG (either DOAG or DPAG) to possess a supersource[2], i.e. a vertex $s \in \text{vert}(\mathcal{D})$ such that every vertex in $\text{vert}(\mathcal{D})$ can be reached by a directed path starting from $s$. Given a DAG $\mathcal{D}$ and $v \in \text{vert}(\mathcal{D})$, we denote by $\text{ch}[v]$ the set of children of $v$, and by $\text{ch}_k[v]$ the k-th child of

[2] If no supersource is present, a new node connected with all the nodes of the graph with null *indegree* can be added.

$v$. We shall denote by $\#^{(c)}$ the class of DAGs with maximum outdegree $c$. The class of all data structures defined over the label universe domain $\mathcal{Y}$ and skeleton in $\#^{(c)}$ will be denoted as $\mathcal{Y}^{\#^{(c)}}$. The void DAG will be denoted by the special symbol $\xi$.

In Fig. 1, we show two examples of visual patterns represented as DOAGs. Each node of the graph represents a different colored component of the visual pattern, while edges are used to encode the concept of "connected with".

Here we are interested in a *feature map* $\widehat{\mathcal{M}} : \mathcal{Y}^{\#^{(c)}} \rightarrow \mathcal{A}$, where the input space is a structured domain with labels in $\mathcal{Y}$. Specifically, we are interested in maps which can be realized by the following recursive definition:

$$\widehat{\mathcal{M}}(\mathcal{D}) = \begin{cases} nil_{\mathcal{A}} & \text{if } \mathcal{D} = \xi \\ \mathring{\mathcal{M}}\left(\boldsymbol{y}_v, \widehat{\mathcal{M}}(\mathcal{D}^{(1)}), \ldots, \widehat{\mathcal{M}}(\mathcal{D}^{(c)})\right) & \text{otherwise} \end{cases}$$

where $nil_{\mathcal{A}}$ is a special void coordinate into the discrete output space $\mathcal{A}$, and $\mathring{\mathcal{M}}$: $\mathcal{Y} \times \mathcal{A}^c \rightarrow \mathcal{A}$ is a SOM, defined on a generic node, which takes as input the label $\boldsymbol{y}$ of the node and the 'encoding' of the subgraphs $\mathcal{D}^{(1)}, \ldots, \mathcal{D}^{(c)}$ according to the $\widehat{\mathcal{M}}$ map. $\widehat{\mathcal{M}}(\mathcal{D})$ can be computed by starting to apply $\mathring{\mathcal{M}}$ to leaf nodes (nodes with null outdegree), and proceeding with the application of $\mathring{\mathcal{M}}$ bottom-up from the frontier to $s \in \text{vert}(\mathcal{D})$

Assuming that each label in $\mathcal{Y}$ is encoded in $\mathcal{U} \subset I\!R^m$, for each node $v$ in vert($\mathcal{D}$), we have a vector $\boldsymbol{u}_v$ of dimension $m$. Moreover, the display output space $\mathcal{A}$, where $\mathcal{A} \equiv [1..n_1] \times [1..n_2] \times \cdots \times [1..n_q]$, is realized through a $q$ dimensional lattice of neurons[3]. Thus, the winning neuron is represented by the coordinates $(i_1, .., i_q)$.

With the above assumptions, we have $\mathring{\mathcal{M}}$: $I\!R^m \times ([1..n_1] \times \cdot\cdot \times [1..n_q])^c \rightarrow [1..n_1] \times \cdot\cdot \times [1..n_q]$, and the $m + cq$ dimensional input vector $\boldsymbol{v}$ to $\mathring{\mathcal{M}}$, representing the information about a generic node $v$, is defined as

$$\boldsymbol{v} = \left[\boldsymbol{u}_v \ \mathcal{D}_{ch_1[v]} \ \mathcal{D}_{ch_2[v]} \ \cdots \ \mathcal{D}_{ch_c[v]}\right],$$

where $\mathcal{D}_{ch_i[v]}$ denotes the coordinates vector[4] of the winning neuron for the subgraph pointed by the i-th pointer of $v$. Each neuron with coordinates vector $\boldsymbol{c}_j$ in the $q$ dimensional lattice has an associated vector weight $\boldsymbol{w}_{\boldsymbol{c}_j} \in I\!R^{m+cq}$. The weights associated with each neuron in the $q$ dimensional lattice $\mathring{\mathcal{M}}$ can be trained using the standard SOM's learning process. The training algorithm for $\widehat{\mathcal{M}}$ is shown in Fig. 2.

The coordinates for the (sub)graphs are stored in $\mathcal{D}_v$, once for each processing of graph $\mathcal{D}$, and then used when needed[5] for the training of $\mathring{\mathcal{M}}$. The

[3] In this paper, experiments are performed by using $q = 2$.

[4] The null pointer $nil_{\mathcal{A}}$ can be defined, for example, by the vector with all components at -1.

[5] Notice that the use of an inverted topological order guarantees that the updating of the coordinate vectors $\mathcal{D}_v$ is performed before the use of $\mathcal{D}_v$ for training.

**Unsupervised Stochastic Training Algorithm for $\widehat{\mathcal{M}}$**

**input:** Set of training DAGs $T = \{\mathcal{D}_i\}_{i=1,\dots,N}$, $c$ maximum outdegree of DAGs in $T$, map $\overset{\circ}{\mathcal{M}}$;
**begin**
randomly set the weights for $\overset{\circ}{\mathcal{M}}$;
**repeat**
randomly select $\mathcal{D} \in T$ with uniform distribution;
`List(`$\mathcal{D}$`)` $\leftarrow$ an inverted topological order for vert($\mathcal{D}$);
**for** $v \leftarrow$`first(List(`$\mathcal{D}$`))` **to** `last(List(`$\mathcal{D}$`))`**do**
train( $\overset{\circ}{\mathcal{M}}$ $\left(\left[\boldsymbol{u}_v\ \mathcal{D}_{ch_1[v]}\ \mathcal{D}_{ch_2[v]}\ \cdots\ \mathcal{D}_{ch_c[v]}\right]\right)$);
$\mathcal{D}_v \leftarrow \overset{\circ}{\mathcal{M}}\left(\left[\boldsymbol{u}_v\ \mathcal{D}_{ch_1[v]}\ \mathcal{D}_{ch_2[v]}\ \cdots\ \mathcal{D}_{ch_c[v]}\right]\right)$;
**end**

**Fig. 2.** The unsupervised SOM-SD algorithm.

vector is an approximation of the true coordinate vector for the graph rooted in $v$, however, if the learning rate $\eta$ is small, this approximation may be negligible. The call *train*() refers to a single step of standard SOM training.

## 3 The Supervised SOM-SD

Given a training set $T = \{(\mathcal{D}_i, t_i)\}_{i=1,..,N}$, the SOM-SD model can be used in a supervised mode as follows. Let us consider the data structure $\mathcal{D}_i$ with desired value $t_i$, the input vector for each node of this data structure is extended by adding a new field containing the information about the target:

$$\boldsymbol{v} = \left[\boldsymbol{u}_v\ \mathcal{D}_{ch_1[v]}\ \mathcal{D}_{ch_2[v]}\ \cdots\ \mathcal{D}_{ch_c[v]}\ \boldsymbol{t}_i\right],$$

where $\boldsymbol{t}_i \in I\!R^r$ is a real valued vector representing the target $t_i$. In this way, the information associated to data structures in the same target class $\tilde{t}$, will have identical last field $\tilde{\boldsymbol{t}}$. Thus, using the unsupervised algorithm on these new representations, the model will partition the neurons in subsets, each one associated with a different target class. We can trivially state that the sSOM-SD computes the a posteriory probability of a data structure given the target class and the training set, i.e., $p(\mathcal{D}_i|t_i, T)$.

After unsupervised training, the neurons in the map are labeled by the most frequent target associated to the data structures for which the neuron is the winner, i.e., given a neuron $n \in \overset{\circ}{\mathcal{M}}$, and the set $Data(n,t) = \{\mathcal{D}|(\mathcal{D}, t) \in T$ and $n$ wins the competition on $supersource(\mathcal{D})\}$, let $WinnerLabel(n) = \arg\max_{t \in C} |Data(n,t)|$. The label associated to each neuron is then used to classify a new data structure $\mathcal{D}$. Specifically, the nodes of $\mathcal{D}$ are presented to $\overset{\circ}{\mathcal{M}}$, having extended the input vector with the null vector (i.e., the target

**Training Algorithm for $\widehat{\mathcal{M}}$**

**input:** $T = \{(\mathcal{D}_i, t_i)\}_{i=1,\dots,N}$,
  $c$ (max. outdegree of DAGs in $T$),
  $\mathring{\mathcal{M}}$;
**begin**
call the unsupervised algorithm on
  the extended data;
    /* Labeling of Nodes */
**foreach** $n \in \mathring{\mathcal{M}}$;
  $label(n) \leftarrow WinnerLabel(n)$;
**end**

**Classification Algorithm for $\widehat{\mathcal{M}}$**

**input:** $\mathcal{D}$;
**begin**
`L` $\leftarrow$ an inverted topological ordered
  list for vert($\mathcal{D}$);
**for** $v \leftarrow$`first(L)` **to** `last(L)`**do**
  $\mathcal{D}_v \leftarrow \mathring{\mathcal{M}}\left(\left[\boldsymbol{u}_v\ \mathcal{D}_{ch_1[v]}\ \mathcal{D}_{ch_2[v]}\ \cdots \right.\right.$
    $\left.\left.\cdots\ \mathcal{D}_{ch_c[v]}\ \mathbf{0}\right]\right)$;
$w \leftarrow winner(supersource(\mathcal{D}))$;
**return** $label(w)$;
**end**

**Fig. 3.** The training and classification algorithms for the supervised mode.

value is unknown), and the label associated with the winning neuron for the supersource of $\mathcal{D}$ is returned. The training and classification algorithms for the supervised mode are described in Fig. 3.

## 4 Experiments

In this section we compare the performance of the uSOM-SD vs. the sSOM-SD model on a classification task with 12 classes. The basic idea is to "measure" the difficulty of the problem via the uSOM-SD model (without using the target information) and then to compare the results with the sSOM-SD model (using the target information).

The data set was taken from the *Policemen Benchmark* [5] which we extended by merging patterns from three different domains (houses, ships, policemen) to increase the complexity of the learning problem, and the number of possible instances (classes). In essence, the *Policemen Benchmark* [5] is a set of labeled DOAGs extracted from images produced by means of a context free attributed plex grammar. Structures are extracted by employing a simple scan line algorithm where images are scanned from bottom to top, and from left to right. The first colored object found constitutes the root node. Objects directly connected form the offsprings. The algorithm proceeds recursively until all objects in an image are considered. Each node created receives a two dimensional label stating the $\{x, y\}$ coordinate of the center of gravity of the represented object.

As a result, the dataset provided graphs with the following properties: Policemen have a maximum outdegree of 3, Houses have 5, and Ships have 6; the minimum (maximum) depth for Policemen is 4 (5), for Houses is 2 (3), for Ships is 1 (2); the minimum (maximum) number of nodes for Policemen is 9 (11), Houses 4 (7), and Ships 3 (13). Hence, policemen patterns produce

Class 'a': policemen with rised left arm ×
Class 'b': policemen with lowered left arm ○
Class 'c': ships with 2 masts ✱
Class 'd': ships with all 3 masts ●
Class 'e': houses without windows ◇
Class 'f': houses with 1 window at L-L corner ▲
Class 'g': houses with 1 window at U-L corner ▼
Class 'h': houses with 1 window at U-R corner □
Class 'i': houses with 2 windows at L-L and U-L corners ▽
Class 'j': houses with 2 windows at L-L and U-R corners +
Class 'k': houses with 2 windows at U-L and U-R corners △
Class 'l': houses with all 3 windows ■

**Fig. 4.** Definition of the 12 classes and their symbols as used later in Fig. 5. L-L means "lower left", U-L "upper left", and U-R "upper right".

deep narrow graphs, ships and houses have a flat and wide data structure. Some graph structures produced by the ships and houses are identical in structure whereas there is no graph structure derived from policemen images that is contained in the domain houses or ships. Thus, some patterns can be distinguished only through features encoded in the labels. For example, when considering policemen, the location of the arm is not encoded in the graph structure, but in the data label, while the graph structure is not affected.

These three domains were combined to form one training set so that the maximum outdegree of all nodes in the training set is 6. As a result (assuming $q = 2$) every input vector $\boldsymbol{v}$ is of dimension $2 + 2 \times 6$. The training (test) set contained $3,750$ $(3,750)$ graphs with a total of $29,808$ $(29,984)$ nodes.

For the supervised task, 12 classes were defined as shown in Fig. 4. The number of examples for each class both in the training and test sets were as in the following table:

| | 'a' | 'b' | 'c' | 'd' | 'e' | 'f' | 'g' | 'h' | 'i' | 'j' | 'k' | 'l' |
|---|---|---|---|---|---|---|---|---|---|---|---|---|
| Training Set | 645 | 605 | 937 | 313 | 28 | 59 | 58 | 172 | 53 | 170 | 188 | 522 |
| Test Set | 670 | 580 | 944 | 306 | 21 | 59 | 64 | 195 | 71 | 173 | 194 | 473 |

Experiments were designed to demonstrate the typical behaviour of sSOM-SD networks in comparison with that of an uSOM-SD network. Networks were trained on the same data set where class memberships formed an additional input when using a supervised training regime – class information was omitted for the unsupervised method. However, to allow the evaluation of the network performance and a comparison with sSOM-SD, we labeled the neurons by a symbolic class label after training the network. All results described in this section were obtained by SOM-SD networks with hexagonal topology and the Gaussian neighborhood function. Networks ranging from a size of $102 \times 118$ down to a size $2 \times 5$ have been trained. Table 1 displays the performance of these networks. It shows that sSOM-SD network perform with 100% correct classification rate down to a very small network size with just 200 neurons whereas the uSOM-SD performs hardly at a level of 90% on a network with more than 60 times as many neurons.

In Fig. 5 we display where root nodes are mapped onto the network. It is observed that an uSOM-SD maps root nodes close to the upper right corner of a network. This is mainly due to the fact that the NULL identifier for missing offsprings for leaf nodes has been chosen to be $-1$ and hence, are values that are nearest to the origin of the map. Thus, leaf nodes tend to be mapped closer to the lower left end of a map. For sSOM-SD networks this is different. Here, the class information has a strong influence on where a root node is mapped. As it can been observed from the plot, some classes are mapped onto a single neuron, whereas others form relatively large clusters. This is due to the fact that some classes are represented through a larger number of different instances than others. For example, there are many different instances of ships (e.g. many ways of arranging the masts, sails, flags) than there are instances for houses which often look very similar. Concerning intermediate and leaf nodes, these activate different regions of the maps (not shown in Fig. 5).

We also observed how a sSOM-SD network represents data on a much smaller output scale. With smaller networks, the training algorithm gradually forces data to be mapped onto smaller clusters until almost every class is mapped onto a single neuron.

## 5 Discussion and Conclusions

In this paper, we demonstrated a new way of addressing supervised learning problem involving structured objects through an extension to the unsupervised SOM-SD algorithm. Specifically, we showed how class memberships of structured data can be incorporated in the learning process to yield improved network performances.

The methodology described in this paper has distinctive advantages over many other supervised learning methods. The algorithm allows to process data with incomplete or missing class labels without further modification. Moreover, it has been demonstrated that a sSOM-SD is able to efficiently represent structured information in a highly reduced manner by mapping it

| Size (w x h = # nodes) | sSOM-SD (train-test) | uSOM-SD (train-test) |
|---|---|---|
| 102 x 118 = 12036 | 100% - 100% | 91.547% - 89.227% |
| 80 x 90 = 7200 | 100% - 100% | 88.347% - 87.36% |
| 60 x 70 = 4200 | 100% - 100% | 86.853% - 85.227% |
| 40 x 50 = 2000 | 100% - 100% | 85.305% - 84.107% |
| 20 x 30 = 600 | 100% - 100% | 77.547% - 74.773% |
| 10 x 20 = 200 | 100% - 100% | 63.067% - 62.933% |
| 5 x 10 = 50 | 93.173% - 93.12% | 54.933% - 54.72% |
| 2 x 5 = 10 | 80.587% - 79.28% | 47.387% - 46.693% |

**Table 1.** Comparison between sSOM-SD and uSOM-SD on the supervised task.

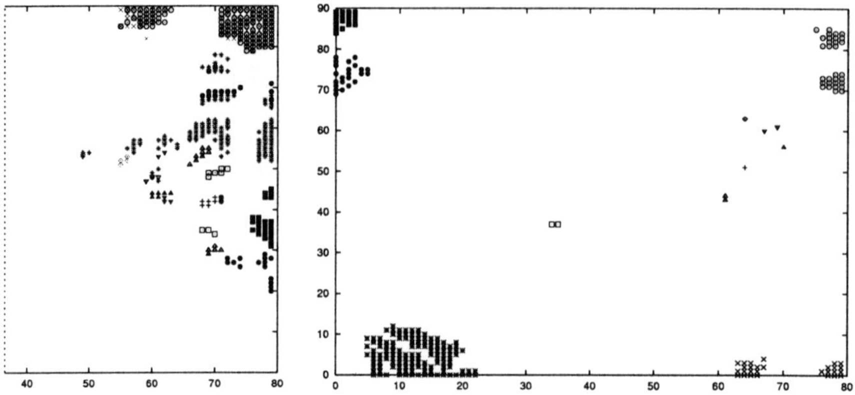


*unsupervised* : *root nodes* *supervised* : *root nodes*

**Fig. 5.** Unsupervised and supervised maps of dimension 80 × 90. Only root nodes are shown; symbols refer to the class assigned to the nodes. For the unsupervised map, only the portion of the map activated by root nodes is shown.

onto a relatively small, often two dimensional output space. It has become evident that through incorporating class information into the learning process, we are able to improve the network performance considerably over the unsupervised counterpart.

## References


1. A.M. Bianucci, A. Micheli, A. Sperduti, and A. Starita. Analysis of the internal representations developed by neural networks for structures applied to qsar studies of benzodiazepines. *Journal of Chemical Information and Computer Sciences*, 41(1):202–218, 2001.
2. Horst Bunke and Alberto Sanfeliu, editors. *Syntactic and Structural Pattern Recognition : Theory and Applications.* Series in computer science ; v. 7. Singapore ; New Jersey : World Scientific, c1990. ISBN-9971505525.
3. Paolo Frasconi, Marco Gori, and Alessandro Sperduti. A general framework for adaptive processing in data structures. In *IEEE Trans on Neural Networks*, volume Vol 9, pages 768–785, 1998.
4. C. Goller. *A Connectionist Approach for Learning Search-Control Heuristics for Automated Deduction Systems.* PhD thesis, Technical University Munich, Computer Science, 1997.
5. Markus Hagenbuchner and Ah Chung Tsoi. The traffic policeman benchmark. In Michel Verleysen, editor, *European Symposium on Artificial Neural Networks*, ISBN 2-9600049-9-X, pages 63–68. D-Facto, April 1999.
6. Teuvo Kohonen. *Self-Organisation and Associative Memory.* Springer, 3rd edition, 1990.
7. A. Sperduti and A. Starita. Supervised neural networks for classification of structures. *IEEE Trans Neural Networks*, Vol. 8(No. 3):714–735, 1997.

# Human Gait Analysis using SOM*

H.M. Lakany

University of Essex, Department of Computer Science
Wivenhoe Park, Colchester CO4 3SQ,
United Kingdom

**Abstract**

This paper addresses the problem of analysing kinematic gait data which has been collected using 3D motion capture equipment that uses IR–reflective markers placed on the joints of the lower extremities of the subject's body. The data comprises motion trajectories of the different joints and it included normal and pathological subjects. The analysis of motion trajectories is done by combining the wavelet transform for feature extraction and a Kohonen self–organising map (SOM) for classification of walking patterns. Rules are then extracted from the SOM after self–organisation to determine the salient features characterising each cluster as well as differentiating it from others. It is shown and experimentally verified that salient features do exist within the internal structure of the kinematic data from which diagnostic signatures are elicited. Existence of such features could be used by clinicians in the orthopaedic field.

## 1 Introduction and Problem Domain

Normal walking depends on a continual interchange between mobility and stability. Any abnormality restricting the normal free mobility of a joint or altering either the timing or intensity of muscle action creates an abnormal gait. Abnormal gait may be due to an injury, disease, pain or problems of motor control. The subject's ability to compensate for the abnormality determines the amount of functionality retained. However, when these compensations introduce penalties in joint strain, muscle overuse, lack of normal muscle growth or soft tissue contracture, then clinical intervention becomes a necessity. In determining appropriate intervention, gait analysis is used to identify gait defects.

Gait analysis instrumentation based on infra–red (IR) cameras and computer–aided systems recording the 3D positions of markers attached to the subject has been used to record gait cycles of the patient and produce patterns and

*The experiments of this paper were carried at the Division of Informatics, University of Edinburgh in Scotland.

plots for clinicians to assess and hence diagnose. To measure kinematics, the subject is filmed while walking using cameras placed on both sides and in front of the walkway such that each marker is seen by at least two cameras at any instant. For kinetics measurements, the subject activates force plates embedded in the walkway. The output of the cameras and the force plates is fed to the computer which estimates the 3D coordinates of the markers and the moments and torques applied on each joint. Trajectories of different gait parameters are plotted – see for example figure 1 showing a summary graph sheet of kinematic measurements of the joints of the left side of an 11 year old child suffering from diplegia. The pathological traces are plotted together with normal ones to show the variations resulting from the impairment and at the upper left corner of the figure, different gait parameters, e.g. cadence, velocity, etc. are estimated for the patient. These graphs are then assessed by the specialists. The number of graphs plotted for gait analysis is immense and extracting useful information from such graphs to accomplish the gait analysis objectives mentioned earlier is a demanding and challenging task. This is due to various reasons that include the complexity of the walking process itself, variability of patients' response to treatment, uncertainty in data quality and the difficulty in distinguishing between primary abnormalities and compensations in the gait pattern.

This research is motivated by the capabilities of humans to perceive gait from reduced spatio–temporal trajectories, and attempts to give machines a similar gait perception capability. It builds upon Murray's [8] ideas of setting standards for normal walking and investigates the existence of a generic feature vector based on kinematic data for both normal and pathological subjects. In addition, it looks into relationships and correlations among features existing in the generic vector. The ultimate objective of this work is to provide clinicians with tools for data reduction, feature extraction and gait profile understanding.

# 2 A Gait Analysis System

A method to extract salient features from hip, knee and ankle angles to characterise normal and pathological cases is proposed. The algorithm is based on transforming the spatio–temporal trajectories of the joints using the continuous wavelet transform (CWT) [7] and then training a Kohonen self–organising map (SOM) as a vector quantiser to cluster the input feature vectors according to their internal structures. Based on the self–organisation of the patterns, rules are extracted for uniquely representing and characterising each cluster of the SOM. Features that contribute in characterising a cluster and in differentiating it from other clusters are selected as *the* significant features of that particular cluster. Results obtained by this research verify the existence of diagnostic signatures that comprise generic and specific features distinguishing between normal and pathological cases, different pathologies and even detecting different sub–groups within a certain pathology. These diagnostic

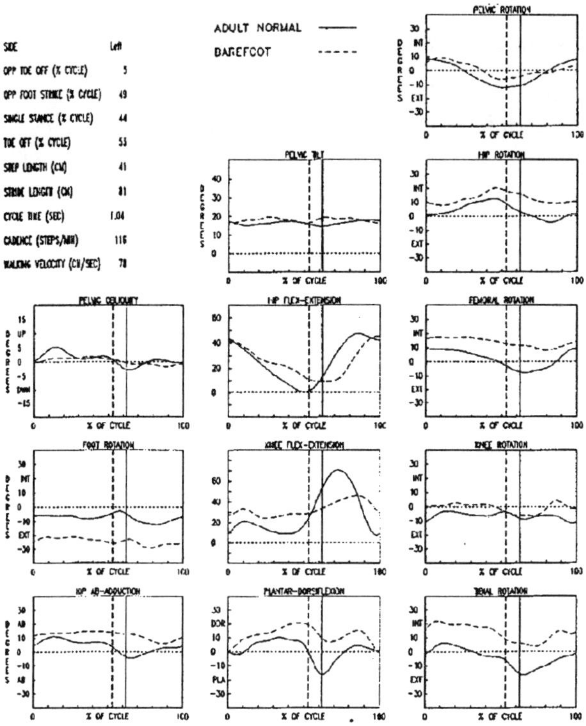


Figure 1: A sample output of different plots of kinematic analysis for an 11 year old child suffering from spastic diplegia. [10]

signatures are automatically extracted.

The algorithm proposed, designed and implemented in this research for gait analysis and recognition comprises the following steps:

- Analysis: this involves spatio–temporal analysis of the trajectories. The type of analysis considered is greatly affected by the application under investigation. Different examples of applications include: recognition of activities [9, 6, 12], of individuals [2, 4], of different injuries [3], of generic gait signatures or gait abnormalities [5].

  Our interest here focuses on recognising different gait signatures for normal and pathological individuals. The wavelet transform is implemented to extract spatio–temporal features in kinematic data.

- Recognition: using the spatio–temporal feature vectors of different subjects, a vector quantiser, namely a Kohonen self–organising map (SOM), is used to classify the different walking profiles into clusters based on the internal structure of the feature vectors. A set of experiments is carried out splitting feature vectors into low and high–scale feature vectors and another set splitting them into low, medium and high–scale features.

- Feature selection: in this step salient features from the original feature vectors of the different classes are selected based on statistical measures of each cluster. These features are considered to be significant and characteristic for the cluster. The mean is used as the statistical measure for choosing the salient features and the standard deviation sets the limits of variation for the mean values estimated in this step assuming Gaussian distribution.

# 3 Experiments and Results

The experiments involved eighty–nine normal subjects – males and females – of 4 to 71 years of age and thirty–two pathological cases including polio, cerebral palsy and spina bifida. In this work, the focus is concentrated on sagittal angles of the hip and knee joints from which the aim is to extract salient features. An example of a plot of the normalised sagittal angles of a normal subject is illustrated in figure 2.

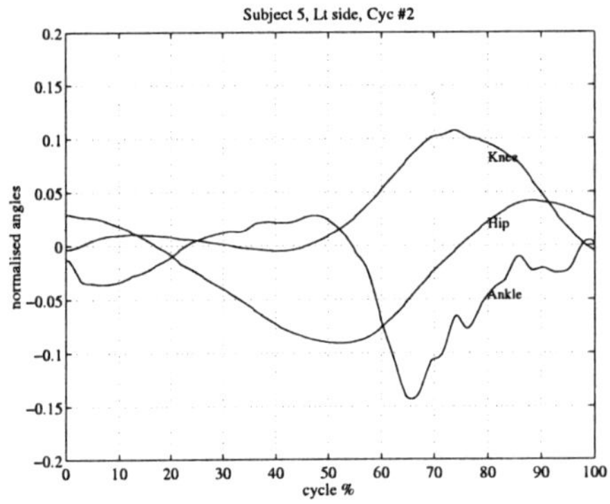


Figure 2: Normalised sagittal angles of the Hip, Knee and Ankle joints of a normal subject.

## 3.1 Spatio–temporal Analysis

The trajectories of the flexion/extension angle of the joints are transformed using the continuous one dimensional wavelet transform (CWT).The wavelet transform is an effective tool for analysis of non–stationary biomedical signals [1]. It provides spectral decomposition of the signal onto a set of basis functions, wavelets, in which representation of temporal dynamics is retained [7]. The continuous wavelet transform (CWT) for a 1D signal $x(t)$ is defined as the inner product of the signal function $x(t)$ and the wavelet family. This is expressed in the following equation:

$$W_x^{\psi}(s,\tau) = \frac{1}{\sqrt{|s|}} \int x(t)\psi^*(\frac{t-\tau}{s})dt \tag{1}$$

where

$\psi^*(t)$ is the conjugate of the transforming function, also called the mother wavelet. $\tau$ is the translation parameter, i.e. representing the shift in the time domain. $s(>0)$ is the scale parameter, and is defined as $1/frequency$.

Figure 3 shows typical scalograms of the Morlet wavelet transform (modulus and phase) of the sagittal angles of the hip (top–most), knee (middle), and the ankle (lower–most) joints of normal (left) and pathological (right) subjects. The ordinate represents the scale which increases downwards and the abscissa is the time; here, time is measured as the percentage of the gait cycle increasing rightwards. Bright areas represent high scalogram values and dark areas low values as indicated by the colour bar.

We focus in our analysis on extracting features from the modulus of the CWT. Phase was not considered as it only adds noise to the signal and carries no useful information on its own. However, in future work, it might be worthwhile investigating the effect of considering *phase differences* between normals and pathological subjects rather than phase only. The stance phase constitutes ≈ 60% of the total cycle, hence we choose to split the scalogram vertically at 60% of gait cycle dividing it into stance phase and swing phase. Furthermore, we look for features in different scale levels, either splitting the scalogram horizontally into two sections (levels) high–scale and low–scale separately, or for some experiments, scale was split into three levels; high, medium and low scale feature vectors. The splitting of the scalogram is shown on the figures by white dotted lines to facilitate the reader's examining them. We therefore handle four regions for each scalogram, shown in figure 3: (1) stance phase, high–scale, (2) swing phase, high–scale, (3) swing phase, low–scale and (4) stance phase, low–scale.

## 3.2 Feature extraction and classification

The results of two sets of experiments are shown in this paper; experiments to examine the features that differentiate normal from pathological cases and experiments covering pathological cases only: to investigate the existence of diagnostic signatures for different pathologies.

The experiments are based only on the sagittal angles, incorporating one or two joints combined together to investigate which joint or combination of joints includes more salient features, and considering one or both sides of the subjects, especially for some pathological cases where there is severe asymmetry. Features are used at different levels, mostly low scale and high scale features. However, for some experiments features are further divided into low, medium and high scale. The two main considerations limiting whether

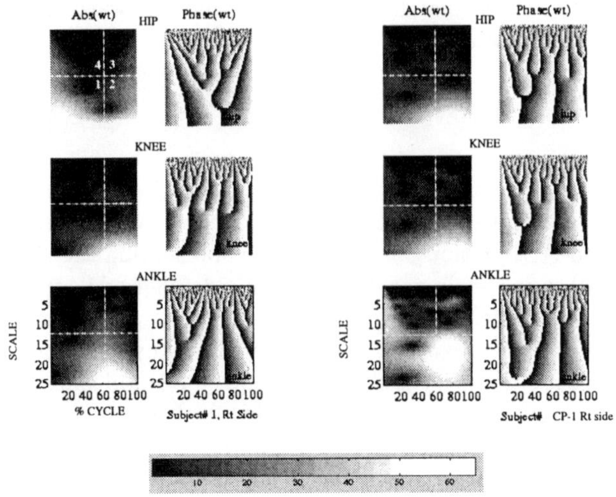


Figure 3: CWT (scalogram) of sagittal angles (hip, knee, ankle joints) of the right side of a normal subject (left) and a pathological subject (right). White dotted lines splits the scalograms into regions for ease of analysis – see text for explanation.

the combination of more than one joint, more than one level or both sides was possible are (1) the huge size of the resulting feature vector to be considered (e.g. 640 features to include both sides and to combine two joints) and (2) the computational time needed for training the SOMs (e.g. three weeks to train a $10 \times 15$ map with 70 input patterns, 160 features each on a Sun Ultra5).

Results are illustrated by two plots, one showing the map after applying UMM algorithm for visualisation [11] to show the different clusters (shown in dark colour) separated by the borders (in bright colour). The vertical axis in the UMM plot represents the average distance between each node and its neighbourhood nodes. The threshold set to discriminate the different clusters depends on the UMM values for every experiment and a quiver and a contour diagram of the map showing the direction of attraction for the different regions.

Figure 4 shows the map trained with high scale CWT of hip joint feature vectors after applying UMM to visualise the different classes. The SOM successfully discriminates normal from pathological subjects. Furthermore, it recognises the two sets of age ranges separately as well as the different pathological cases based on the global information (high scale) of the hip joint. The quiver diagram shows the direction of attraction for each class of the SOM.

Figure 5 shows the resulting SOM/UMM for pathological cases only to investigate differences between different pathologies. This experiment involves symmetrical diplegias, left and right asymmetrical diplegias and left and right

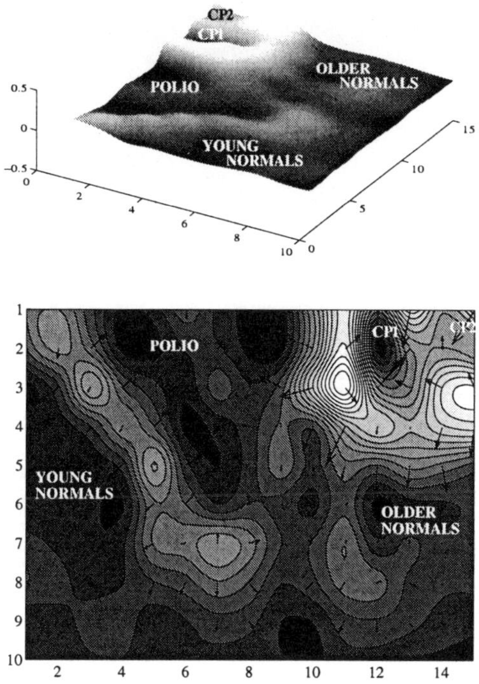


Figure 4: SOM/UMM (above) trained using HIGH scale feature vectors of the sagittal angles of the hip joint of both normal and pathological subjects and a contour and quiver diagram (below) showing the different classes and clustering attraction regions.

hemiplegias. The map was trained with the high scale features of the right and left hip joints combined together. The map self–organised as follows: right asymmetrical diplegias and right hemiplegias clustered together, and so did the left asymmetrical diplegias and left hemiplegias while the symmetrical diplegias grouped between these clusters pulled by their attraction regions in different directions. This result was discussed with the clinicians, who stated that some of the asymmetric cases are diplegics with a superimposed hemiplegia and often clinicians face a difficulty in accurately diagnosing these cases.

## 3.3 Interpretation and rule extraction from SOMs

After training the SOM and applying the UMM, all vectors/cases that lie in the same cluster are considered to have a strong similarity, i.e. have similar internal structure. Single neurons in deep valleys indicate possible outliers.

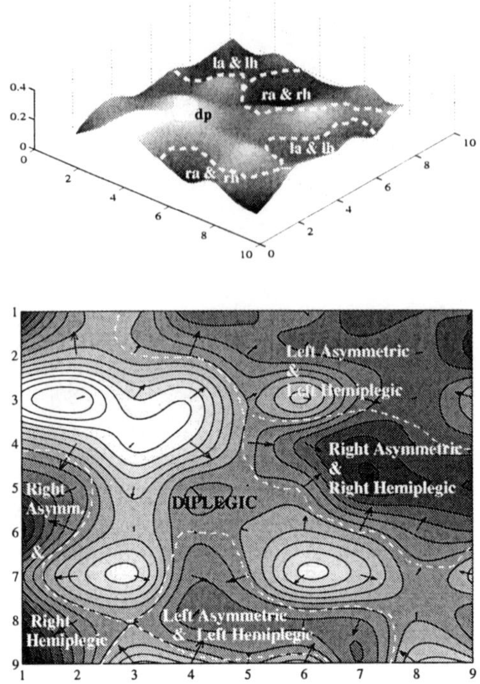


Figure 5: SOM/UMM (above) trained using HIGH scale features of the sagittal angles of combining right and left HIP joints of pathological subjects and a contour and quiver diagram (below).

Hills separate different clusters. it is required to find generic salient features or prevailing conditions that would characterise the different clusters. These conditions should not only characterise the cluster but also differentiate it from the others. The SIG algorithm suggested by [11] is implemented to extract rules. The rules are symbolic and determine the variance of the most significant features for each cluster.

Due to shortage of space, it is difficult to show the results of this module of the algorithm

# 4 Discussion and Conclusion

A continuous wavelet transform was used to convey the spatio–temporal characteristics of motion trajectories. It proved to be convenient because it maintains the temporal dynamics of the motion signal. Furthermore, it analyses the signal on an adaptable time–scale. The scalograms produced were split into different levels – high, medium and low scale from which feature vectors were selected.

A Kohonen SOM was used as the vector quantiser required for classification for two main reasons, visualisation and rule extraction.

Different experiments were carried out to investigate the walking signatures in normal and pathological subjects using different levels of scale. Maps trained with high scale feature vectors (i.e. low frequency) self–organised discriminated between normal and pathological subjects, between normal subjects in different age ranges and between different pathologies. They also succeeded in detecting different gender, symmetry of both sides and in identifying different classifications within CP.

The classifier proved promising in detecting cases of CP that were initially misclassified by experts. Such cases are considered complicated and hard to diagnose.

SOMs showed that it was difficult to find consistency within the internal structure of the low scale features (i.e. high frequency) due to high signal to noise ratios.

Rules are extracted after training the SOMs to: (1) select features that are *specific* and *characteristic* for each cluster and (2) set conditions for each selected feature by making use of the distribution properties of the features of a cluster. The rules extracted are simple yet they provide clinical gait analysis with a quantifiable measure for accurate comparison between different cases. The resulting set of salient features is a subset of the spatio–temporal features extracted from CWT. Reduction ratios varied from 15% to 80%.

In conclusion, the algorithm exhibited its ability to detect and distinguish between normal and pathological subjects, different pathologies and different categories within a specific pathology (CP).

The procedure correctly classified some difficult pathological cases which were initially misclassified by specialists. This demonstrated its potential as an alternative method of gait analysis.

The procedure results in a set of quantifiable rules relating to salient features extracted – significant and characteristic – which provide a degree of accuracy and a measure for comparison.

# References

[1] M. Akay. Wavelets in biomedical engineering. *Annals of Biomedical Engineering*, 23(5):531–542, 1995.

[2] A. Birbilis. Recognising walking people. Master's thesis, Dept. of Artificial Intelligence, Univ. of Edinburgh, Scotland, 1995.

[3] M. Köhle and D. Merkl. Identification of gait patterns with self–organizing maps based on ground reaction forces. In *Proceedings of the European Symposium on Artificial Neural Networks (ESANN'96), Bruges, Belgium.*, 1996.

[4] H. Lakany and G. Hayes. An algorithm for recognising walkers. In *First International Conference on Audio and Video based Biometric Person Authentication (AVBPA) - Crans–Montana, Switzerland*, Lecture Notes in Computer Science. Springer Verlag, Mar. 1997.

[5] H. M. Lakany and G. Hayes. A generic kinematic pattern for human walking. In *Proceedings of World MultiConference on Systemics, Cybernetics and Informatics (SCI'98), Orlando, Florida.*, 1998.

[6] D. Meyer, J. Denzler, and H. Niemann. Model based extraction of articulated objects in image sequences for gait analysis. In *Fourth International Confrerence on Image processing (ICIP'97), Santa Barbara, USA.*, volume 3, pages 78–81, 1997.

[7] Y. Meyer. *Wavelets.* Springer Verlag, Berlin, 1989.

[8] M.-P. Murray, A. Drought, and R. C. Kory. Walking patterns of normal men. *The Journal of Bone and Joint Surgery*, 46–A(2):335–360, Mar. 1964.

[9] R. Polana and R. Nelson. Low level recognition of human motion. In *Proceedings of the 1994 IEEE Workshop on Motion of Non–rigid and articulated objects, Nov. 11–12, 1994, Austin , Texas*, pages 77–82, Los Alamitos, California, USA, 1994. IEEE Computer Society Press.

[10] J. Rose and J. G. Gamble. *Human Walking.* Williams & Wilkins, 2 edition, 1994.

[11] A. Ultsch and H. Siemon. Kohonen's self organizing feature maps for exploratory data analysis. In *Proceedings of International Neural Network Conference*, pages 305–308, Dordrecht, Netherlands, 1990. Kluwer.

[12] Y. Yacoob and M. J. Black. Parameterized modeling and recognition of activities. In *Proceedings of the International conference on computer vision ICCV'98, Bombay 1998.*, pages 120–127, 1998.

# Exploring Financial Crises Data with Self-Organizing Maps (SOM).

Ismael Arciniegas[a], Betty Daniel[a], and Mark J Embrechts[b]
[a]Unversity At Albany, [b]Rensselaer Polytechnic Institute
Albany, Troy, USA

Abstract

The recent Asian financial crises showed the need for a better understanding of the determinants of financial crises' aftermath. We used the SOM to explore the multidimensional properties of a data set of recent speculative attacks in search of potential associations between economic descriptors, and the size and duration of speculative attack's real effects. We found that speculative attack's real effects were associated with: the health of the banking system; the law origin; the level of economic development; the terms of trade; and the financial development. Interestingly, we did not find evidence of the beneficial roles played by the IMF and the high interest rate policies in solving financial crises[1].

## 1 Introduction

A speculative attack is an economic event where a country's fixed exchange rate is under pressure by speculators. Often the monetary authorities try to defend the currency at the cost of heavy reserve losses and high interest rates. Nasty recessions following speculative attacks occurred in the recent Asian crises as well as in Mexico 1995, Chile 1982, among others. Empirical research on speculative attacks has mostly focused on the factors associated with the occurrence of the event however; research on the aftermath is scarce. The aim of this paper is to use the power of the SOM to better understand the aftermath that follows a speculative attack. We used the tool of the SOM to explore the empirical evidence in search of relevant associations between descriptors of the external, legal and financial environment, and the response of the economic activity to an attack.

Exploratory data analysis in empirical economics is important in detecting stylized facts and in model formulation. Typically in speculative attacks' economics exploratory data analysis is done by univariate techniques because they do not impose parametric structures and require fewer assumptions than other statistical methods. A speculative attack is an economic event whose impact on the economic activity depends on complex interactions between traditional economic

---

[1] We benefited from very helpful comments from Guido Deboeck. Of course remaining errors are just ours.

fundamentals and several other descriptors of the country's financial, legal, and political structure. Traditional univariate approaches may not be adequate for the exploration of speculative attacks' real effects given the multidimensional non-linear relations among its descriptors. The SOM is a better technique to explore speculative attacks' data because it provides an efficient non-linear description of the multidimensional data distribution, allowing good data reduction.

In this paper we collected data on 59 speculative attacks occurred in small open economies between 1970 and 1997. We selected a list of 11 variables that may describe the crisis' aftermath based on the review of the economic literature and the use of descriptive statistics. After analyzing the component planes of the feature map, we found that speculative attack's real effects were associated with: the health of the banking system; the law origin; the level of economic development; the terms of trade; and the financial development. Interestingly, we did not find evidence of the beneficial roles played by the IMF assistance and the high interest rate policies in solving modern financial crises.

# 2 Some Economics

A speculative attack on the domestic currency may end up in devaluation or not as the central bank may be able to use monetary policy to support the exchange rate. After a successful attack the exchange rates are floated and not just devalued, restricting severely the government's capabilities to adopt policies consistent with the alignment in the exchange rate. Thus, a speculative attack is a more complex economic event than devaluation. Theoretically, shocks to the exchange rate may cause increases in interest rates, which increases the opportunity cost of consumption and therefore reduces output. Empirically, World Economic Outlook (WEO) estimated the average cumulative output loss per speculative attack in 1.3 % (see [1]).

The economic literature suggests many ways, in which the real effects caused by speculative attacks may be amplified by specific characteristics of the economic environment at the onset of the crisis or by government policies in the aftermath. The banking system may amplify output effects as exchange rate shocks can also trigger credit crunches. Poor creditor's protection may cause capital outflows in the aftermath of the crisis that would reduce the current account deficit causing further contractions in aggregate demand. The effect of the demand reduction on output depends on the ability to export the excess of output, which depends on the country's terms of trade. The recovery from the crisis would be speed up by the return of the foreign capital, which depends on the behavior of foreign interest rates. The IMF's rescue packages may also boost the investor's confidence on the economy in distress, but at the same time those packages may enforce contractionary policies, which may exacerbate the recessions. Other variables that may influence the country's recovery from a speculative attack are budget deficit; capital controls; stock market liquidity; and Deposit Insurance Schemes (DIS).

# 3 The Database

Historical evidence suggests that speculative attacks' aftermath can be short-lived events; hence, in the construction of our database we used the highest available frequency of the data (monthly). Following [2], we identified an attack when the reading of an index of speculative pressure *I* was above three standard deviations above the mean (see Table 1). This index considers not only changes in the exchange rate but also, in the foreign reserves.

$$I = \left(\frac{\Delta_e}{e}\right) - \left(\frac{\sigma_e}{\sigma_R}\right) * \left(\frac{\Delta_R}{R}\right)$$

$\Delta_e/e$ = Rate of change of the exchange rate.

$\Delta_R/R$= Rate of change of the reserves.

$\sigma_e$= Standard deviation of the rate of change in the exchange rate.

$\sigma_R$= Standard deviation of the rate of change in the reserves.

Previous the SOM exploration, we selected among all the "potentially influential" economic descriptors in the literature, the ones with sounder empirical and theoretical support. Then based on data availability we measured for each episode the variables' empirical proxies. Our data set included for each speculative attack's aftermath measures of: real effects' size; real effects' duration; financial development; size of financial intermediation; legal origin; reserve losses; depreciation of real exchange rate; health of the banking system; budget deficit; terms of trade; foreign interest rate; level of economic development; and the IMF rescue assistance. The first two indicators measured the response of the real economic activity to the speculative attack. We measured more than one proxy for several variables (e.g. real effects, financial intermediation, and terms of trade). The data source was the International Financial Statistics database of the IMF.

# 4 Method, Results, and Discussion

The SOM is a powerful tool in data exploration; however, a preprocessing of the raw data may facilitate the work of the SOM. We preprocessed the raw data in two ways: standardization and data transformation. The first one was justified given the wide variation in range among the descriptors used as inputs in the construction of the feature map. We scaled all the data by variance in order to limit the effect of the range's wide in the SOM. Second, a Kernel estimation of each descriptor's density distribution indicated that some of them had wide tails (e.g. financial development, and real effect's size). The descriptors with these characteristics were transformed with a logarithmic function to obtain more uniform distributions. We computed several feature maps using different training parameters and selected the ones that were better organized. We tested the robustness of the "best" map using alternative empirical proxies for the descriptors. The plates showed were computed using the training parameters that led to the most organized and robust feature map. The plates presented here are edited version of the ones computed with Viscovery SoMine.

## 4.1 Response Variables

The real effect's size and duration measured the economy's response to the speculative attack. The real effect' size was measured as the accumulated output loss between the point of the attack and the point, in which output reached its trend (level). Output was measured by the seasonally adjusted industrial production and the trend was computed using the Hodrick-Prescott filter. Duration was defined as the counted number of months between the onset of the attack and the point, in which output reached its trend[2]. Map 1 and Map 2 present the component plates for real effect's size and duration. We noted that both component maps were organized so that attack episodes with similar responses were allocated together. The arrow indicates an ordered transition from strong to mild real effects[3]. Based on the *"topology preserving"* property of the feature map, these plates suggest that similar episodes were characterized by similar response variables. Our measure of similarity among episodes was the euclidean distance. A comparison between map 1 and 2 show that they have *similar* direction of organization, with stronger effects in the left side and milder effects in the right side, which implies that both indicators are closely correlated.

## 4.2 Legal Origin

Comparative Law scholars classify the countries' legal systems in four legal families: French; English; German; and Scandinavian. Researchers had shown that the level of investor protection is statistically different among countries with different legal origins [3]. They showed that English Law countries had the best investor's rights protection among the legal families. We included legal origin as a proxy of investor's rights as was not possible to build indicators of investor's rights for all the attack's episodes. The component plane for law origin (Map 3) suggests that countries with French legal origin are more prone to speculative attacks than countries belonging to other legal families. Interestingly, the episodes associated with English law were grouped within the same patch, suggesting that the aftermath's descriptors of countries with an English Legal origin were similar. Comparing Map 1, Map 2, and Map 3 together, we see that the English legal origin

---

[2] We defined a speculative attack, as contractionay if within the 18-*month* window immediately after the speculative attack, there was a detected negative accumulated output gap between the potential and the actual output. Otherwise, we assigned a *zero* to the output losses associated with that attack.

[3] There were few episodes, in which the color transition was not smooth (e.g. Spain 92 and Denmark 83), suggesting that there were potentially important variables not included in the list of descriptors.

was associated with mild real effects, but the other legal origins were associated with mild, and strong effects. English legal origin differentiates itself from the other two legal origins in its association with the aftermath's size.

## 4.3 Twin Crisis

We identified a twin crisis when the beginning of a banking crisis was followed by a speculative attack within 48 months. In Map 4 we noted that the twin crisis descriptor was not randomly scattered on the map, instead it was grouped in three separated areas, suggesting some order in the map. Comparing Map 1 and 4, we see that almost all the episodes associated with strong real effects were also associated with a twin crisis descriptor. This finding indicates that episodes with banking crises at the onset of the attack had in general stronger real effects. For instance, Finland 91 and Finland 82 are located in opposite sides of the real effect's size map. We remarked that in 1991 Finland was in a twin crisis. The dark patch on the left of Map 4 also suggests that the association between the twin crisis descriptor and the real effects depends on the values of the other descriptors.

## 4.4 Financial Development

We used as proxy for financial development the liquid liabilities of the financial system to GDP, which represents the size of the formal financial intermediary sector relative to the economic activity. Map 5 shows a topological order, in which the transition of the financial development is Southwest-Northeast; the transition in the response map is East-West, suggesting a non-linear association between these two descriptors. There is a big white group at the top of the map that includes the episodes with the lowest financial development, moving along the frontier of this group does not change the color transition on the duration component maps. On the other hand, there is a darker patch at the bottom of the map that includes the episodes with the highest financial development. Moving along this frontier we can see changes on the color transition in the real effects' size map for high response values. We noted that the strongest crises were out of the borders that define the patch for high financial development and those twin crises with higher financial development had relatively milder real effect's size.

## 4.5 IMF

Map 6 shows the component map for the IMF's assistance. We compared the IMF component map with the other descriptors component maps. An interesting remark is that the frontier (strong–mild) of the real effects size map did not coincide with the IMF-non IMF frontier. For IMF countries there was presence of both strong (left patch) and mild (right patch) real effects, suggesting that in general the IMF failed as much as succeeded dealing with financial crisis.

# 5 Final Remarks

For lack of space we do not present here the plates for level of development, budget deficit, terms of trade, foreign interest rate, financial intermediation efficiency, and speculative attack's severity[4]. We did find topological ordering in the OECD, the terms of trade, and the severity component maps, but not in the budget deficit, the foreign interest rate, and the financial intermediation efficiency.

We showed in the analysis above how the SOM can be used to explore speculative attacks' data. We uncovered interesting associations of attacks' aftermath with: the health of the banking system; the law origin; the level of economic development; the terms of trade; and the financial development. We did not find evidence of the beneficial roles played by the IMF's assistance and the high interest rate policies in solving modern financial crises.

## References

1. International Monetary Fund. World Economic Outlook 1998.
2. Kaminsky G, and Reinhart C. The Twin Crises: The Causes of Banking and Balance-of-Payments Problems. American Econ. Rev. 1999; 89:473-500.
3. La Porta R, Lopez-de-Silanes F, Schleifer A, and Vishny R. Law and Finance. J. Political Econ. 1998; 106-6:1113-1155.

**Table 1. Speculative Attacks Episodes***

| | | | | |
|---|---|---|---|---|
| Argentina (81) | Chile (72) | Finland (91) | Mexico (94) | Spain (92) |
| Argentina (86) | Chile (73) | Indonesia (78) | Norway (73) | Sweden (77) |
| Argentina (89) | Chile (74) | Indonesia (83) | Norway (78) | Sweden (81) |
| Argentine (90) | Chile (76) | Indonesia (86) | Norway (86) | Sweden (92) |
| Bolivia (82) | Chile (82) | Israel (74) | Norway (92) | Turkey (94) |
| Bolivia (83) | Chile (84) | Israel (77) | Peru (87) | Uruguay (82) |
| Bolivia (85) | Colombia (83) | Israel (83) | Philippines (83) | Venezuela (84) |
| Brazil (86) | Colombia (85) | Malaysia (75) | Philippines (84) | Venezuela (86) |
| Brazil (89) | Denmark (79) | Malaysia (97) | Philippines (86) | Venezuela (89) |
| Brazil (90) | Denmark (83) | Mexico (76) | Spain (76) | Venezuela (94) |
| Brazil (91) | Finland (73) | Mexico (82) | Spain (77) | Venezuela (95) |
| Chile (71) | Finland (82) | Mexico (83) | Spain (82) | |

* The numbers between parenthesis represent the year

[4] Level of development was defined by the membership or not to the OECD. Budget deficit was defined as the average fiscal deficit within the duration of the crisis. Terms of trade were defined as the average ratio of unit value of exports over imports within the crisis duration. The financial intermediation efficiency was measured as the ratio of deposit money bank domestic assets to deposit money bank domestic assets plus central bank domestic assets. Severity was measured as the unweighted average between ex-ante six-month loss in reserves and the ex-post six-month real exchange rate depreciation.

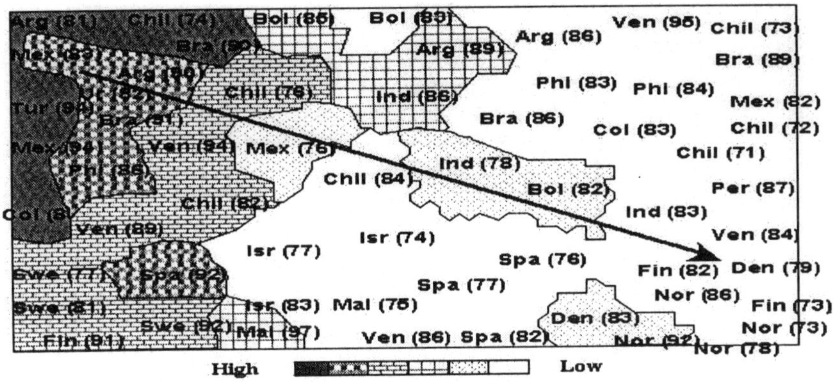


**Map 1. Real Effects' Size**

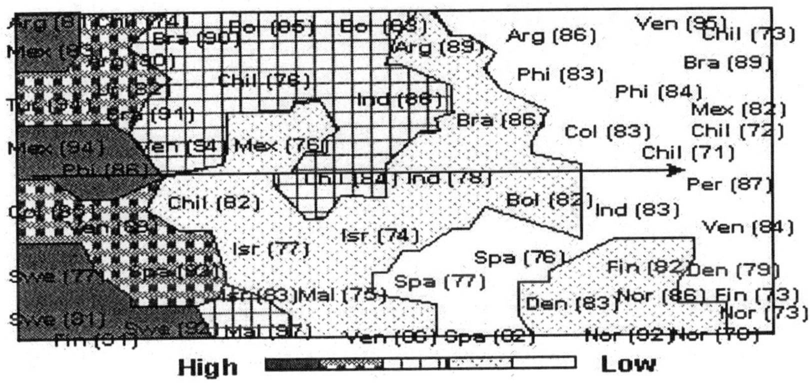


**Map2. Real Effect's Duration**

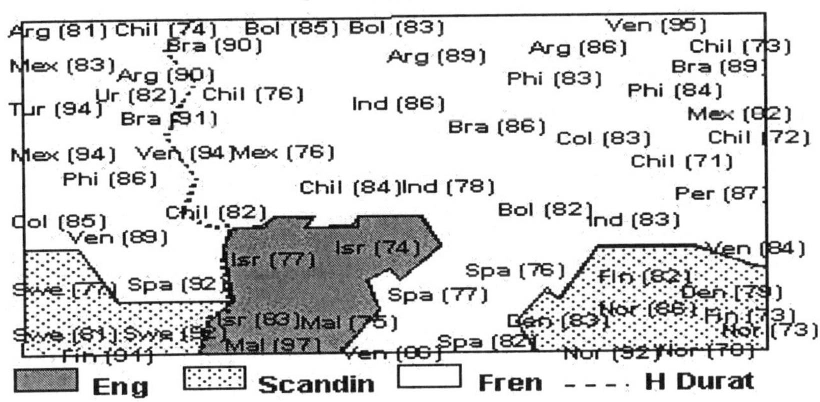


**Map 3. Law Origin**

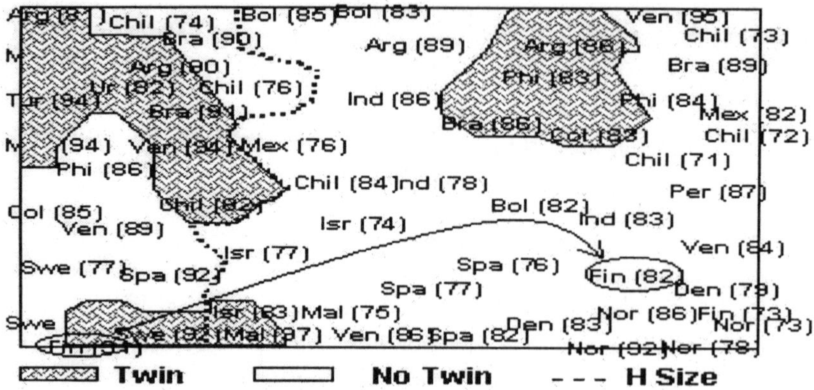


**Map 4. Twin Crises**

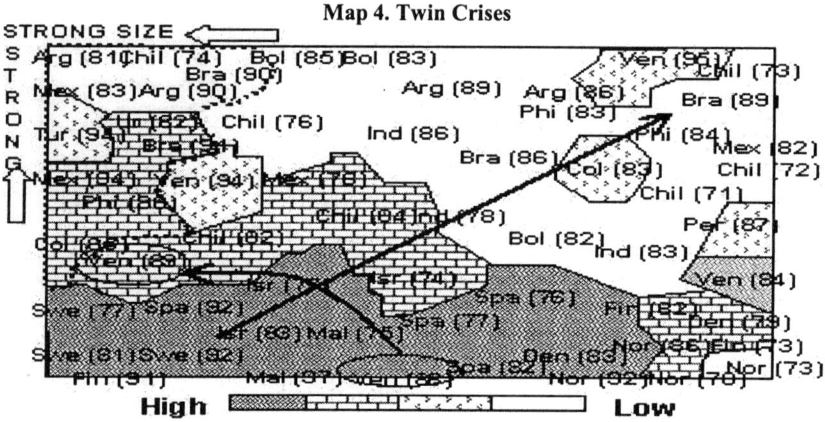


**Map 5. Financial Development**

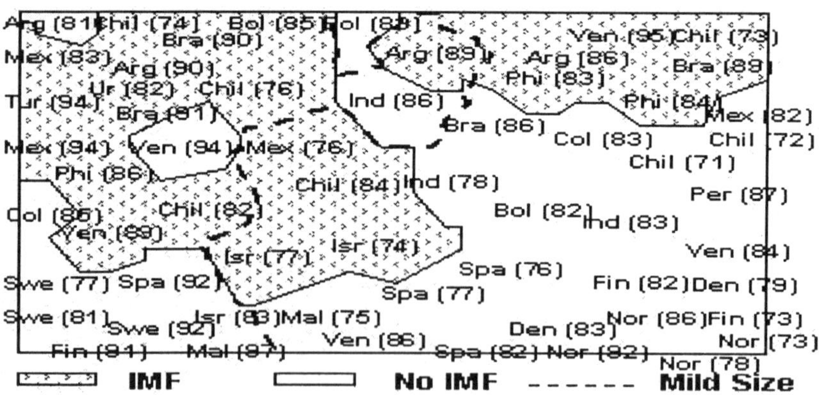


**Map6. IMF**

# Analysing Health Inequalities Using SOM

Zheng Rong Yang
Dept. of Computer Science, Exeter University, UK
Email: Z.R.Yang@exeter.ac.uk

## Abstract

As proposed by WHO that global health promotion in this century should be towards reducing health inequalities and improving health status. While how to identify health inequality structure is critical. It is believed that health inequality analysis is a complex problem. Self-organisation mapping is therefore employed in this paper to analyse health inequalities based on a data set provided by the Centre for Disease Control and Prevention, USA (CDC).

# 1 Introduction

In order to promote human health in this century, the World Health Organisation (WHO) has identified five goals: improving health status, reducing health inequalities, enhancing responsiveness to legitimate expectations, increasing efficiency and protecting individuals, families and communities from financial loss [2]. An important step to achieve the first two goals is to identify health inequalities (the difference in health status).

Health inequalities are commonly identified by dividing a population into several groups using one of the socio-economic health indicators, such as mortality, dietary and socio-economic position through univariate statistical analysis [3, 4, 5, 7]. However relations between health inequalities and health indicators are complicated and cannot be fully accounted for through such analysis so reducing the accuracy with which health inequalities may be identified.

This paper presents a method of applying self-organisation mapping (SOM) [6] to health inequalities analysis based on multiple health indicators. A local area with a number of health indicators is regarded as a pattern. The SOM will map the patterns from a high dimensional space to a low dimensional space to visualise the spatial relationship among them. The original topological structure among patterns will be well reserved during mapping. However each output node on the SOM map is forced to stay in a fixed place as a grid and there will be four corners at the SOM output map. While quantitative health inequalities analysis is desirable [1, 8]. This makes it difficult to analyse health inequalities quantitatively using the SOM output map. This paper therefore uses the weight space for quantifying health inequalities quantitatively.

A data bank from Centre for Disease Control and Prevention, USA (CDC) is analysed. The data bank contains 50 states, Washington DC as well as the national average from 1992 to 1996. Each of them is measured by 22 health indicators. The analysis pointed out that the health inequalities among these 50 states and Washington DC are generally uniformly distributed.

# 2 Method

Each state as well as Washington DC with 22 health indicators at a given year is regarded as a pattern. The national average with 22 health indicators is also regarded as a pattern. There are 52 patterns in each year. Two patterns will be mapped to the close output nodes on the SOM map after mapping if the states of these two patterns have a similar health status. The distance between the weight vectors from the closest output nodes should be relatively small. However the magnitude of the distance between a pair of weight vectors corresponding to a pair of the closest output nodes is not necessary the same as the other pair of weight vectors. This is the basis for using the SOM for quantitatively analysing health inequalities in this paper. The method of using the SOM for quantitatively analysis health inequalities is described as following three steps

*Step 1.* To refer 50 states, Washington DC and the national average to 52 patterns. Name $p_0$ for the pattern of the national average and name $p_{name}$ for any other states as well as Washington DC.

*Step 2.* To map the patterns using the SOM and to locate the best and worst health status on the weight vector map. Name the state with the worst health status as $p_{worst}$.

*Step 3.* To quantify the health inequalities for each status using $E(|p_{name} - p_0|)$ and $E(|p_{name} - p_{worst}|)$.

# 3 Result

The data from CDC comprises 50 states, Washington DC and the national average from 1992 to 1996. Each pattern is a collection of 22 health indicators. The four health indicators used for a presentation in this paper in 1992 are: prevalence of low birth weight as measured by the percentage of live born infants weighing under 2,500 grams at birth (LOW), births to adolescents (ages 10-17 years) as a percentage of total live births (ADO), prenatal care as measured by the percentage of mothers delivering live infants who did not receive care during the first trimester of pregnancy (PRE), and childhood poverty, as measured by the proportion of children under 15 years of age living in families at or below the poverty level (5-17 years) (POV).

Each state will be associated with one output node on the SOM map or be associated with one of the weight vectors after mapping. Some weight vectors may have more than one states associated. Different states will stay at different output nodes on the SOM map after mapping. The health inequalities among them are quantified by the distances measured from weight space corresponding to these nodes. The larger the distance between two weight vectors, the larger the health inequality among the states associated with these two weight vectors. The national average in each year is also regarded as a special pattern for training the SOM and it will appear near the centre somewhere on the SOM map.

Figure 1 (left) shows the mapping result for two indicators (LOW and ADO). In the map, we see that the pattern of the national average marked with "US" stays exactly at the centre of the map. At two extremes of the map are "AR" and "AK". It can be seen from Table 1 that all the health indicators occurring to "AK" show smaller values compared with the national average while all the health indicators occurring to "AR" show larger values compared with the national average. In particular, all the health indicators occurring to "DC", which is mapped to the same node as "AR", demonstrate much larger values compared with the national average, see Table 1. The node associated with "AR" and "DC" therefore represents the worst health

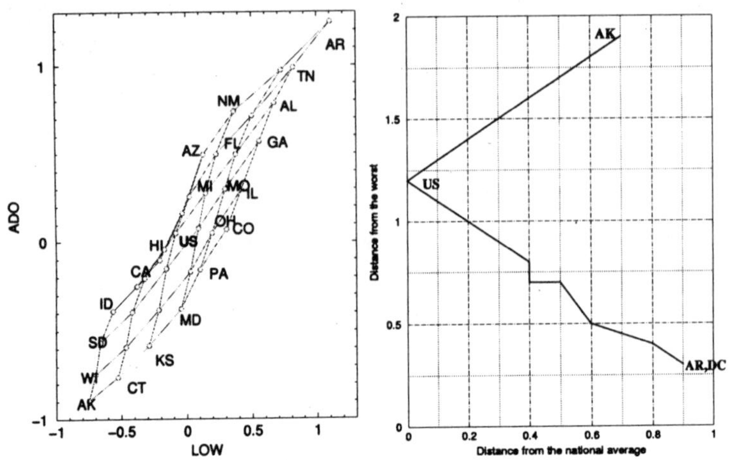


Figure 1: Pattern distribution (left) and Combination distance (right)

status and the node associated with "AK" illustrates the best health status in USA in 1992. The national average is a boundary to divide all 50 states as well as Washington DC into two parts. The states whose health status are worse than the national average (the right-top part in Figure 1: left) and the states whose health status are better than the national average (the left-bottom part in Figure 1: left).

Table 1. A comparison between extream nodes

| | US | AK | AR | DC |
|---|---|---|---|---|
| LOW | 7.1 | 4.9 | 8.2 | 14.3 |
| ADO | 4.9 | 3.7 | 7.4 | 7.1 |
| PRE | 22.3 | 16.9 | 27.7 | 43.1 |
| POV | 20.8 | 9.5 | 25.4 | 49.3 |

Table 2 gives the details of the expected distance. From this, health status of each state can be ranked. A simple rank is listed in Table 2 by considering the two expected distances. A complicated rank scheme can be derived by considering social science knowledge.

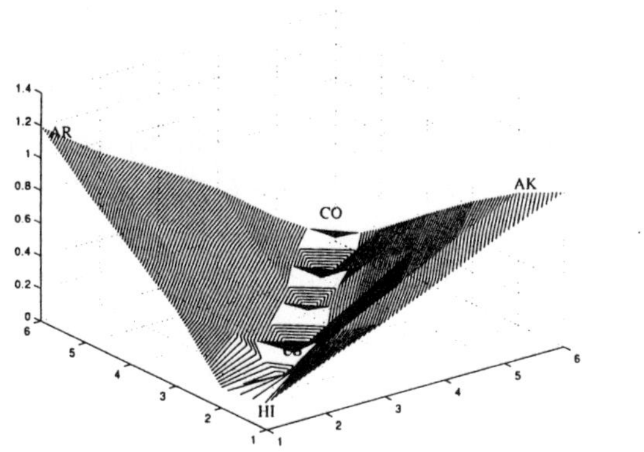


Figure 2: Distance from the national average (ADO)

Table 2. The expected distance from the national average and the worst status

| | $E(\|p_{name} - p_0\|)$ | $E(\|p_{name} - p_{worst}\|)$ | Rank |
|---|---|---|---|
| AK,IA,ME,MA,MN | 0.6657271 | 1.855246 | 1 |
| NE,NH,ND,UT,VT | 0.6657271 | 1.855246 | 1 |
| WI | 0.6123794 | 1.801898 | 1 |
| SD,WA | 0.5744173 | 1.763936 | 2 |
| ID,MT,OR | 0.4776766 | 1.667195 | 3 |
| CT,RI | 0.4417522 | 1.631271 | 3 |
| KS,NJ | 0.2039547 | 1.393473 | 4 |
| HI,NV | 0.1219920 | 1.311510 | 4.5 |
| CA,NY | 0.0882726 | 1.277791 | 5 |
| US | 0.0000000 | 1.189518 | 6 |
| MD,VA | 0.0392950 | 1.150223 | 6.5 |
| PA | 0.1133030 | 1.076215 | 7 |
| IN,WY | 0.1882460 | 1.001272 | 7 |
| AZ,OK | 0.2173181 | 0.972200 | 8 |
| MI | 0.2325044 | 0.957014 | 8 |
| OH | 0.2830011 | 0.906517 | 8 |
| FL,KY | 0.3172156 | 0.872303 | 9 |
| MO | 0.3833002 | 0.806218 | 9 |
| CO,DE | 0.3928336 | 0.796685 | 9.5 |
| NM,TX | 0.4532142 | 0.736304 | 10 |
| IL | 0.5045063 | 0.685012 | 11 |
| GA,NC | 0.6423831 | 0.547135 | 12 |
| AL | 0.7552295 | 0.434289 | 13 |
| TN,WV | 0.9034563 | 0.286062 | 14 |
| AR,DC,LA,MS,SC | 1.1895188 | 0.000000 | 15 |

Figure 2 draws one interesting graph, in which we can see the health status distributions of a health indicator displayed in contours. The contours are generated by the distance between any weight vector and the weight vector associated with the national average. The graph can be enlarged to display all the patterns of states to show their health status distribution. The advantage of such a method is that the health status from all the states can be visualised in a graph.

Using the weight vectors, we can calculate: (1) the expected distance between any pattern of a state and the pattern of the national average; (2) the expected distance between any pattern of a state and the pattern of "AR", which is associated with the worst health status. They quantify the health inequalities among 50 states and Washington DC. From this, health promotion officers can make their decision of how to promote health status. A curve can be drawn in Figure 1 (right) based on these distances. The curve shows that the health status in USA in 1992 had an approximately uniform distribution. This can also be proved by the two expected distances from Table 2 that the two distance has a bi-linear relationship.

# Summary

This paper has presented a method for analysing health inequalities using the SOM. With the SOM the health inequalities can be well visualised and then an analysis of health inequality structure can be completed by making an expectation of the distance between any state and national average and the distance between any state and the worst state. By combining these two distances, we are able to rank the health status for each state in USA. From this, the health promotion officers are able to make an efficient health promotion strategy for improving the health status. It has also been found that the health status in USA in 1992 had an approximately uniform distribution.

# References

[1] Anand S. and Hanson K. Disability-adjusted life years: a critical review. Journal of Health Economics 1997; 16: 685-702.

[2] Brundtland G.H. Making A Difference. The World Health Report 1999: World Health Organization.

[3] Catford D. Health promotion in the market place- constraints and opportunities. Health Promotion International 1995; 1: 41-50.

[4] Charlton B.G. Is inequality bad for the national health?. The Lancet 1994; 343: 221-222.

[5] Charlton B.G. Living on the margin- a salutogenic model for socioeconomic differentials in health. Public Health 1995; 4: 235-243.

[6] Kohonen T. The self-organizing. Proceedings of the IEEE 1990; 78: 1464-1480.

[7] Turrel A. Socioeconomic differences in food preference and their influence on healthy food purchasing choices. Journal of Human Nutrition and Dietetics 1998; 11: 135-149.

[8] Murray C.J.L. and Lopez A.D. The Global Burden of Disease. Harvard School of Public Health 1996.

# Integrating Contextual Information into Text Document Clustering with Self-Organizing Maps

Daniel Pullwitt and Ralf Der

University of Leipzig, Department of Computer Science
04109 Leipzig, Germany
{pullwitt, der}@informatik.uni-leipzig.de

**Summary.** Exploration of large text collections requires suitable methods of presentation. The Self-Organizing Map has shown promising results over the past years. The topographic map approaches usually use the common vector space model for text document representation. We present here a new two stage representation which uses sentences as intermediate information units. In this way contextual information is preserved and influences the process of self-organization. We demonstrate that presence of the contextual information improves the quality of the resulting document maps. The procedure is computationally more expensive but we present modifications of the algorithm which cope with this problem.

## 1. Introduction

Unsupervised neural networks, in particular Self-Organizing Maps (SOMs) [3], have been repeatedly used in text document clustering. Several successful examples can be found like *WEBSOM* [4] or *SOMLib* [6].

Most of the research done in regard to but not limited to text document classification is concerned with the classification methodology itself. Computations on the basis of large sets of documents are made feasible by inclusion of dimensional reduction and fast winner search [4] or hierarchical clustering [5].

While clustering methodology has evolved, the representation model used has not been changed from the simple vector space model based on keywords [7]. However the resulting quality that can be obtained depends on the representation used.

In this paper we present a new model based on the common vector space model. We introduce an intermediate step based on single sentences to include contextual information. It is demonstrated that the additional information can replace more refined indexing used in other approaches with regard to clustering quality. We are currently evaluating the model in comparison to common representation models.

## 2. A Two Stage Vector Space Model for Text Representation

### 2.1 The Common Vector Space Model

The vector space model [7] is a common method for text document representation. It is widely used in information retrieval and SOM based clustering [1, 2, 5, 6]. In an indexing process text documents are analysed and important keywords are extracted. Each keyword is represented by a distinct vector component. The value of a component depends on the frequency of the word occurrence in the document.

The simplicity of the vector space model is not without drawbacks. A lot of information contained within a document is lost including everything provided by context. It is not possible to disambiguate distinct word meanings based on context. Sections of different topics are mixed together without distinction.

## 2.2 Motivation for the Two Stage Vector Space Model

In this paper we propose two stage model which reduces the loss of contextual information. It introduces an intermediate separation of documents into smaller units taking the context of words into account. A set of categories is identified on the basis of similarities in these units (e.g. word usage). In a second step document representations are created using the categories.

The representation quality depends on the selected type of intermediate unit. For example, early versions of *WEBSOM* [2] used groups of 3 words to form word category maps. These categories are used to map groups of words to single symbols, reducing the dimension of input space. Contextual information is used only to determine word mapping but the document clustering itself is based on words. This approach was less successful than the random projection used later [4].

Considering the structure of languages it appears to be a natural choice to use sentences as intermediate units. Sentences are relatively self-contained units within a text which include most of the contextual information of their words. Documents containing multiple topics can be split into sets of sentences for each topic. Therefore separating a document into its sentences allows differentiation of both distinct word meanings of single keywords and distinct topics.

Despite the fact that sentence separation appears to be a natural way to process textual information it has not been used in vector space based clustering before. This may be due to the fact that at first impression it adds a huge amount of additional data making effective clustering impossible. In this paper we will show how the computation can be made feasible by using a suitable representation model and adapted algorithms.

## 2.3 Description of the Model

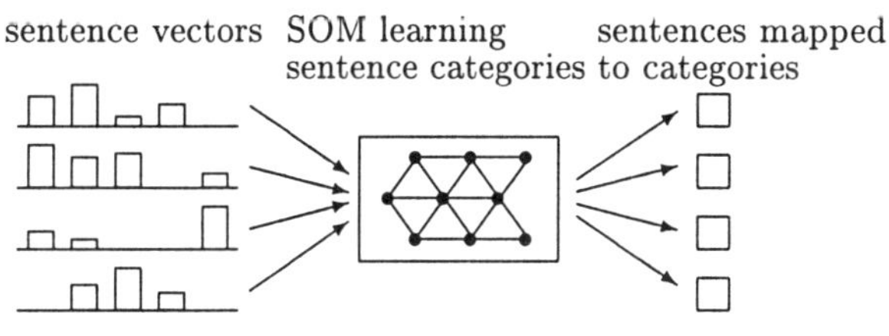


**Fig. 2.1.** SOM based mapping of sentences to categories

**2.3.1 Step 1: Sentence Vectors.** According to the common vector space model each sentence is represented by a single vector. Almost all vector components are zero due to the small number of words within a sentence and relevance filtering applied, including the removal of common stop words. Varying with the type of documents the average number of non-zero components in a sentence vector is often smaller than 10. This allows compressed storage and efficient distance calculation (section 2.4).

A SOM is trained with the complete set of sentence vectors of all documents. The resulting prototype vectors each represent a category of sentences similar in word usage (including context) and therefore similar in content. They also allow to map sentences to their respective categories (figure 2.1) without further distinction within a category.

Experiments indicate that usually a number of 400 to 600 categories is sufficient for detailed modelling, although the number may vary depending on the actual document collection. The level of detail of the representation grows with the number of categories. A higher number leads to a more detailed modelling thus emphasising differences, while reducing the number too far prevents any useful distinction between documents. A strategy for automatically determining the appropriate number of categories has yet to be found.

If only categories are to be obtained topological ordering gained by SOMs is not necessary. Structural information emphasises important features of the documents and therefore allows to easily determine the quality of sentence mapping. It is helpful in analysing different mappings dependent on different document collections and on the number of categories. If ordering is of no concern, an arbitrary vector quantiser (e.g. k-means) can be used alternatively, reducing computational costs significantly.

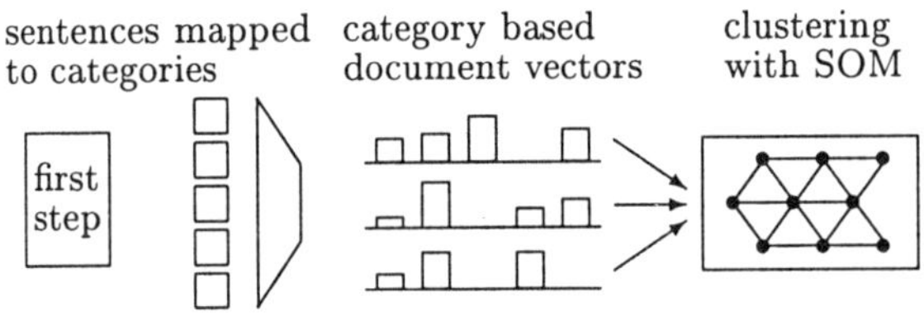


**Fig. 2.2.** Clustering documents using sentence categories

**2.3.2 Step 2: Document Vectors.** In our model the document representation is based on sentences instead of keywords as in the common vector space model. Using the sentence vectors and categories obtained in step 1, sentences are mapped to categories. Document vectors are formed of the frequency distribution of sentence categories (figure 2.2). A SOM is trained with the document vectors. The resulting map is used to visualise the structure of the document collection.

It can be noticed that the sensitivity of a document vector to errors in indexing depends on the number of sentences. If the document contains only a few sentences small changes in one sentence can affect its category mapping leading to a noticeable different document vector. Robustness of representation with regard to indexing errors grows with the length of documents. If only short documents are used, the sentence clustering stage is rather similar to document clustering using the common vector space model. Therefore the two stage model is more suitable for long documents while short ones can be clustered more efficiently using common methods.

## 2.4 Optimisations

Clustering using the proposed model requires a larger amount of vector calculations as compared to the common vector space model. In particular clustering of sentences

($10^5$-$10^7$ vectors) is computational intensive. The structure of data used allows optimisations which affect efficiency:

As mentioned before sentences are represented by extremely sparse vectors of high dimension. By using a cosine based distance measure

$$d(\mathbf{x}, \mathbf{y}) = 1 - cos(\mathbf{x}, \mathbf{y}) = 1 - \frac{< \mathbf{x}, \mathbf{y} >}{||\mathbf{x}|| ||\mathbf{y}||} \tag{2.1}$$

only the components present in both vectors need to be considered. Therefore the number of necessary floating point operations is linearly proportional to the number of evaluated vector elements.

To be more precise: By using precalculated length of both vectors and $N$ shared vector components distance calculation (2.1) requires $N + 2$ floating point multiplications and $N$ additions. On prototype updating only the shared components and the stored length need to be changed, latter can be recalculated efficiently using length and cosine calculated in distance measuring.

Computational costs and achieved quality of the document SOM depend mainly on the dimension of document vectors and therefore on the number of sentence categories. The number of categories affects how detailed the documents are modelled and how distinguished the resulting clusters will be. A low number of categories leads to low dimensional vectors and a more general representation allowing fast clustering but only vague distinctions. Results more precise require more categories leading to higher computational costs.

Computational costs can also be reduced in a variety of methods: As mentioned before in most cases determining sentence categories does not require topological ordering. Using a simpler algorithm than SOM causes a significant speed up. Further reduction of complexity can be obtained by determining categories on a subset of documents instead of the whole collection. The difficulty lies in determining a representative subset. If adequately chosen computation is sped up noticeably, otherwise the categories are suitable to model just a few documents. Sentences not represented by the training set will be mapped to seemingly arbitrary categories resulting in useless or misleading representations.

# 3. Experimental Results

The described two stage representation model has been implemented and used with several test sets. All results described are obtained by indexing using only a simple stop word list but no word stemming. Better pre-processing is supposed to improve quality of the representation while also reducing dimension of the feature space thus reducing complexity of computation.

Keywords relevant for representation are filtered and weighted using an entropy based measure. The entropy $E(i)$ of word $i$ is calculated as

$$E(i) = -\sum_{j=1}^{n} d_{ij} log d_{ij} \tag{3.1}$$

using the distribution

$$d_{ij} = f_{ij} / \sum_{k=1}^{n} f_{ik} \tag{3.2}$$

of word frequencies $f_{ij}$ over documents $j = 1, \ldots, n$. Extreme values of entropy $E(i)$ are unsuitable for classification and therefore omitted: Small values indicate usage only in a fraction of the documents and therefore words too special in meaning allowing no generalisation. Keywords with high entropy are distributed evenly over large subsets of the collection and are too general to allow any meaningful differentiation. It is possible to reduce the set of keywords to less than one third of all occurring words without a noticeable loss of quality.

All experimental sentence classifications have been done using SOMs to allow easier identification of important categories. Using a simpler algorithm reduces computation time significantly.

## 3.1 CIA World Factbook 1998

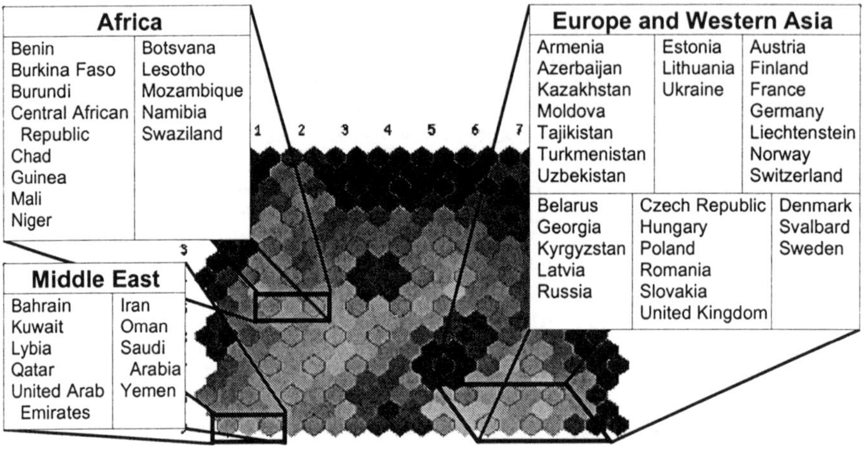


**Fig. 3.1.** Results of clustering the *CIA World Factbook 1998*. The level of darkness is proportional to the distances between prototypes. A few important document groups have been marked and labelled. Most of the dark area (low similarity) is occupied by descriptions of islands. Full results can be found at http://www.informatik.uni-leipzig.de/~pullwitt/wfb98/

The *CIA World Factbook*[1] consists of 278 documents of homogenous structure. Its documents do not consist of sentences but can be divided in sections describing distinct features of countries like geography, politics or economics. These 'sentences' have been mapped to 400 categories by a $20 \times 20$ SOM. The categories found reflect topics present in the documents: economical aspects like available resources, industry or infrastructure, aspects of climate and environment, even descriptions of the appearance of flags.

The resulting 400 dimensional document vectors are mapped on a $10 \times 10$ SOM which reveals meaningful structures (figure 3.1). Within the map it is possible to locate geographical regions like the Middle East or Europe (which can easily be separated in Eastern and Western Europe) or types of entries like oceans and islands.

The results are similar to those obtained by using the common vector space model with more refined pre-processing including word stemming [5]. It can be

[1] documents available at http://www.gutenberg.net

concluded that the additional contextual information in our model allows the SOM to generate information similar to the one usually provided in indexing.

### 3.2 Alex Catalogue of Electronic Texts

This set of 119 documents concerning philosophy[2] contains more than 200.000 sentences. It has been used mainly to test the model with larger sets of data and analyse its robustness and performance. Sentence categories were calculated by an $20 \times 25$ SOM. They reflect different aspects of the documents like government, religion, meaning and understanding or physical and mathematical properties. Computation of the sentence SOM has been extremely time consuming[3] thus indicating the need of a training algorithm adapted to the problem. The resulting document map is ordered mainly by authors and epochs but also reflects thematic aspects of the collection.

## 4. Conclusions

The proposed two stage vector space model for text representation has been successfully applied to text collections. Inclusion of contextual information by analysing sentences allows distinction between different word meanings and distinct topics within single documents, both unconsidered in the common vector space model.

However, effectivity varies with the length of the documents. For very small documents ($< 10$ sentences) clustering sentences is rather similar to clustering documents using the common vector space model which shows to be more appropriate.

Level of detail in document representation can be adjusted by means of the number of sentence categories. If a general map for initial orientation lacking detailed structure is needed, using only a few categories allows relatively fast computation.

Additional costs caused by the larger data sets necessary in clustering can be reduced by modified storage and algorithms. Using a suitable configuration of sentence categories and adapted training algorithms clustering can even be faster than one based on the common vector space model.

In all experiments so far we used only simple indexing. Other published results used in comparison are based on a more refined pre-processing which includes word stemming. Document maps obtained by our model are at least similar in quality despite the limited pre-processing. Using the same indexing our model usually shows better results. It can be concluded that our two stage representation is well suited for the self-organized mapping. It allows to determine information of similar quality with regard to document differentiation as usually provided in indexing.

Our model has to be evaluated further, in particular systematic comparisons to the common vector space model needs to be done on different text collections.

---

[2] documents available at http://www.infomotions.com/alex/

[3] The training process took nearly one week on a 550Mhz Pentium III while a fast k-means clustering (which also requires fewer training steps) could process the same data within a few hours.

# References


1. H. Chen, C. Schuffels, and R. Orwig. Internet categorization and search: A machine learning approach. In *Journal of Visual Communication and Image Representation*, volume 7 of *Special Issue on Digital Libraries*, pages 88–102. 1996.
2. T. Honkela. *Self-Organizing Maps in Natural Language Processing.* PhD thesis, Helsinki University of Technology, Espoo, Finland, 1997.
3. T. Kohonen. *Self-Organizing Maps.* Springer, 1995.
4. T. Kohonen, S. Kaski, K. Lagus, J. Salojärvi, V. Paatero, and A. Saarela. Organization of a massive document collection. *IEEE Transactions on Neural Networks, Special Issue on Neural Networks for Data Mining and Knowledge Discovery*, 11(3):574–585, May 2000.
5. D. Merkl and A. Rauber. CIA's view of the world and what neural networks learn from it: A comparison of geographical document space representation metaphors. In *Proceedings of the 9th International Conference on Database and Expert Systems Applications (DEXA'98), Vienna, Austria*, Lecture Notes In Computer Science. Springer, 1998.
6. A. Rauber and D. Merkl. The SOMLib digital library system. In *Proceedings of the 3rd European Conference on Research and Advanced Technology for Digital Libraries (ECDL'99), Paris, France, September 22. - 24. 1999*, Lecture Notes In Computer Science (LNCS 1696). Springer, 1999.
7. G. Salton. *Automatic Text Processing: The Transformation, Analysis and Retrieval of Information by Computer.* Addison-Wesley, 1989.

# Self-Organizing Internet Semantic Network

Sergey Shumsky
NeurOK LLC
Moscow, RUSSIA

Abstract

This paper presents a concept of global distributed information network comprising millions of personal agents acting on the Internet on behalf of their owners.

## 1 Introduction

Semantic networks are well known knowledge representation models, where a set of nodes, associated with *notions* are interconnected by links, associated with *relations*. The largest and the best known semantic network in the world is the World Wide Web comprising at the end of the year 2000 about $3 \cdot 10^9$ hyper-text pages connected by $\sim 5 \cdot 10^{10}$ hyper-links [1]. The structure of those links accumulates the knowledge of millions of people, who have deliberately put into their web-pages hyperlinks associated with a discussed topic. Contemporary search engines, pioneered by Google, make use of this distributed world-wide knowledge to greatly enhance the quality of information retrieval in WWW [2]. In order to take advantage of the network topology these search engines have to create their own copy of the whole WWW. Taking into account the size of the latter, this task requires power of supercomputer to monitor a dynamically changing and constantly growing Web. Google is planning to increase the number of its computers to 10,000 in the near future [3]. But sooner or later centralized search engines will fail to keep pace with the exponentially growing knowledge network. In any case it is difficult to imagine a unique centralized index of all databases available on the Internet with the estimated volume of $\sim 10^{16}$ bytes [1]. That or another kind of a distributed global search system will soon be in great demand.

This article supports the claim that constantly growing network may be indexed only by another network growing at the same pace. We seek such a solution in a realm of self-organizing semantic networks – quite a natural approach for a neural network community [4-5]. On the previous WSOM'99 we presented Semantic Explorer, a self-learning semantic agent, using self-organizing semantic categories to organize documents in regard to their topics [6]. Here we describe the network of such semantic agents, with self-organizing interconnections.

Agents in such network have two indices: of documents and of other agents, treating both of them in a similar manner – based on semantics of the content. If an agent is unable find a document on a given topic, he re-directs such a query to the agents that are mostly likely to find it. The connections between the agents adapt themselves based on the history of successfully served queries. Altogether those agents form a distributed self-organizing search engine. Running on the corporate servers as well as on PCs of Internet users, such information retrieval network will be capable of indexing on the fly all the documents available in the Net. It aims to automatically organize the content-based access to all textual information arrays.

# 2 Semantic Agent

The self-organizing search network is based on the semantic agents. Each such agent has its own semantic space, formed by a set of semantic categories, corresponding to clusters of documents with similar content. The link to a document or other agent is represented by a vector in the agent's semantic space. (Fig. 1). This space may result e.g. from the learning procedure described below.

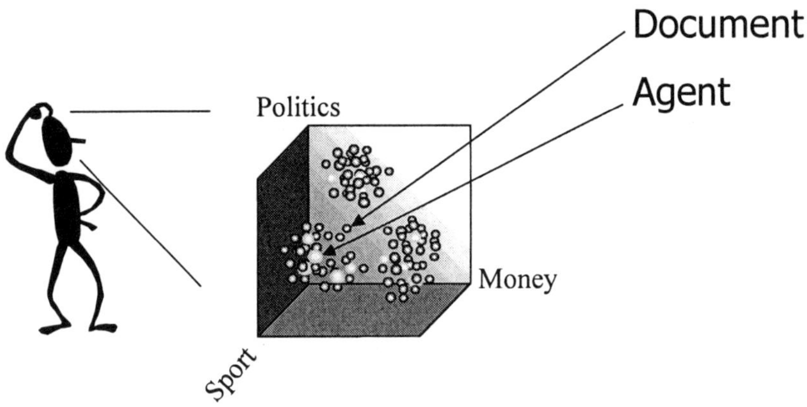


Fig. 1 Documents and other agents are represented by vectors in the agent's semantic space

Let K-dimensional semantic vectors represent the meanings of words. Initial value of these vectors are randomly distributed. Suppose, that agent has some document collection to learn from. The learning then proceeds along the following lines:

- Compute semantic vectors of documents as the centroids of all their words
- Find K clusters of documents
- Compute new semantic vector of each word with components being proportional to the words count in the found clusters
- Proceed to the next iteration with new semantic vectors of words

The above learning procedure belongs to a family of converging EM-algorithms, extracting semantic categories using an information bottleneck approach [7,8]. Learning may also proceed in on-line manner and with the growing number of clusters-categories.

Resulting semantic vectors of words retain only general information about the related topics. Various word-forms or close synonyms have almost identical semantic vectors, as far as they are typically used in the same context. Thus, documents with a similar content will have a close semantic representation regardless of the exact choice of used terms. It allows to construct semantic filters fine-tuned by a small set of related terms or documents, thus greatly facilitating learning the profiles of the agents' owners (based on their queries) and of other agents, treated as documents sources. Each agent then becomes a kind of electronic secretary filtering out news and documents based on users' preferences.

# 3 Semantic Agents Network

Consider now a distributed search system comprising numerous semantic agents, each having its own index of documents based on his owner's interests. The semantic index of each such agent comprises not only links to the documents, but also links to other agents. The corresponding semantic vectors are computed as centroids of all documents, received so far from a given agent.

In a way such network resembles the WWW, where numerous documents are interconnected by associative hyperlinks. In our network numerous search agents are also interconnected by associative links. Like in the WWW one may use those hyperlinks when he needs additional information on the related subject.

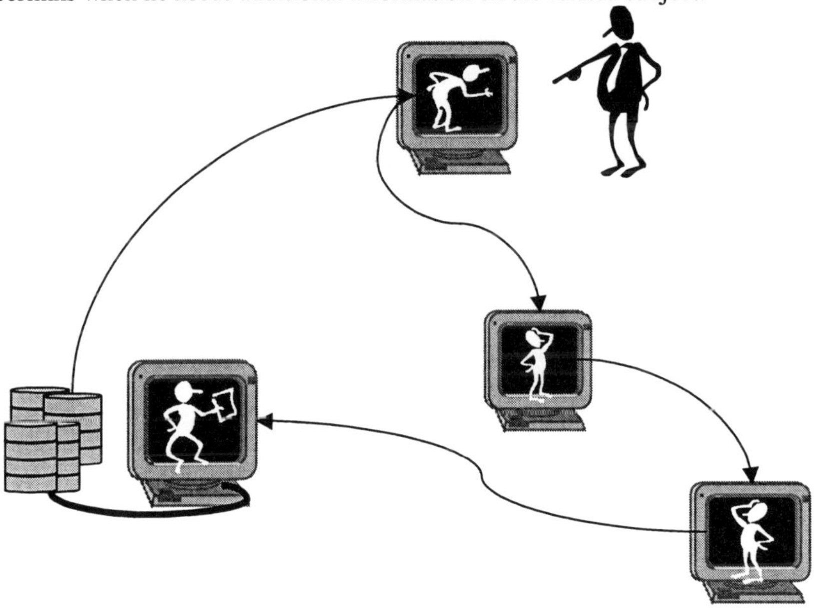

Fig. 2 Query redirection in the agents network

Namely, if one agent can't find a required number of documents on a given topic in his local index, he redirects the query to another agents he knows, related to this subject (Fig. 2).

As a result, such agents network forms a distributed search engine, where each node, though possessing only limited amount of local information, can handle global queries.

The key feature of such distributed search engine is content dependent query redirection, based on semantic indexing. Queries are sent to other agents depending on the previous experience about the nature of information, received from them so far. Each piece of information received from an agent corrects the coordinates of its representation in the semantic index of the recipient. One can use, e.g. the Robbins-Monro learning algorithm [9] to update the semantic vectors of agents:

$$\mathbf{a}_n = \frac{n-1}{n}\mathbf{a}_{n-1} + \frac{1}{n}\mathbf{d}_n,$$

where $\mathbf{a}_n$ is the semantic vector of some agent after receiving from him the $n$-th document with semantic vector $\mathbf{d}_n$.

Each link to an agent has also its own utility based rating. Those ratings are used for choosing right candidates for queries redirection and pushing news, ads and announcements in the network. Rating adaptation procedure discussed below is designed to optimize network efficiency.

# 4 Network Economy

A well-known drawback of network-based search systems, like Freenet [10] or Gnutella [11], is exponential queries expansion. To limit query reproduction, its lifetime is usually limited to some constant value.

We chose original "free-market" based approach. Namely, agents in our network provide chargeable search services to each other. Each query has some limited amount of network currency, *neuro*, dissipating in the course of query processing in the network. Neuro circulating through the network is used by the agents to fine-tune their connections with the other agents, based on their *utility*, much as the money flow in the real economy organizes the structure of business relationships.

To this end, we introduce the following rules of simplified network economy:

- Each query processing costs one neuro
- Each document (query) transaction costs one neuro
- Agents aim to minimize their expenditures

Agent then becomes a bookkeeper, counting the balance of transactions with each known agent. Economical rationality forces agents to address the real experts in that or another topic, thus minimizing effective cost of search in the network.

Let us illustrate the impact of economical behavior by a very simple example (**Fig. 3**). Suppose agent **A** has to bring his owner 10 documents on some specified topic. Agent **A** does not have links to such documents in his local index, but he knows agent **B**, related to the topic of query. Agent **A** redirects the query to agent

**B**, supplying his request with 11 neuro (1 for query processing and 10 for transfer of the found documents). Agent **B** appears to have only 5 documents on the subject. He takes one neuro for query processing, 5 neuro accompany 5 found documents returning to **A**, and the rest 5 neuro supply the query redirected to agent **C**, which **B** considers most relevant. The latter appears to be a real expert, and returns all 4 prepaid documents to **A** (taking 1 neuro for processing query). Overall **A** paid 2 neuro for 9 documents. The mean cost of the document received from **B** is (11-5)/5 = 6/5. Documents received from **C** have negative cost (0-4)/4 = -1.

Next time **A** will address a similar query directly to **C** (**Fig. 4**). Since **C** is supposed to be an expert, he returns all 10 documents. And this time **A** pays only 1 neuro for 10 documents. Thus in this particular example economical rationality allowed to find a route more than 2 times cheaper than the previous one!

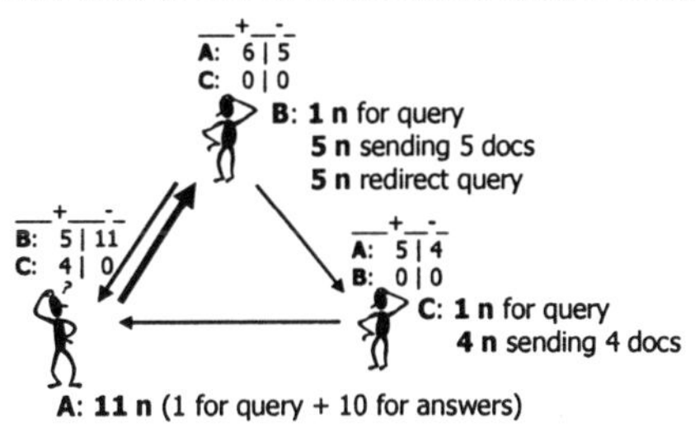


**Fig. 3 Agent** B **redirects query from agent** A **to agent** C

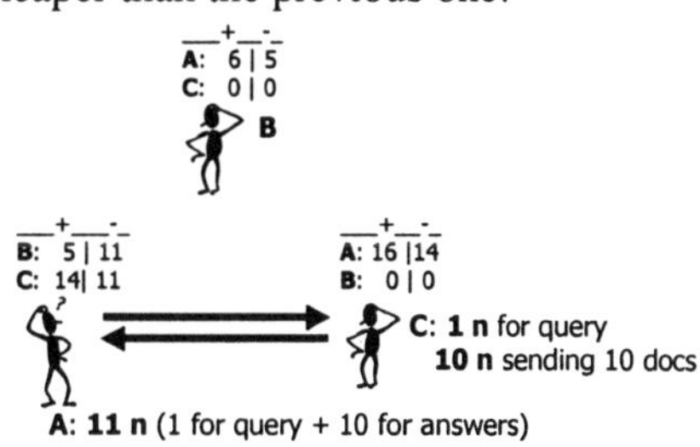


**Fig. 4 Agent** A **addresses directly agent** C**, who is better expert than** B

In general, bookkeeping allows of pointing out not only the sources of supply, but also the sources of demand on that or another topic. Our network thus provides an ideal infrastructure for information pushing. Ads, news or announcements will readily find the right addressees.

Accounting may be used not only for price calculation, but also for finding a related probability of receiving requested document from an agent. This allows of combining semantics with economics. Namely, it is natural to consider the overall relevance of an agent to a query as a product of the Bayessian probability of query relevance to an agent, and the probability of receiving from him relevant documents.

# 5 Discussion

Self-organizing network of agents, constantly adapting to each other and to their owners, provides a new quality of communication in Internet. One can submit not only queries, but also any kind of announcement or request, which will be readily delivered by the agents to the addressees with related interests. Internet is a programmable media, which is now underestimated by portals, utilizing primitive broadcasting mass media model. Let communications program themselves as the neural networks do, and each user will automatically be connected to the ones with similar interests. Without any additional efforts users will suddenly find themselves embedded in numerous self-organized virtual communities of like-minded persons, whether they are music-lovers, dog-fanciers or stamp collectors. Job hunting will also be greatly facilitated: anyone who has something to offer might hope to satisfy someone's demand.

This article gives only the outline of self-organizing agents network, developed by NeurOK LLC, an agentware company founded in 1998. For additional information please visit company's site www.neurok.com.

## References


1. http://www.sims.berkeley.edu/how-much-info/internet.html
2. www.google.com
3. Burrett M. Google: The Search Engine That Could http://www.linux.com/jobs/newsitem.phtml?sid=72&aid=8848
4. Kohonen T. Self-Organizing Maps, 2-nd ed. Springer-Verlag, Berlin, 1997
5. WEBSOM - Self-Organizing Maps for Internet Exploration http://websom.hut.fi/websom/
6. Shumsky S. Navigation in databases using Self-Organizing Maps. In: E.Oja and S.Kaski (ed) Kohonen Maps. Elsevier, Amsterdam, 1999, pp 197-206
7. Tishby N., Pereira F., and Bialek W. The Information Bottleneck Method. Proceedings of the 37th Annual Allerton Conference on Communication Control and Computing , 1999
8. Hofmann T. Probabilistic Latent Semantic Indexing. Proceedings of the 22nd Annual International ACM SIGIR Conference on Research and Development in Information Retrieval, 1999
9. Robbins H., and Monro S. A stochastic approximation method. Annals of Mathematical Statistics 1951; 22:400-407
10. http://freenet.sourceforge.net/
11. http://gnutella.wego.com/

# Recursive learning rules for SOMs

John A. Lee, Nicolas Donckers, and Michel Verleysen

Université catholique de Louvain, Laboratoire de Microélectronique,
Bâtiment Maxwell, Place du Levant, 3, B-1348 Louvain-la-Neuve, Belgium

**Summary.** Three extensions of the traditional learning rule for Self-Organizing Maps are presented. They are based on geometrical considerations and explore various possibilities regarding the norm and the direction of the adaptation vectors. The performance and convergence of each rule is evaluated by two criteria: topology preservation and quantization error.

## 1 Introduction

Self-Organizing Maps (SOMs, [1,2]) are well known in the domain of Vector Quantization (VQ). Unlike other VQ methods, the neurons (or prototypes) used for the quantization are given a position in a grid, which is often one- or two-dimensional. This predefined geometrical organization, combined with a well chosen learning rule, generates a self-organizing behavior, useful in numerous areas like nonlinear projection and data representation. More technically, learning rules for VQ can be classified into two sets, according to the number of neurons which are adapted at each stimulation of the network:

- 'Winner Takes All' (WTA) rules, like for Competitive Learning (CL, [3]), where only one neuron is adapted;
- 'Winner Takes Most' (WTM) rules, like for Neural Gas (NG, [4]), where all neurons are adapted.

In order to observe self-organization, learning rules for SOMs have to fulfill two conditions: they belong to the WTM set and they use information given by the position of each prototype in the grid. Other characteristics of the rules are more or less unconstrained. Section 2 studies these degrees of freedom and presents four different learning rules (three alternative rules in addition to the traditional one). Next, section 3 describes some experiments and criteria to test the four rules. Section 4 gathers results and discusses them. Finally, conclusion (section 5) shows that one of the proposed rules improves convergence with respect to the traditional SOM rule.

## 2 Four learning rules

First of all, let's write the traditional learning rule for SOMs. Suppose that the map is defined by:

1. matrix $\boldsymbol{W}$, of which rows $\boldsymbol{w}_r$ give the weights in the feature space;
2. function $d(q, r)$, measuring the distance between neurons $q$ and $r$ in the grid space.

At learning time $t$, a vector $\boldsymbol{x}_i$ of the feature space stimulates the map (see also Figure 2, left):

$$\boldsymbol{w}_r^{t+1} = \boldsymbol{w}_r^t + \Delta \boldsymbol{w}_r^t = \boldsymbol{w}_r^t + \gamma_r^t(\boldsymbol{x}_i - \boldsymbol{w}_r^t). \tag{1}$$

In this equation, $\gamma_r^t = \alpha^t e^{-0.5(d(r,*)/\lambda^t)^2}$ is the learning rate for neuron $r$, computed as the product of the global learning rate $\alpha^t$ (between 0 and 1, decreasing in time) and the neighborhood factor depending on the neighborhood width $\lambda^t$ (between 0 and $\infty$, decreasing in time). Finally, $*$ is the index of the 'winning' neuron, such that $\|\boldsymbol{x}_i - \boldsymbol{w}_*\| \leq \|\boldsymbol{x}_i - \boldsymbol{w}_s\|$ for all neurons $s$.

From a geometrical point of view, the traditional SOM rule adapts neurons radially around the stimulating vector $\boldsymbol{x}_i$. Now, consider that the self-organizing map is like a fishing net crumpled on the ground and that the neurons are like small balls linked together by short pieces of elastic cord. Consider also that the stimulating vector is like the hand of the fisherman, taking a node of the net between two fingers. In this case, the neurons do not move directly towards the hand, i.e. radially, but instead they are pulled by neighboring neurons. Formally, it means that the neurons have to be adapted towards neuron $q$, where index $q$ indicates the neuron preceding neuron $r$ on the shortest path [5] between neuron $r$ and winning neuron $*$. This idea is illustrated on the right of Figure 2 and leads to the following learning rule:

$$\Delta \boldsymbol{w}_r^t = \gamma_r^t(\boldsymbol{w}_q^{t+1} - \boldsymbol{w}_r^t), \tag{2}$$

where neuron $r$ has to be adapted *after* neuron $q$. Note that the neighborhood factor cannot be suppressed, although the new rule presents an interesting property without the neighborhood factor. Indeed, suppose that the map is reduced to a string in a one-dimensional feature space and that its current state is such that:

- $\boldsymbol{w}_r^t = r$, with $\boldsymbol{w}_*^t = * = 0$;
- $\boldsymbol{w}_r^t$ is linked with $\boldsymbol{w}_{r-1}^t$;
- the stimulating vector is $\boldsymbol{x}_i = 0$.

If $\lambda^t$ is set to $+\infty$, then the neighborhood has no limit, $\gamma_r^t$ degenerates to $\alpha^t$ and one can rewrite the traditional rule into:

$$\boldsymbol{w}_r^{t+1} = \boldsymbol{w}_r^t + \alpha^t(\boldsymbol{x}_i - \boldsymbol{w}_r^t) = (1 - \alpha^t)r, \tag{3}$$

while the fisherman's rule leads to a recurrence:

$$\boldsymbol{w}_r^{t+1} = \boldsymbol{w}_r^t + \alpha^t(\boldsymbol{w}_{r-1}^{t+1} - \boldsymbol{w}_r^t) = (1 - \alpha^t)r + \alpha^t \boldsymbol{w}_{r-1}^{t+1}, \tag{4}$$

solved by $\boldsymbol{w}_r^{t+1} = (1 - \alpha^t) \sum_{i=0}^{r} i(\alpha^t)^{r-i}$.

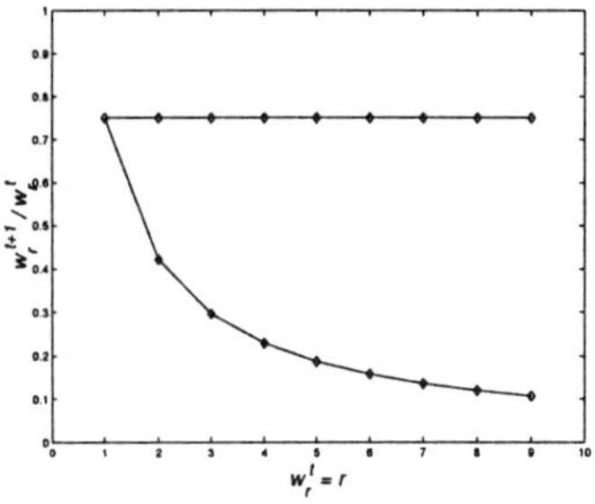

**Fig. 1.** Weight ratio after and before adaptation, with $\alpha^t = 0.25$, for the traditional SOM rule (constant horizontal line) and for the fisherman's rule (decreasing curve)

Figure 1 is a plot of the weight ratio $\boldsymbol{w}_r^{t+1}/\boldsymbol{w}_r^t$, showing that even without the neighborhood factor, the fisherman's rule induces an attenuation of the adaptation, as the distance to the winning neurons grows. This constatation does not imply that the neighborhood factor is useless in the fisherman's rule. Actually, it would be like running the traditional rule with $\lambda^t$ set as constant: convergence is not guaranteed.

## 3 Description of the experiments

Comparing the traditional learning rule and the fisherman's one is almost impossible because their nature is totally different. In order to make them comparable, one has to list their differences one by one and try each combination. Actually, there are two differences between both learning rules: the traditional rule is non-recursive and purely radial, while the fisherman's one is recursive but not radial. This leads to four combinations showed in equations 5a to 5d, table 1 and figures 2 and 3. The four rules can be written more or less the same way:

$$\Delta \boldsymbol{w}_r^t = \gamma_r^t \ \|\boldsymbol{x}_i - \boldsymbol{w}_r^t\| \ \frac{\boldsymbol{x}_i - \boldsymbol{w}_r^t}{\|\boldsymbol{x}_i - \boldsymbol{w}_r^t\|}, \tag{5a}$$

$$\Delta \boldsymbol{w}_r^t = \gamma_r^t \ \|\boldsymbol{w}_q^{t+1} - \boldsymbol{w}_r^t\| \ \frac{\boldsymbol{x}_i - \boldsymbol{w}_r^t}{\|\boldsymbol{x}_i - \boldsymbol{w}_r^t\|}, \tag{5b}$$

$$\Delta \boldsymbol{w}_r^t = \gamma_r^t \ \|\boldsymbol{x}_i - \boldsymbol{w}_r^t\| \ \frac{\boldsymbol{w}_q^{t+1} - \boldsymbol{w}_r^t}{\|\boldsymbol{w}_q^{t+1} - \boldsymbol{w}_r^t\|}, \tag{5c}$$

$$\Delta \boldsymbol{w}_r^t = \gamma_r^t \ \|\boldsymbol{w}_q^{t+1} - \boldsymbol{w}_r^t\| \ \frac{\boldsymbol{w}_q^{t+1} - \boldsymbol{w}_r^t}{\|\boldsymbol{w}_q^{t+1} - \boldsymbol{w}_r^t\|}. \tag{5d}$$

In each of these four equations, one finds from left to right the learning rate, the norm of the adaptation and its direction. To evaluate these four rules, two criteria are defined. The first one is the relative quantization error, written as:

$$E_{VQ} = \sqrt{\sum_i \|\boldsymbol{x}_i - \boldsymbol{w}_*\|^2} \Big/ \sqrt{\sum_i \|\boldsymbol{x}_i - \mu \boldsymbol{x}_i\|^2}, \tag{6}$$

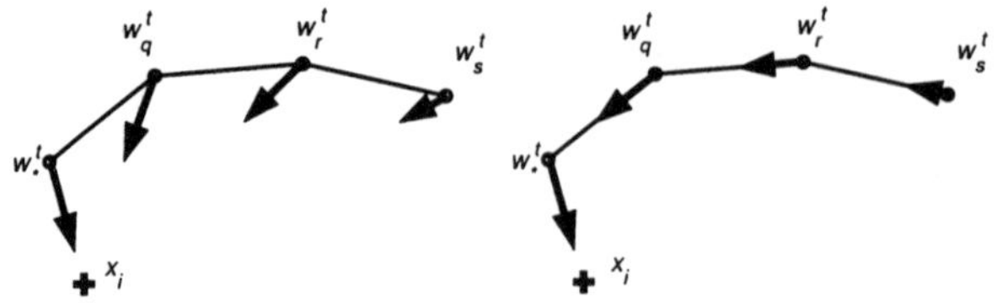

**Fig. 2.** Traditional SOM rule (left) and fisherman's rule (right)

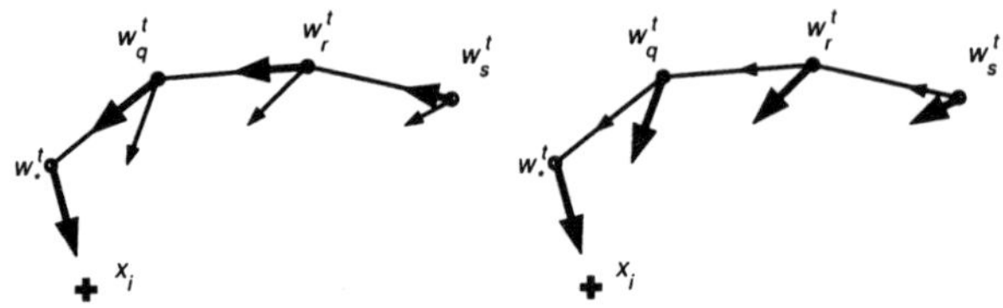

**Fig. 3.** Hybrid rules: non radial non recursive (left) and recursive radial (right)

**Table 1.** Rules

| | Non-recursive | Recursive |
|---|---|---|
| Radial | Rule 5a (Figure 2, left)<br>Traditional SOM Rule | Rule 5b (Figure 3, right)<br>Hybrid |
| Non-radial | Rule 5c (Figure 3, left)<br>Hybrid | Rule 5d (Figure 2, right)<br>Fisherman's rule |

where index $i$ traverses the whole learning set, $\boldsymbol{w}_*$ without time index $t$ is the winning neuron for stimulus $\boldsymbol{x}_i$ (after learning) and $\mu\boldsymbol{x}_i$ is the average of all stimuli. In the equation of $E_{VQ}$, the divisor is the quantization error with one prototype placed on the average of all stimuli. The second criterion measures the topology preservation:

$$E_T = \sum_s \sharp \left\{ r | r \neq s, r \notin N(s), \|\boldsymbol{w}_r - \boldsymbol{w}_s\| < \max_{q \in N(k)} \|\boldsymbol{w}_q - \boldsymbol{w}_s\| \right\}, \quad (7)$$

where $\sharp$ is the set cardinality operator and $N(k)$ is the set of direct neighbors of neuron $s$. This error criterion is specially designed for honeycombed maps and counts the number of prototypes which are badly positioned on the map. More precisely, $E_T$ approximates each hexagonal cell with a circle and counts all prototypes that are inside the circle while they should be outside. If the map is thoroughly unfolded, then the circles approximate very well the hexagonal cells and $E_T$ gives a value close to zero. On the contrary, when the map is crumpled, hexagonal cells are distorted, circles grow and so does $E_T$. Unlike other criteria [6], $E_T$ easily detects when maps are twisted, as shown in Figure 4.

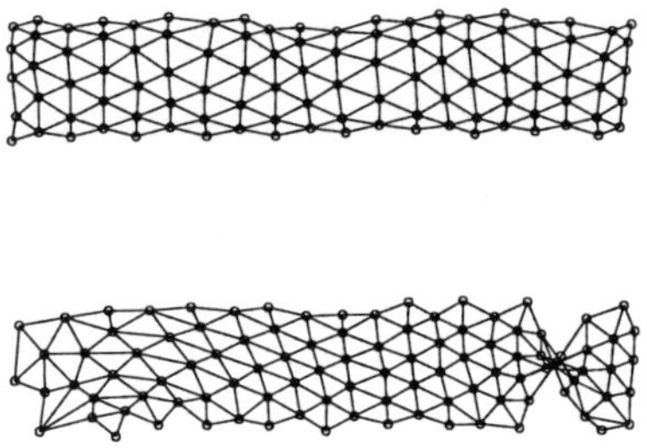

**Fig. 4.** Rectangular honeycombed maps with 100 neurons: if parameters are well chosen, then convergence is fast (first map), and $E_T$ is below 20; on the contrary, if $\lambda^t$ decreases too fast, then convergence is slower, twist effects may occur (second map) and $E_T$ around 100

The experiments were conducted on rectangular honeycombed maps (i.e. with hexagonal neighborhood structure), with 100 neurons, like shown in Figure 4). The training set was always the same: 2500 samples drawn from a uniform distribution of width 1 and length 4, so that the map can fit perfectly. The maps were randomly initialized with 100 of the 2500 samples and convergence was stopped after 5 epochs on the data set. Different values were tested for parameters $\alpha^t$ and $\lambda^t$, the initial value being always ten times larger than the fifth one (last epoch); the three intermediate values are computed with an exponential decrease from the initial to the final value. Each parameter configuration was repeated 300 times.

# 4 Results and Discussion

Table 2 and 3 summarizes some results for the four rules ($\lambda^t$ is given relatively to the largest grid distance). A first good result is that all four rules perform well when $\alpha^t$ and $\lambda^t$ values are well chosen. But, when parameters are too low or too high, some differences appear. The main difference exists between radial rules and non-radial ones, which are rapidly outperformed. Two rules remain: the traditional one and the recursive radial. When $\lambda^t$ is low, the traditional rule wins, but shortly. On the contrary, when $\lambda^t$ is high, the recursive rule is clearly better. Generally, when $\lambda^t$ is high, twist effects tend to disappear with the recursive rule. An intuitive explanation for this unexpected but good performance may be the following: after initialization, the map is strongly crumpled and therefore the path along the links from any neuron $r$ to the winning neuron $*$ can be much longer than the distance between stimulus and neuron $r$. This leads to stronger adaptation norms and better unfolding. Moreover, the recursive rule can move neurons beyond the stimulus (this occurs when the last neuron on the shortest path is further than stimulus). The poor performance of non radial rules is probably also due to initialization: when the map is crumpled, the adaptation vectors have a strong tangential part and a weak radial component, so that neurons move a lot, but without coherence in their direction.

**Table 2.** Mean values of $E_{VQ}$ for various parameter settings

| $\alpha^1$ | 0.5 | 0.5 | 0.5 | 0.5 | 0.7 | 0.7 | 0.7 | 0.7 |
|---|---|---|---|---|---|---|---|---|
| $\lambda^1$ | 0.3 | 0.5 | 0.7 | 0.9 | 0.3 | 0.5 | 0.7 | 0.9 |
| 5a | 0.0732 | 0.0871 | 0.1076 | 0.1317 | 0.0725 | 0.0869 | 0.1076 | 0.1320 |
| 5b | 0.0736 | 0.0851 | 0.0988 | 0.1145 | 0.0727 | 0.0843 | 0.0981 | 0.1132 |
| 5c | 0.0800 | 0.1118 | 0.1569 | 0.1957 | 0.0774 | 0.1093 | 0.1516 | 0.1918 |
| 5d | 0.0790 | 0.1016 | 0.1265 | 0.1495 | 0.0772 | 0.0984 | 0.1240 | 0.1480 |

**Table 3.** Mean values of $E_T$ for various parameter settings

| $\alpha^1$ | 0.5 | 0.5 | 0.5 | 0.5 | 0.7 | 0.7 | 0.7 | 0.7 |
|---|---|---|---|---|---|---|---|---|
| $\lambda^1$ | 0.3 | 0.5 | 0.7 | 0.9 | 0.3 | 0.5 | 0.7 | 0.9 |
| 5a | 0.88 | 5.67 | 16.96 | 35.07 | 3.27 | 9.53 | 24.67 | 44.89 |
| 5b | 1.51 | 2.95 | 5.71 | 16.46 | 0.70 | 5.36 | 7.37 | 16.44 |
| 5c | 62.22 | 137.64 | 264.68 | 421.46 | 42.53 | 131.15 | 238.25 | 398.88 |
| 5d | 32.46 | 79.94 | 152.38 | 236.06 | 24.79 | 64.36 | 137.83 | 223.80 |

# 5 Conclusion

This study has shown that the recursive radial rule is more robust than the traditional SOM rule when the neighborhood is wide. Actually, recursiveness fastens the convergence during the first epochs. Therefore, an idea for future work consists in combining both rules: the recursive rule for map initialization and rough unfolding, and the traditional one for fine tuning.

*Acknowledgement.* This work was realized with the support of the 'Ministère de la Région wallonne', under the 'Programme de Formation et d'Impulsion à la Recherche Scientifique et Technologique'. Michel Verleysen is Research Associate of the Belgian National Fund for Scientific Research (FNRS).

# References

1. T. Kohonen. Self-organization of topologically correct feature maps. *Biological Cybernetics*, 43:59-69, 1982.
2. T. Kohonen. Self-organization and Associaative Memory. Springer-Verlag, Berlin, 3rd edition, 1989.
3. A. Ahalt, A.K. Krishnamurthy, P. Chen and D.E. Melton. Competitive learning algorithms for vector quantization. *Neural Networks*, 3:277-290, 1990.
4. T. Martinetz and K. Schulten. A neural gas network learns topologies. In T. Kohonen *et al.*, editor, *IEEE International Conference on Artificial Neural Networks, Espoo, Finland*, volume 1, 397-407. Elsevier, 1991.
5. E.W. Dijkstra. A note on two problems in connection with graphs. *Numerical Math.*,1:69-271, 1959.
6. H.U. Bauer and K.R. Pawelzik. Quantifying the neighborhood preservation of self-organizing maps. *IEEE Transactions on Neural Networks*, 3:570-579, 1992.

# Induced Voronoï Kernels for Principal Manifolds Approximation

Michaël Aupetit Pierre Couturier Pierre Massotte

LGI2P – EMA – Site EERIE
Parc Scientifique Georges Besse
30035 Nîmes

Abstract

We present a new interpolation technique allowing to build an approximation of *a priori* unknown principal manifolds of a data set which may be non-linear, non-connected and of various intrinsic dimension. This technique is based onto the Induced Delaunay Triangulation of the data built by a Topology Representing Network. It opens the way to a new field of research and applications in data analysis, data modeling and forecasting.

# 1 Introduction

The notion of Principal Manifold has been introduced by Hastie and Stuetzle [1]. The idea is to represent a data distribution by a nonparametric model which has "to pass through the middle of the data set" generalizing the self-consistency property of principal components to non-linear manifolds. In [2], Mulier and Cherkassky pointed out that Kohonen Self-Organizing Maps (SOMs) may be viewed as discrete equivalent to Principal Manifolds, one-dimensional and two-dimensional maps corresponding to Principal Curves and Principal Surfaces respectively. Some authors tried to get continuity in SOMs as Ritter [3], Walter [4], Göppert [5] and in our previous works [6]. However, to use these approaches to approximate Principal Manifolds, the dimensionality of the model (i.e. the dimension of the SOM) has to be specified *a priori*. As far as we know, no existing technique is able to deal with possibly non-connected distributions with *a priori* unknown varying intrinsic dimension. We present a way to come over these limits.
Our approach is based on Topology Representing Networks (TRNs) presented by Martinetz in [7]. A TRN is a set of units which represents the data set using a Vector Quantization algorithm (e.g. the "Neural-Gas" [8] ) and which at the same time, learns the topology of this data set, dynamically creating and deleting links between the units thanks to a Hebb-like rule (This is called a Competitive Hebbian Learning or CHL). The resulting graph structure is a subpart of the Delaunay triangulation of the units in the data space, which covers only the regions where the data fell into, and which is called the "Induced Delaunay Triangulation" (IDT). The IDT is made of simplices whose dimension approximates the local intrinsic dimension of the data distribution. Here following the idea of Mulier and Cherkassky about the close connections between SOMs and Principal Curves, we assume that a TRN gives a discrete equivalent to the *collection* of Principal

Manifolds of a possibly non-connected and non-linear distribution which may also have a varying intrinsic dimension. We show that TRNs are an appealing basis to construct smooth Principal Manifolds.

In our approach, the TRN gives a skeleton to the approximating manifolds of the data distribution. Each point inside a simplex of the TRN is associated to an "output" point of the same space through an interpolating function which create a non-linear manifold homeomorphic to that simplex.

When a datum is presented, it is first projected onto the "closest" simplex of the IDT (the definition of this closeness is given hereafter). Then, the closest point (in the sense of the Euclidean norm) of the approximating manifold to the input datum is found using an iterative gradient descent. At last, a gradient descent step is performed to adapt the parameters of the model and a new datum is presented. The detailed algorithm is presented in Section 3.

The interpolation technique presented allows to get a possibly non-linear manifold and to decide onto which simplex the datum has to be projected. This technique leans on bases functions that we call "Induced Voronoï Kernels" (IVK) and which have the orthonormality property, i.e. each unit projects on itself. It makes easier to get the self-consistency property, i.e. any point of the approximating manifold projects on itself. Note that basis functions like gaussian kernels of RBFs networks [9] cannot ensure orthonormality, while interpolation techniques such as B-splines [10], Lagrangian interpolant [4] or Natural Neighbors [11] ensure it but suffer from their inadequacy with the IDT and their high memory or computing time complexity especially in dimension higher than 3.

IVKs avoid nasty computation using geometrical properties of Voronoï regions.

## 2 Induced Voronoï Kernels

The Induced Voronoï Kernels are based onto the Induced Voronoï Tessellation of the units that we define as the dual structure of the Induced Delaunay Triangulation.

Considering $W=(\mathbf{w}_1;\ldots;\mathbf{w}_n)$ a set of n vectors in $R^d$ that we call units in the following, an edge of the IDT is created between the two closest units (in the sense of the euclidean norm) to the current input datum $\mathbf{v}$ of $R^d$. The IDT is a subpart of the Delaunay Triangulation (DT) of W in $R^d$ (More details may be found in [7]). Let N(i) be the set of indices of all the units in W, connected to the unit i through an edge of the Induced Delaunay Triangulation. We define the Induced Voronoï Tessellation (IVT) as the collection of Induced Voronoï regions attached to each unit i :

$$V_i^{\text{induced}} = \{\mathbf{v} \in R^d \,|\, \forall j \in N(i), \|\mathbf{v} - \mathbf{w}_i\| \le \|\mathbf{v} - \mathbf{w}_j\| \}$$

Note that Induced Voronoï regions of different units may overlap.

Now we define $H_{(\mathbf{w}_i,2)}(\mathbf{v})$ ,the homothety of center $\mathbf{w}_i$ and factor 2. It is easy to show that the region obtained through the homothety H of the Induced Voronoï region of the unit i is such that any neighboring unit N(i) of i is located on the border of that region. We exploit this property to create Induced Voronoï Kernels (IVKs) based on these magnified Induced Voronoï regions (Figure 1a).

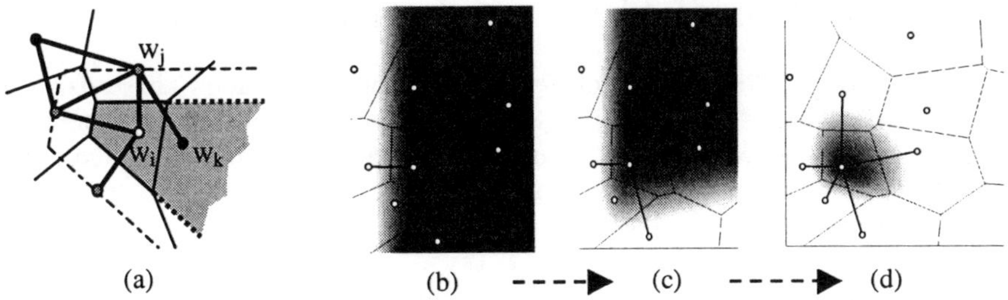


**Figure 1**: (a) In gray, the Induced Voronoï region of the unit i (white circle) depending on the position of its neighbors (gray circles). Thin plain lines are the edges of the Voronoï regions, bold plain lines the edges of the Induced Delaunay Triangulation and dash-dotted lines the edges of the magnified Induced Voronoï region of i the Induced Voronoï Kernel leans onto. (b)(c)(d) The way the Induced Voronoï Kernel of a unit is built using a tensor product of cubic splines (white is 0 and black is 1). All of this generalizes to any dimension.

A set of cubic spline functions is used to defined the IVKs. Considering an input datum $\mathbf{v}$, we compute the IVK associated to the unit i by projecting $\mathbf{v}$ in $\mathbf{v_p}$ orthogonally onto each neighboring link with vertices $\mathbf{w_i}$ and $\mathbf{w_j}$ (j in N(i)).
The position of $\mathbf{v_p}$ onto the line ($\mathbf{w_i w_j}$) gives an input value to the cubic spline attached to that link. While $\mathbf{v_p}$ passes continuously from $\mathbf{w_i}$ to $\mathbf{w_j}$, the output of the spline passes continuously (at the first order) from 1 to 0, is 0 beyond $\mathbf{w_j}$ and 1 before $\mathbf{w_i}$. The IVK of the unit i is the tensor product of all such spline functions (Figure 1b-c-d). Hence, the IVK is equal to 0 outside the magnified Induced Voronoï region of $\mathbf{w_i}$ and equal to 1 at $\mathbf{w_i}$.
The formal definition of the IVK of a unit i for a given vector $\mathbf{v}$ of $R^d$ is:

$$H_i(\mathbf{v}, W) = \prod_{q \in N(i)} h(u_{iq}(\mathbf{v})) \text{ with } h(t) = \begin{cases} 1 & \text{if } t < 0 \\ (2t+1)(t-1)^2 & \text{if } t \in [0,1] \\ 0 & \text{if } t > 1 \end{cases}$$

$$\text{and } u_{iq}(\mathbf{v}) = \frac{(\mathbf{v} - \mathbf{w}_i)^T (\mathbf{w}_q - \mathbf{w}_i)}{(\mathbf{w}_q - \mathbf{w}_i)^2}$$

In practice, *only* the IVKs of the closest unit i to $\mathbf{v}$ and of its neighbors N(i) are computed. This allows to take care about the IDT and to ensure the continuity of the non-linear manifold at the edges of the Voronoï Tesselation, as far as the IDT is not "pathological". An example of "pathology" is presented on the Figure 1a where IVKs of units i and k overlap in such a way that if $\mathbf{v}$ passes continuously from the Voronoï region of $\mathbf{w_j}$ to that of $\mathbf{w_k}$, discontinuities may appear at the border. However, such "pathologies" are unlikely to occur because the units i and k being closer to each other than to the third unit j, they are very likely to be linked if such a $\mathbf{v}$ is presented during the learning phase of the IDT.
Thus computed, IVKs ensure the orthonormality property (hence the self-consistency) such that the closest unit to $\mathbf{v}$ is the only one activated if $\mathbf{v}$ merges with it, and it is so even in "pathological" cases.
We can show that the units activated by $\mathbf{v}$ are a subset of the *natural neighbors* of $\mathbf{v}$, i.e. $\mathbf{v}$ would be connected to them if the DT of the units W *and* $\mathbf{v}$ were built. More precisely, these activated units are all 0.5-Observable neighbors of $\mathbf{v}$ (see [12] for the definition of the $\gamma$-Observability).

# 3 Principal manifolds: principle and algorithm

The first part of the training consists in a Vector Quantization phase using the "Neural-Gas" and CHLs algorithms. It makes the units respecting the distribution and learning its topology. Then, the units are frozen and the following algorithm is applied to project the input datum onto the approximating manifold and tune its parameters.

1. The activation $H_{win}$ (i.e. the IVK) of the closest "winning" unit win to the input datum **v** and those of its neighbors N(win) are computed. We call the set of activated units $S_{activ}$ (Figure 2a). The winning unit is defined as:
$$\mathrm{win} = \arg \min_{j \in \{1..n\}} (\|\mathbf{v} - \mathbf{w}_j\|)$$

2. The units of $S_{activ}$ are sorted in the decreasing order of their activation and the "winning" unit is set as the first element of $S_{simplex}$. If the next unit j in the ordered list is connected to all the units already in $S_{simplex}$, then j is added to $S_{simplex}$ and the next unit of the list is considered. If the considered unit of the ordered list is not connected to *all* those of $S_{simplex}$, then the loop stops. $S_{simplex}$ defines the simplex onto which **v** should be projected (Figure 2b).

3. This simplex gives implicitly the dimension of the parametric model (i.e. the local dimension of the approximating manifold). Starting from the winning unit $\mathbf{w_{win}}$ as the "origin" unit $\mathbf{w_{orig}}$, the aim is to find the set of coordinates K of the vector $\mathbf{v_p}$ in the basis formed by the simplex $\{(\mathbf{w_j}-\mathbf{w_{orig}})\}_{j \text{ in } S_{simplex}\setminus\{orig\}}$, such that $\mathbf{w}(\mathbf{v_p}^*)$ is the closest point of the approximating manifold to **v** (Figure 3a):

$$\mathbf{w}(\mathbf{v}_p^*) = \arg \min_K (\|\mathbf{v} - \mathbf{w}(\mathbf{v}_p)\|) \text{ with } \mathbf{v}_p = \mathbf{w}_{orig} + \sum_{\substack{j \in S_{simplex} \\ j \neq orig}} k_j \cdot (\mathbf{w}_j - \mathbf{w}_{orig})$$

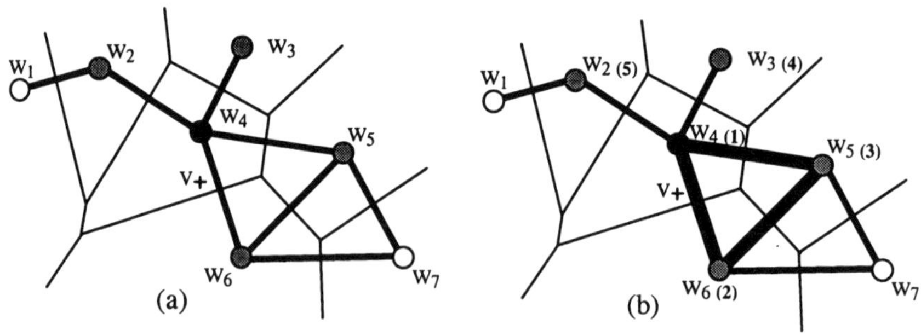


**Figure 2**: (a) the winning unit (black circle) is the closest unit to the datum **v**. The IDT (bold lines) defines the neighbors of the winning unit (gray circles). The IVK of all black and gray circles is computed and they all belong to $S_{activ}$. (b) The activated units are then ranked (number between parentheses) in the decreasing order of their activation. $S_{simplex}$ the simplex onto which **v** should be projected, is determined as the largest simplex of the IDT which contains successive ordered units starting from the winning unit (large bold lines).

where $\mathbf{w}_{orig}$, $S_{simplex}$ and $S_{activ}$ which were first initialized according to the position of $\mathbf{v}$, are dynamically found in the same way according to $\mathbf{v_p}$ such that $\mathbf{v_p}$ remains inside the current simplex during the minimization phase. $\mathbf{w}(\mathbf{v_p})$ is defined as:

$$\mathbf{w}(\mathbf{v_p}) = \frac{\sum_{i\in S_{activ}} H_i(\mathbf{v_p}, W)\cdot \mathbf{L}_i(\mathbf{v_p})}{\sum_{i\in S_{activ}} H_i(\mathbf{v_p}, W)} \quad \text{with } \mathbf{L}_i(\mathbf{v_p}) = A_i * (\mathbf{v_p} - \mathbf{w}_i) + \mathbf{w}_i$$

$\mathbf{A}_i$ is a gradient matrix coding the linear deformation of the space around the unit i. The initial value of $\mathbf{A}_i$ is the identity matrix. $\mathbf{w}(.)$ is an interpolation function between *all* the units of $S_{activ}$ and not only those of $S_{simplex}$. This allows to ensure the continuity at the first order between neighboring approximating manifolds of the same dimension (e.g. a set of neighboring triangles creates a continuous surface which passes through all their vertices). The normalization in $\mathbf{w}(.)$ gives the partition-of-unity property which makes easier the approximation of linear manifolds (e.g. if $\mathbf{A}_i$ is the identity matrix, then $\mathbf{w}(\mathbf{v_p})=\mathbf{v_p}$ and the approximating manifold is directly the simplex).
The minimization to find $\mathbf{w}(\mathbf{v_p}^*)$ consists in an iterative gradient descent onto the cost function: $E=0.5(\mathbf{v}-\mathbf{w}(\mathbf{v_p}))^2$.
The rule to find the coordinates of $\mathbf{v_p}^*$ is: $\Delta k_j = \alpha \cdot \frac{\partial \mathbf{w}(\mathbf{v_p})}{\partial k_j} \cdot (\mathbf{v} - \mathbf{w}(\mathbf{v_p}))$

with

$$\frac{\partial \mathbf{w}(\mathbf{v_p})}{\partial k_j} = \frac{\sum_{i\in S_{activ}} H_i(\mathbf{v_p}, W)\cdot \left(\mathbf{A}_i * (\mathbf{w}_j - \mathbf{w}_{orig}) + (\mathbf{L}_i(\mathbf{v_p}) - \mathbf{w}(\mathbf{v_p}))\cdot \sum_{q\in N(i)} \frac{s_{q,j,i,orig}\cdot h(u_{iq}(\mathbf{v_p}))}{h'(u_{iq}(\mathbf{v_p}))}\right)}{\sum_{i\in S_{activ}} H_i(\mathbf{v_p}, W)}$$

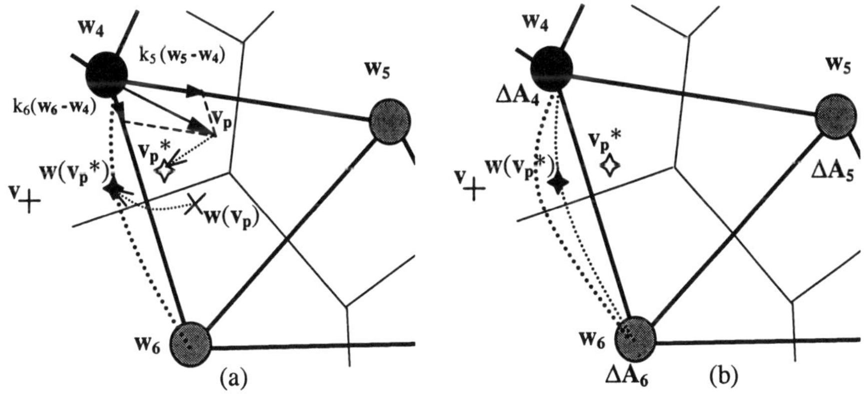


**Figure 3**: (a) $\mathbf{v_p}$ is located in the basis $(\mathbf{w}_5\text{-}\mathbf{w}_4, \mathbf{w}_6\text{-}\mathbf{w}_4)$ formed by the simplex with the coordinates $(k_5,k_6)$. $\mathbf{v_p}^*$ is found minimizing (dotted arrows) the distance between $\mathbf{w}(\mathbf{v_p})$ and $\mathbf{v}$, by tuning the coordinates $(k_5,k_6)$ of $\mathbf{v_p}$. $\mathbf{w}(\mathbf{v_p}^*)$ is then the projection of $\mathbf{v}$ onto the approximating manifold (dotted line). (b) The parameters of the approximating manifold ($\mathbf{A}_i$ matrices) are tuned to make it closer to the datum (bold dotted line).

where

$$s_{q,j,i,orig} = \frac{(\mathbf{w}_j - \mathbf{w}_{orig})^T (\mathbf{w}_q - \mathbf{w}_i)}{(\mathbf{w}_q - \mathbf{w}_i)^2} \text{ and } h'(t) = \begin{cases} 0 & \text{if } t \notin [0,1] \\ 6t(t-1) & \text{if } t \in [0,1] \end{cases}$$

The iterative process to find the set of coordinates K stops when the variation of the norm $\|\mathbf{v}-\mathbf{w}(\mathbf{v_p})\|$ between two successive steps is below a threshold T. We chose this set of coordinates for $\mathbf{v_p}$ instead of the euclidean one, because it ensures in a very straight and simple way that $\mathbf{v_p}$ remains *inside* the current simplex during the minimization process (i.e $k_i$ must be between 0 and 0.5 else a new simplex is considered according to $\mathbf{v_p}$).

**4**. when $\mathbf{v_p}^*$ has been found, the approximating manifold is tuned to get closer to the data. Thus, the matrix $\mathbf{A}_i$ of each unit of $S_{activ}$ (i.e. the units activated by $\mathbf{v_p}^*$) is adapted following an iterative gradient descent onto E (Figure 3b):

$$\Delta a_i^{[r,c]} = \alpha \cdot \frac{H_i(\mathbf{v_p^*}, W)}{\sum_{j \in S_{activ}} H_j(\mathbf{v_p^*}, W)} \cdot (v_p^{*[c]} - w_i^{[c]}) \cdot (v^{[r]} - w^{[r]}(\mathbf{v_p^*})) \text{ with } \mathbf{A}_i = \begin{pmatrix} a_i^{[1,1]} & \cdots & a_i^{[1,c]} & \cdots & a_i^{[1,d]} \\ \vdots & \ddots & & & \vdots \\ a_i^{[r,1]} & & a_i^{[r,c]} & & a_i^{[r,d]} \\ \vdots & & & \ddots & \vdots \\ a_i^{[d,1]} & \cdots & a_i^{[d,c]} & \cdots & a_i^{[d,d]} \end{pmatrix}$$

where $\mathbf{x}^{[m]}$ denotes the $m^{th}$ coordinate of $\mathbf{x}$. Note that the optimal configuration of the $a_i$ is attainable simply using a gradient descent because E is quadratic in $\mathbf{A}_i$.

**5**. return to step **1**.

# 4 Experiments

We use a Neural-Gas of 20 units during 100000 learning steps to quantify a noisy non-connected distribution with various intrinsic dimensions in $R^3$ (Figure 4a). The topology of the distribution is learnt using Competitive Hebbian Learning. Then we apply the algorithm described above and adapt the gradient term of each unit for 5000 steps with the noisy distribution. Figure 4b shows the noisy set and the organization of the units after the quantization. We set $\{\alpha,T\}=\{0.1,0.0001\}$.
Figure 4c shows the approximating manifolds after the adaptation of the gradient terms. This shows the ability of this approach to approximate non-linear and possibly non-connected manifolds of *a priori* unknown intrinsic dimension.
Units in the "flag" part of the distribution are not very well distributed to give a good model of the local manifold. This shows that the quality of the model strongly depends on the number of units and their location. This is part of our future work to improve the placement of the units onto the distribution, for example by tuning the $\mathbf{w}_i$ following an iterative gradient descent onto E at the step 4 of the algorithm. The IDT giving the dimension of the approximating manifold, needs also fine tuning to respect as well as possible the topology of the underlying distribution. We also investigate the use of the Levenberg-Marquardt optimization algorithm [13] to find $\mathbf{v_p}^*$ in a more efficient way.

# 5 Conclusion

Induced Voronoï Kernels are a way to interpolate in spaces of possibly high dimension. They are built from the Induced Delaunay Triangulation which is learnt and reflects closely the data structure. We present how IVKs may be used to approximate *a priori* unknown Principal Manifolds possibly non-linear, non-connected and of various intrinsic dimension while representing their topology. The next step is to use such models to realize a continuous associative memory: presenting to the model an incomplete vector to get its missing components. This approach has been studied with PSOMs and used successfully by Walter [4]. IVKs may be also used in place of Gaussian kernels in RBF networks applied to function approximation, coming over the problem of setting the width of the kernels which would be set automatically according to the position of the neighboring units.
At last, we emphasize that IVKs give a solution to a problem untractable with existing methods so far. The very aim of this approach is to give a model of a data distribution as complete as possible, i.e. being continuous and preserving its topological properties (connectivity and intrinsic dimension). IVKs in the spirit of SOMs, are based on topological links binding the units together. However this approach is in no way a tool to visualize high-dimensional data distribution, although it might be used to help their analysis.
IVKs let's foresee new applications in data modeling (3D reconstruction, function approximation), in data analysis and forecasting (Principal Manifolds, continuous associative memory, continuous projection...) and in system control (Q-learning, continuous path planning in robotics based on the IDT).

### References


[1] Hastie T, Stuetzle W. Principal curves. Journal of the American Statistical Association 1989, vol. 84, no. 406, pp. 502-516.

[2] Mulier F, Cherkassky V. Self-organization as an iterative kernel smoothing process. Neural Computation 1995, vol. 7, pp. 1165-1177.

[3] Ritter H. Parametrized self-organizing maps. In S. Gielen and B. Kappen Eds., Proc. of the Int. Conf. on Art. Neural Networks 1993, pp. 568-575, Springer Verlag, Berlin.

[4] Walter J. Rapid learning in robotics. Cuvillicr Vcrlag, Göttingen, Germany. Url:http://www.techfak.uni-bielefeld.de/~walter/

[5] Göppert J, Rosenstiel W. Interpolation in SOM: Improved generalization by iterative methods. In EC2&Cie eds., Proc. of the Int. Conf. on Artificial Neural Networks 1995, vol. 10, Paris, France.

[6] Aupetit M, Couturier P, Massotte P. Function approximation with continuous self-organizing maps using neighboring influence interpolation. Proc. of NC'2000, Berlin.

[7] Martinetz T, Schulten K. Topology Representing Networks. Neural Networks 1994, vol. 7, no. 3, pp. 507-522, Elsevier Science.

[8] Martinetz T, Berkovich S, Schulten K. "Neural-Gas" network for vector quantization and its application to time-series prediction. IEEE Trans. on Neural Networks 1993, vol. 4, no. 4, pp. 558-569.

[9] Moody J, Darken C. Fast learning in networks of locally-tuned processing units. Neural Computation 1989, vol. 1, pp. 281-294, MIT.

[10] Bartels R, Beaty J, Barsky B. B-splines. Mathématiques et CAO, vol.4, Hermès 1988.

[11] Sibson R. A brief description of natural neighbour interpolation. Interpreting Multivariate Data, pp. 21-36, V. Barnet eds., Wiley, Chichester 1981.

[12] Aupetit M, Couturier P, Massotte P. Vector Quantization with γ-Observable Neighbors. Workshop on Self-Organizing Maps (WSOM01), Lincoln, UK, June 2001.

[13] Marquardt D. J. Soc Appl. Math 1963, vol. 11, pp.431-441

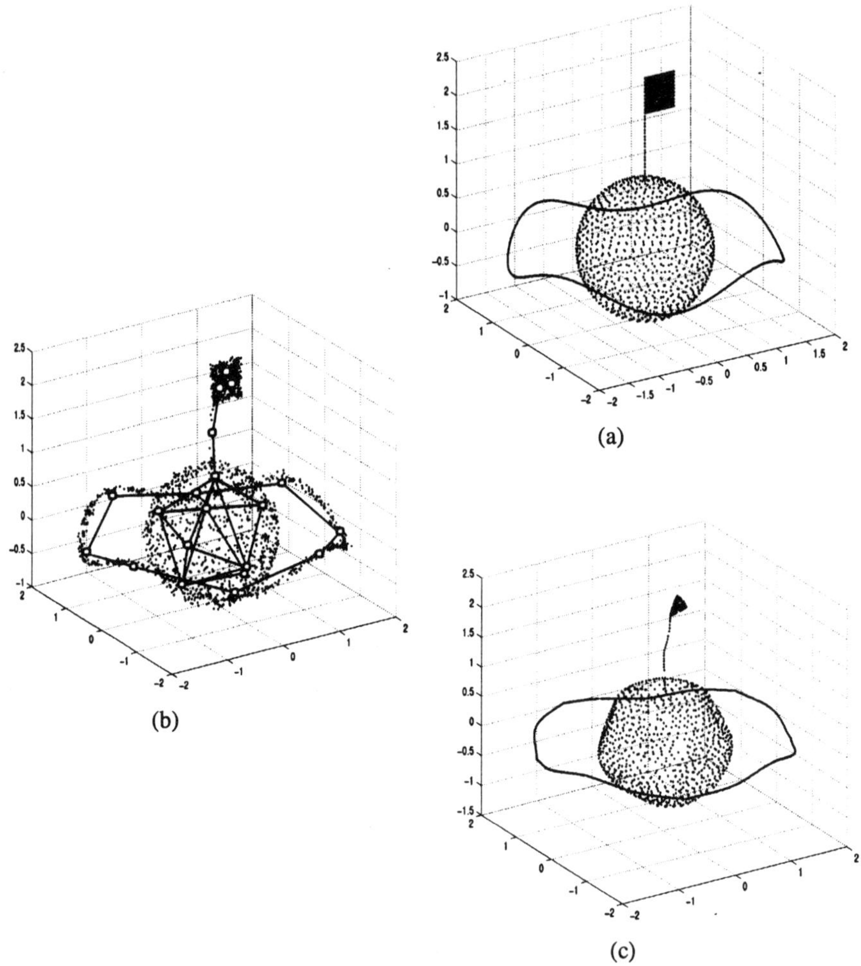


**Figure 4**: (a) The principal manifolds to approximate. (b) The 20 units of the TRN representing the topology of the noisy data set. (c) The resulting non-linear non-connected approximating manifolds using IVKs.

# Visualisation Induced SOM (ViSOM)

Hujun Yin
Dept. of Electrical Engineering and Electronics, UMIST
Manchester, M60 1QD, UK
h.yin@umist.ac.uk

Abstract

When used for visualisation of high dimensional data, the self-organising map (SOM) requires a colouring scheme such as U-matrix to mark the distances between neurons. Even so, the structures of the data clusters may not be apparent and their shapes are often distorted. In this paper, a visualisation-induced SOM (ViSOM) is proposed as a new tool for data visualisation. The algorithm constrains the lateral contraction forces between a winning neuron and its neighbouring ones and hence regularises the inter-neuron distances. The mapping preserves directly the inter-neuron distances on the map along with the topology. It produces a graded mesh in the data space and can accommodate both training data and new arrivals. The ViSOM represents a class of discrete principal curves and surfaces.

## 1. Introduction

Sammon [1] pioneered the nonlinear mapping for visualisation of multivariate data, when the linear principal component analysis (PCA) was the primary tool for dimension reduction. The objective of the Sammon mapping is to minimise the differences between inter-pattern distances in the original space and those in the projected space. The projection of data from a high-dimensional space to a low one (usually two dimensions) can reveal the data structures and cluster tendency. The Sammon mapping has been shown to be superior to PCA for data structure analysis [1]. However, the Sammon algorithm is point-to-point mapping, which does not provide the explicit mapping function and cannot accommodate new data points [1,2]. For any additional data, the projection has to be re-calculated from scratch based on all data points. This may prove difficult or even impossible for many practical applications where data arrives sequentially and memory space for the data is limited.

Neural networks open up an alternative approach to the nonlinear data analysis. They are biological inspired adaptive mapping methods and can learn complex nonlinear relationships of variables in a system from data samples. Mao and Jain have given an overview on this topic [2]. The Kohonen's SOM is a simplified mathematical model of the topological mapping in the cerebral cortex [3]. The inputs to the SOM are often drawn from high dimensional space, while the output, i.e. the map, is often a 2-D grid. The algorithm has been used as a tool for dimensional reduction [3-5]. One of greatest properties of the SOM is its topological preservation. That is, close points in the input space will be mapped to nearby neurons in the map space. Such properties can be employed to visualise *relative* or *qualitative* mutual relationships among inputs. However, as the inter-neuron distances are not directly visible from the map, one has to use a colouring scheme such as U-matrix [4,5] or interpolations [6] to mark relative distances between the weights of nearby neurons referred in the input space. Even so, the structures of the data clusters may not be apparent or appear often distorted.

In this paper, a new variant, namely the visualisation induced self-organising map (ViSOM), is proposed. The ViSOM projects the high dimensional data in an unsupervised manor as does the SOM, but constrains the lateral contraction force and hence regularises the inter-neuron distance to a parameter, which defines and controls the resolution of the projection. It preserves the data structure as well as the topological order as faithfully as possible. The ViSOM is a nonlinear projection but of a simple computational structure. The analysis and experimental results show that the ViSOM has great advantages over the commonly used SOM, PCA and Sammon methods. It represents a discrete (more accurately, uniformly quantised) principal curves or surfaces [7].

# 2 Data Projection Methods

Searching for better and suitable data projection methods has always been an integral aim of pattern recognition and data analysis. Such a method will enable us to observe and detect underlying data distributions and pattern structures. A great deal of research has been devoted to this subject and a number of useful methods have been proposed. Some key methods and algorithms are briefly reviewed next.

## 2.1 PCA and Principal Curve/Surface

PCA is a classic linear data analysis method aiming at finding orthogonal principal directions from a set of data, along which the data exhibit the largest variances. By discarding the minor components, the PCA can effectively reduce data variables and display the dominant ones in a linear low dimension. It is the optimal linear projection in the mean-square error sense.

The extension to nonlinear PCA, however, has not been straightforward, due to the lack of a unified mathematical structure and an efficient or reliable algorithm, and in some cases due to excessive freedom in selection of representative basis functions [8]. Principal curves and surfaces [7,9] are primary nonlinear extension of PCA. They define a smooth self-consistent curve/surface, i.e. each point on the curve/surface is the mean of the points mapped onto it. However, a valid algorithm is required for a good implementation.

## 2.2 Sammon Mapping

A traditional subject related to dimension reduction and data projection is multidimensional scaling (MDS), which includes the PCA as one of the projection methods. The MDS searches for a mapping to a low (usually 2) dimensional plot, on which inter-point distances of projected points resemble those inter-point distances in the original space [9,10]. Sammon proposed a recursive learning algorithm using the Newton optimisation method for the optimal configuration that minimises the so-called *Sammon stress* [1].

In addition to being computationally costly and not adaptive, one major drawback of the MDS methods including the Sammon mapping is lack of an explicit projection function. Thus for any new input data, the mapping has to be recalculated based on all available data. These methhods become impractical for applications where data arrive consecutively and/or for predictions on new data points. Mao and Jain have proposed a feedforward neural network to parameterise the Sammon mapping function and a back-propagation training algorithm has been derived for minimising the Sammon stress [2].

### 2.3 SOM

The SOM is an unsupervised learning algorithm that uses a finite grid of neurons to frame the input space. The SOM tends to cluster or quantise the data space, find the representatives or features, and minimise the mean-square errors between features and the data points they represent [11],

In the SOM, a neighbourhood conscience learning is adopted to form topological ordering among the neurons in the map, thus the map can be used to show the *relative* relationships among the data points. However, the SOM does not show the inter-neuron distances on the map. A colouring scheme such as U-matrix is needed to imprint the inter-neuron distances so that the boundaries between clusters can be marked. Even so, the cluster structures often appear in distorted and unnatural forms. A faithful display of structure shapes and a direct display of distributions of the data set are more desirable for visualisation.

## 3. Visualisation induced SOM

In order to capture the data structure on the map, the distance quantities need to be preserved on the map. The key of the ViSOM is to constrain the lateral contraction forces between a winning neuron and its neighbouring ones and hence to regularise the inter-neuron distances.

### 3.1 ViSOM Structure and Derivation

The ViSOM uses a similar grid structure of neurons to the SOM. It consists of an array of nodes arranged in a low dimensional rectangular or hexagonal grid. Each node, indexed $c$ (index $c=(i,j)$ for a 2-D map), has an associated weight vector, $\mathbf{w}_c=[w_{c1}, w_{c2}, \ldots\ w_{cn}]^T$, where $n$ is the input dimensions.

At time step $t$, an input $\mathbf{x}(t)$ is drawn randomly from the data set or data space. A winning neuron $v$ can be found according to its distance to the input, i.e.,

$$v = \arg\min_{c\in\Omega} \left\| \mathbf{x}(t) - \mathbf{w}_c \right\| \tag{1}$$

where $\Omega$ denotes the map space, i.e. the grid.

Then the SOM updates the weight of the winning neuron according to,

$$\mathbf{w}_v(t+1) = \mathbf{w}_v(t) + \alpha(t)[\mathbf{x}(t) - \mathbf{w}_v(t)] \tag{2}$$

where $\alpha(t)$ is the learning rate at time $t$.

The weights of the neurons in the neighbourhood, in the SOM, are updated as,

$$\mathbf{w}_k(t+1) = \mathbf{w}_k(t) + \alpha(t)\eta(v,k,t)[\mathbf{x}(t) - \mathbf{w}_k(t)] \tag{3}$$

where $\eta(v,k,t)$ is the neighbourhood function, which is monotonically decreasing with $\|v\text{-}k\|$. The second term in (3), especially the updating force, can be rearranged and decomposed into two components,

$$\mathcal{F}_{kx} \equiv \mathbf{x}(t) - \mathbf{w}_k(t) = [\mathbf{x}(t) - \mathbf{w}_v(t)] + [\mathbf{w}_v(t) - \mathbf{w}_k(t)] \equiv \mathcal{F}_{vx} + \mathcal{F}_{kv} \tag{4}$$

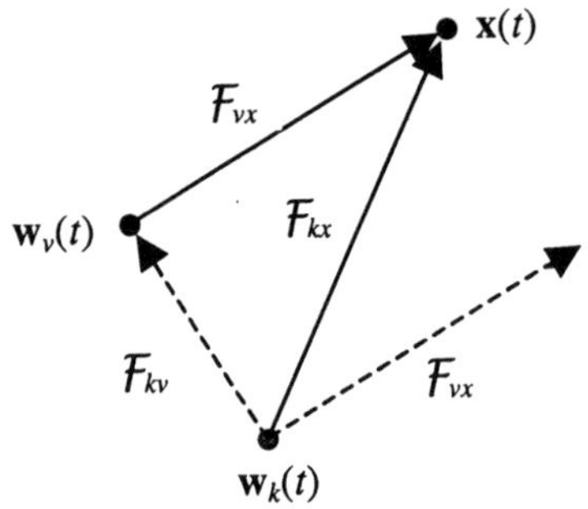


Figure 1. Decomposition of the updating force.

This can be shown in Fig. 1. The first force, $\mathcal{F}_{vx}$, represents the updating force bringing the winner $v$ to the input $\mathbf{x}$, which is the same as that used by the winner in Eqn. (2). While the second force, $\mathcal{F}_{kv}$, represents a lateral force bringing neuron $k$ to the winner $v$, i.e. a contraction force. It is this contraction force that brings neurons in the neighbourhood towards the winner and thus forms a contraction around the winner on the map at each time step. The ViSOM constrains this lateral contraction force by regularising the distance between a neighbouring neuron to the winner. If the $\mathbf{w}_k$ is far from $\mathbf{w}_v$ under a pre-specified resolution, this force remains, otherwise an opposite force applies. The scale of this force is controlled by a normalised distance between these two weights, see Eqn (5) in the next subsection. As each neuron represent the mean of the data points mapped onto it, and the inter-neuron distances are uniformly quantised or digitised. The ViSOM represent a uniformly quantised principal curve/surface, while the common SOM represent an entropy quantised principal curve/surface.

## 3.2 ViSOM Algorithm

The details of the derived ViSOM algorithm are given below.

1) Finding the winner from Eqn. (1),
2) Updating the winner according the Eqn. (2),

3) Updating the neighbourhood according to,

$$\mathbf{w}_k(t+1) = \mathbf{w}_k(t) + \alpha(t)\eta(v,k,t)\{[\mathbf{x}(t) - \mathbf{w}_v(t)] + [\mathbf{w}_v(t) - \mathbf{w}_k(t)]\frac{(d_{vk} - \Delta_{vk}\lambda)}{\Delta_{vk}\lambda}\} \quad (5)$$

4) where $d_{vk}$ and $\Delta_{vk}$ are the distances between neurons $v$ and $k$ in the data space and on the unit grid or map respectively, and $\lambda$ is a pre-specified resolution parameter. The smaller $\lambda$ is, the higher resolution the map can provide. $\lambda$ can be made adaptive according to the data variance and required resolution.
5) Refreshing the map by choosing randomly the weights of the neurons as inputs at a small percentage of updating times (e.g. 10% iterations). This step can be optional.

A refreshing phase is added to help the map's smooth expansion to the areas where there are few or none data points as well as to regularise the distances between them. It enables the map to extrapolate smoothly and uniformly beyond the data regions. Then the map is able to position and display any outliers in less density regions on the map. It is particularly useful in applications such as novelty detection, where a SOM can only overlay the outliers to the existing data regions, often at the borders of the grid.

The key feature of the ViSOM is that the distances between the neurons on the map reflect the corresponding distances in the data space. That is, the distant measure is faithfully preserved on the map (at least locally). When the map is trained and data points are mapped on the map, the distances between mapped data points on the map will resemble approximately those in the original space (subject to the resolution of the map). This makes visualisation more direct and quantitatively measurable. The size of the neighbourhood function should decrease from initial large value to a final value, which should not be zero, however. The rigidity of the map is controlled by this ultimate size. The larger this size is the flatter of the final map in the data space will be.

# 4 Experimental Results

## 4.1 Illustrative Data Set

To demonstrate the applications of the ViSOM and its structure preserving properties and in comparison to the SOM, an illustrative data set of 2 dimensions was used. 100 points were randomly drawn from each of two Gaussian sources. Their mean vectors are $[1.2, 1.0]^T$ and $[0.5, 0.5]^T$, and covariance matrices are $\begin{bmatrix} 4.0 & -0.9 \\ -0.9 & 0.3 \end{bmatrix}$ and $\begin{bmatrix} 2.0 & 0.2 \\ 0.2 & 3.0 \end{bmatrix}$ respectively. The scatter of these points is shown in both Fig. 2 (a) and (b) as circles. A SOM and a ViSOM, both of 20×20, were applied to the data. Both networks used random initial states and exponential neighbourhood functions, $\exp(-dis(c,v)^2/2\sigma^2)$, where $c$ and $v$ denote node $c$ and winner $v$ respectively, $dis(.)$ denotes the index distance between these two nodes, and $\sigma$ represents the neighbourhood size decreasing from a large size (e.g. 10 for this case) to eventually 1 for the SOM and 4 for the ViSOM. The maps after 10,000 iterations are shown in Fig. 2 (a) and (b). The ViSOM places a fine mesh grid through data points and extrapolates smoothly. The projections of the data points on both maps are shown in Fig. 2 (c) and (d). They are the maps used for the visualisation purpose. It shows clearly that the advantages of the ViSOM in capturing and showing the data distribution and structure. The mapped data points on the SOM do not reveal the distribution and structure of the data. The map's resolution parameter $\lambda$ was set 0.5 in this example. A larger map and smaller $\lambda$ should be used for a finer projection.

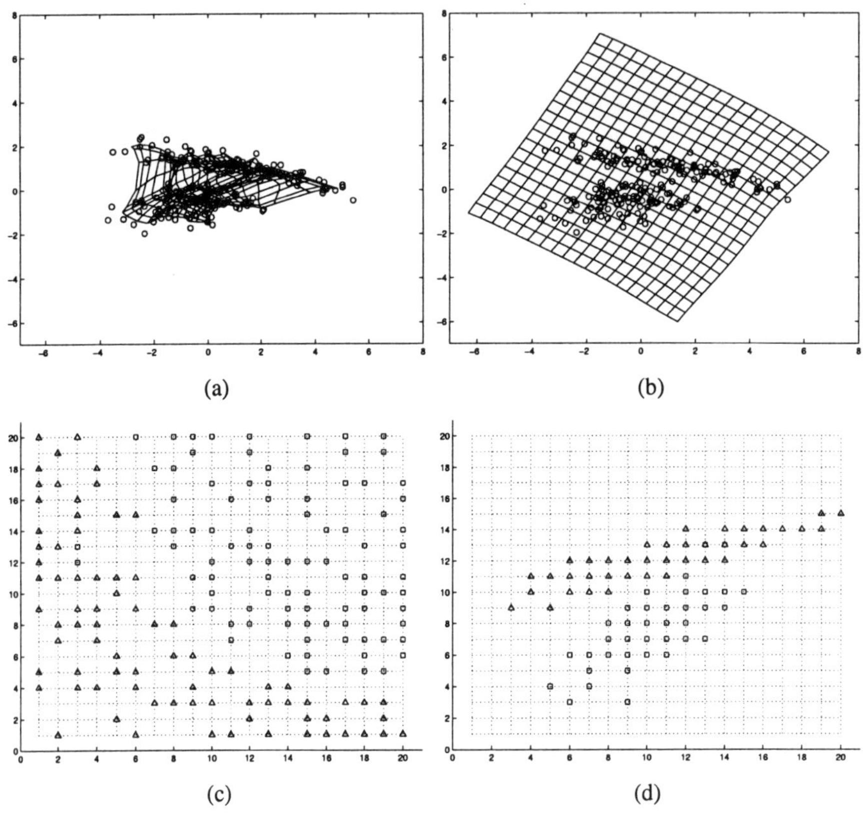

Figure 2. (a) A trained 20×20 SOM in data space, (b) A trained 20×20 ViSOM in data space, (c) Data projected on the SOM, (d) Data projecteded on the ViSOM, the squares and triangles in (c) and (d) indicate data points from two Gaussians.

## 4.2 Iris Data Set

The Iris data set is made of 150 4-D points from three categories, each of which has 50 points [12]. Many data clustering and visualisation experiments have used this data set as a benchmark. A 100×100 hexagonal ViSOM has been applied to the data set and the result, i.e. the projected data on the map, is shown in Fig. 3 (d). For comparison, a SOM of the same size and structure has also been used to map the data and the result is presented in Fig. 3 (c), together with the results of the PCA and Sammon mapping in Fig. 3 (a) and (b). The initial states of both SOM and ViSOM were placed on a plan spanned by the first two principal components of the data. The results shown were after 10,000 iterations. The final neighbourhood size for the ViSOM was set to 4, and resolution $\lambda$=0.075.

From the results, it can be seen that the Sammon method is better than the linear PCA, as it reveals better structure details. The ViSOM has preserved both inter- and intra-cluster structures on the map. Futhermore, the ViSOM can provide the projection function, while the Sammon mapping cannot. The SOM can also form the mapping function. However, the SOM preserves directly the topology only, i.e. it can only provide relative relationships among data points. It is hard for the SOM to reveal both inter-cluster and intra-cluster distribution.

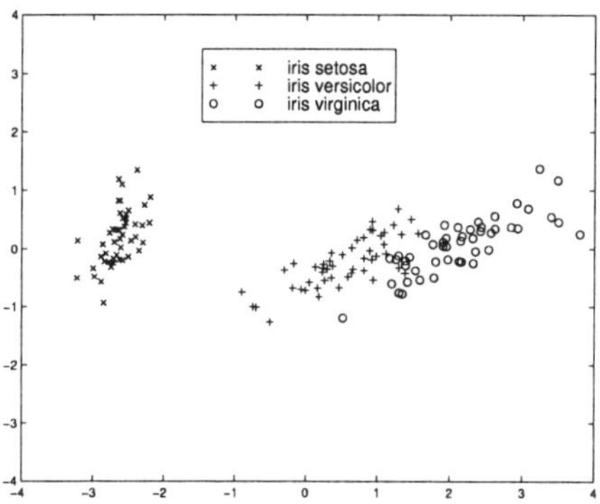


(a) PCA

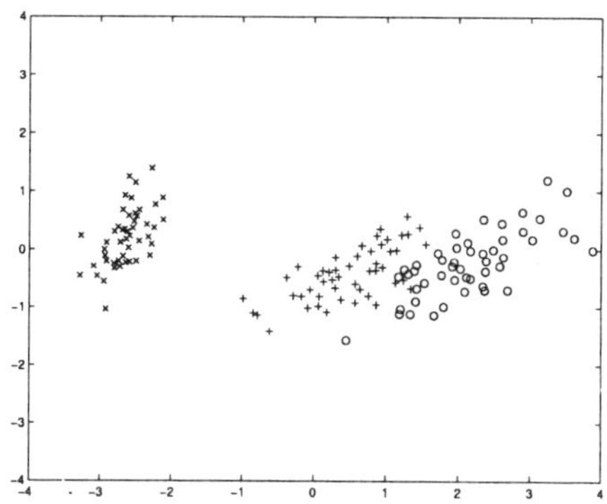

(b) Sammon Mapping

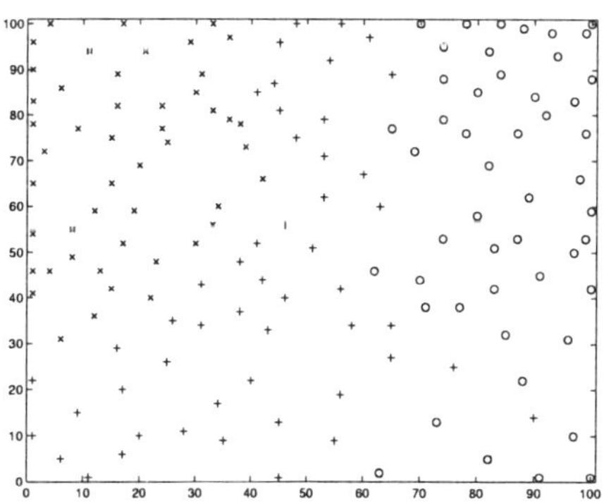

(c) SOM

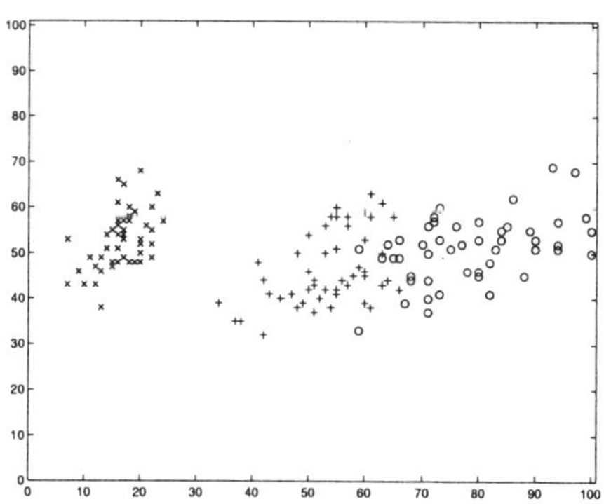

(d) ViSOM

Figure 3. Mappings of Iris data set: (a) PCA, (b) Sammon Mapping, (c) SOM, (d) ViSOM.

# 5 Conclusions

In this paper, a new variant SOM, namely the ViSOM, is proposed for visualisation and projection of high dimensional data. It is similar in structure to the SOM, but constrains the *lateral contraction force* within the updating neighbourhood. As a result, the map preserves the inter-neuron distances as well as topology as faithfully as possible. The ViSOM produces a smooth and regularly graded mesh through the data points and enables a quantitative, direct and visually appealing measure of inter-point distances on the map. Thus it provides a better visual exhibition of data points and more importantly their structure and distribution on the map. The ViSOM is as a simple algorithm as is the SOM.

## References

[1]. Sammon, J.W., "A nonlinear mapping for data structure analysis," *IEEE Trans on Computer*, vol. C-18(5), pp. 401-409, 1969.
[2]. Mao, J. and Jain, A.K., "Artificial Neural Networks for Feature Extraction and Multivariate Data Projection," *IEEE Trans. on Neural Networks*, vol. 6, pp. 296-317, 1995.
[3]. Kohonen, T., *Self-Organising Maps,* Springer: Berlin, 1995.
[4]. Ultsch, A., "Self-organising neural networks for visualisation and classification," in *Information and Classification*, O. Opitz, B. Lausen and R. Klar eds., pp. 864-867, 1993
[5]. Kraaijveld, M.A., Mao, J. and Jain, A.K., "A nonlinear projection method based on Kohonen's topology preserving maps," *IEEE Trans. Neural Networks*, vol. 6, pp. 548-559, 1995.
[6]. Yin, H. and Allinson, N.M., "Interpolating self-organisng map (iSOM)," *Electronics Letters*, vol. 35, pp. 1649-1650, 1999.
[7]. Hastie, T. and Stuetzle, W., "Principal curves," *Journal of the American Statistical Association*, vol. 84, pp. 502-516, 1989.
[8]. Malthouse, E.C., "Limitations of nonlinear PCA as performed with generic neural networks," *IEEE Trans. Neural Networks*, vol. 9, pp. 165-173, 1998.
[9]. Ripley, B.D., *Pattern Recognition and Neural Networks*, Cambridge University Press: Cambridge, UK, 1996.
[10]. Cox, T.F. and Cox, M.A.A., *Multidimensional Scaling*, Chapman & Hall: London, 1994.
[11]. Yin, H. and Allinson, N.M., "On the distribution and convergence of feature space in self-organising maps," *Neural Computation*, vol. 7, pp. 1178-1187, 1995.
[12]. Fisher, R.A., "The use of multiple measurements in taxonomic problems," *Annual Eugenics*, vol. 7, pp. 178-188, 1936.

# An approach to automated interpretation of SOM

Markus Siponen[1,2], Juha Vesanto[1], Olli Simula[1] and Petri Vasara[2]

(1) Neural Networks Research Centre,
Helsinki University of Technology,
P.O.Box 5400, 02015 HUT, Finland
{Juha.Vesanto,Olli.Simula}@hut.fi
(2) Jaakko Pöyry Consulting
P.O.Box 4, 01621, Finland
Petri.Vasara@poyry.fi

**Abstract** The objective of this work was to develop automatic tools for post-processing of SOMs, especially in the context of hierarchical data — data where each higher level object consists of a varying number of lower level objects. Both low and high level data is available and needs to be utilized. The information from lower levels is transferred to higher level using data histograms of lower level clusters. The clusters are formed and interpreted automatically so as to summarize the information given by the SOM, and to produce meaningful indicators that are useful also to problem domain experts. The results show that the approach works well at least in the case study of pulp and paper mills technology data.

**Keywords:** interpretation, self-organizing map, clustering, data mining

## 1 Introduction

The self-organizing map (SOM) [6] is frequently being utilized in data analysis, including analysis of hierarchical data [13]. An example of such data is forest industry data used as a case study in this project [10][1]. The data consists of paper mills where each mill contains one or more pulp and paper machines. On a higher level, pulp and paper companies own varying number of mills. An earlier study concerned a very similar data set [13]. In that study, at least, one of the major workphases was interpretation of the resulting SOM: the data clearly consisted of a number of clusters, but quantifying their significant properties required a great deal of manual work.

In this paper, an approach is proposed which automates much of the whole analysis process. An automatic clustering procedure is utilized, and the resulting clusters are analysed with post-processing methods which facilitate interpretation. In hierarchical context, data histograms based on the lower level clusters are used as features on the upper level. Because each cluster has an interpretation, the new features are meaningful also to problem domain expers.

[1] This work has been carried out in the "Enterity" project in Helsinki University of Technology. The financing of TEKES and Jaakko Pöyry Consulting is gratefully acknowledged.

# 2 Methodology

The data analysis approach described in this section consists of four phases. The first phase is training a SOM based on the data. This involves several choices, such as normalization method and training parameters. However, these are not discussed in this paper. The second phase is clustering of the SOM (Section 2.1), and the third phase is interpretation of the clusters (Section 2.2). If the data is hierarchical and the lower level information needs to be transferred to a higher level, an additional fourth phase is needed to construct the frequency components for each higher level object (Section 2.3).

## 2.1 Clustering

There are several ways to perform clustering of the SOM [12]. Frequently it has been based on visual cues and manual assignment of map units to clusters. In order to automate the analysis process this approach is not applicable. In this paper, the clustering procedure is based on $k$-means algorithm and Davies-Bouldin index [4,2,12]. Several different partitions of the SOM prototypes are made with $k$-means algorithm using a range of $k$-values. The best of these is the one that minimizes Davies-Bouldin index: $I_{DB} = \frac{1}{C}\sum_{i=1}^{C} \max_j(\frac{S_i+S_j}{M_{ij}})$, where $C$ is the number of clusters, $S_i$ is the dispersion (e.g. mean squared distance from center) of cluster $i$, and $M_{ij}$ is the distance between centers of clusters $i$ and $j$. This clustering procedure tries to find spherical clusters which are internally compact but widely separated.

Note that the clustering method could be applied directly to the data instead of the SOM. The SOM is used as an intermediate phase because this reduces the computational complexity of clustering, and because the trained SOM can be efficiently used for visualization, both important properties in practice [12].

## 2.2 Interpretation of clusters

Methods which aid in interpretation of SOMs is a topic which has been gaining interest recently with the use of the SOM for data mining [3]. The aim of interpretation is to create an understanding of which components are important with respect to each cluster; both within the cluster and with respect to other clusters.

A recently favoured approach is to derive some measure of significance and rank the components using it [7,9,5]. The data in the case study consisted mostly of sparse variables where a big value is much more significant than a small value. The measures take this into account by comparing how big the relative values of the components are in each cluster:

$$s_v(i,k) = \frac{m_{ik} - \min_k}{\max_k - \min_k}, \tag{1}$$

where $m_{ik}$ is the mean value for component $k$ in cluster $i$, and $\min_k$, $\max_k$ are minimum and maximum values of component $k$. This measures the relative importance of the components within each cluster, but it does not take other clusters into account at all. For this purpose, a second measure can be used:

$$s_{\hat{v}}(i,k) = \frac{s_v(i,k)}{\frac{1}{C-1}\sum_{j\neq i} s_v(j,k)}, \tag{2}$$

which measures how big value component $k$ has with respect to its value in other clusters. Both $s_v$ and $s_{\hat{v}}$ are maximized to find the most significant component. The measures are very closely related to those defined in [7]: their product $s_v s_{\hat{v}}$ corresponds to the $G^2$ measure for keyword selection for areas in document maps.

While definitions of $s_v$ and $s_{\hat{v}}$ can be easily changed to accomodate the situation where small values are significant, they fail in more general cases. A simple refinement is to use standard deviation $\sigma_k$ instead of the component value. This approach is closely related to the LabelSOM method [9].

Earlier, the characterization of SOM clusters has been done using rules [8,11]. Also this approach was applied: for each cluster, a set of rules of type $r$ `is true, if` $x_{lk} > \alpha_k$ was generated. The rules were evaluated using the product of two measures: $P(r|i) = n_{r\&i}/n_i$ which measures the confidence of rule $r$ being true in cluster $i$ (internal significance), and $P(i|r) = n_{r\&i}/n_r$ which measures confidence in a data sample being in the cluster $i$ if rule $r$ is true (external significance). Their product measures the significance of the rule:

$$S_r(i,r) = P(i|r)P(r|i) = n_{r\&i}^2/(n_r n_i), \tag{3}$$

where $n_r$ is the number of samples for which the rule $r$ is true, $n_i$ is the number of samples in cluster $i$, and $n_{r\&i}$ is the number of samples in cluster $i$ for which the rule $r$ is true. The measure $S_r(i,r)$ reaches its maximum value 1 if the rule and cluster correspond to each other perfectly.

### 2.3 Frequency components

After lower level maps have been clustered and interpreted, the information can be transferred to an upper level. This is accomplished through the use of frequency components, each of which corresponds to one cluster on the lower level map.

To each lower level vector $\mathbf{x}_l$, a vector $\mathbf{p}_l$ is associated which indicates which cluster $\mathbf{x}_l$ belongs to. At most rudimentary level, this can be a binary vector $\mathbf{p}_l = [p_{l1}, ..., p_{lC}]$, with $p_{li} = 1$ if the best-matching unit of $\mathbf{x}_l$ belongs to cluster $i$ and zero otherwise. In this work, a more sophisticated version was used based on a gaussian mixture model estimated on top of the SOM [1]. The probabilities of each data vector to belong to each map unit were calculated based on the mixture model, and these probabilities were averaged over each cluster, giving a non-binary vector $\mathbf{p}_l$ for each low level data vector.

For each higher level object (such as a forest industry company) a set of new features was constructed by aggregating the $\mathbf{p}_l$ vectors of all lower level objects (such as pulp and paper mills) associated with it. In the case study, the aggregation was done as an average of the vectors.

# 3 Case study

A database of pulp and paper mills was investigated [10]. The database contained data about various technical aspects of pulp and paper mills around the world. Analysis presented here contains two levels, low level map of the mills ($n = 4205$) and a higher level map of the forest industry companies ($n = 279$).

The mill data consisted of 47 variables most of which were production capacities of various pulp and paper types. All variables were scaled to have unit ([0,1]) range. A map with 22 x 15 units was trained (using default training parameters as defined in SOM Toolbox[2]). The map was then clustered using $k$-means (see previous section) resulting in 15 clusters, and analysed using $s_v$ as the measure of significance.

Figure 1 shows U-matrix of the map, the 15 automatically determined clusters, and the most signicant component (and its value) in each cluster. In addition, some rules were constructed for each of the clusters and tested using the data belonging to that cluster, see Table 1.

Each of the 15 clusters formed one feature used in the higher level company map. The feature vectors for each company were formed with the procedure described in the previous section, such that each feature indicated the relative frequency of mills of that particular type in the company. A 16th component was added with the total number of mills in each company. No scaling was performed on the frequency components, but the 16th component was scaled between [0,1]. The same training and interpretation procedure as for the lower level map was applied, except for the fact that $s_{\hat{v}}$ was used as the significancy measure instead of $s_v$.

The clusters, most significant components and one component plane ("Integrated news") of the company level map are shown in Figure 2. The clusters are not quite as clear as in the mill map, but they still give a coherent picture of the forest industry companies.

# 4 Conclusions

In this paper, a framework has been presented for interpretation of the cluster structure and contents of SOMs, and for generation of new meaningful components for higher level data in hierarchical data sets. In the case study, the methodology worked well, and the domain experts were very satisfied with the results. It was not investigated whether the achieved clusters and interpretations were optimal in some sense, but they gave the experts a sensible and coherent picture of the data.

On the company level map, the cluster based components offered a vast improvement in interpretability over the projection (ie. SOM unit-)coordinates utilized in earlier work [13]. Apart from selecting the significancy measures, giving actual names to the frequency components is the only part of the procedure

[2] A SOM library for Matlab: `http://www.cis.hut.fi/projects/somtoolbox/`

which makes it less than totally automatic. The list of most significant components gives a starting point for the naming, but the data analyst must apply his or her own insight to crystallize the automatically produced information.

## References


1. Esa Alhoniemi, Johan Himberg, and Juha Vesanto. Probabilistic Measures for Responses of Self-Organizing Map Units. In H. Bothe, E. Oja, E. Massad, and C. Haefke, editors, *Proceeding of the International ICSC Congress on Computational Intelligence Methods and Applications (CIMA'99)*, pages 286–290. ICSC Academic Press, 1999.
2. David L. Davies and Donald W. Bouldin. A Cluster Separation Measure. *IEEE Trans. on Pattern Analysis and Machine Intelligence*, PAMI-1(2):224–227, April 1979.
3. Guido Deboeck and Teuvo Kohonen, editors. *Visual explorations in Finance using Self-Organizing Maps*. Springer-Verlag, London, 1998.
4. Robert M. Gray. Vector quantization. *IEEE ASSP Magazine*, pages 4–29, April 1984.
5. Samuel Kaski, Janne Nikkilä, and Teuvo Kohonen. Methods for interpreting a self-organized map in data analysis. In Michel Verleysen, editor, *Proceedings of ESANN'98, 6th European Symposium on Artificial Neural Networks, Bruges, April 22-24*, pages 185–190. D-Facto, Brussels, Belgium, 1998.
6. Teuvo Kohonen. *Self-Organizing Maps*, volume 30 of *Springer Series in Information Sciences*. Springer, Berlin, Heidelberg, 1995.
7. Krista Lagus and Samuel Kaski. Keyword selection method for characterizing text document maps. In *Proceedings of ICANN99, Ninth International Conference on Artificial Neural Networks*, volume 1, pages 371–376. IEE, London, 1999.
8. W. Pedrycz and H. C. Card. Linguistic interpretation of self-organizing maps. In *Proceedings of International Conference on Fuzzy Systems '92*, pages 371 – 378, 1992.
9. Andreas Rauber and Dieter Merkl. Automatic labeling of self-organizing maps: Making a treasure-map reveal its secrets. In *Proceedings of the 3rd Pasific-Area Conference on Knowledge Discovery and Data Mining (PAKDD'99)*, 1999.
10. Markus Siponen. Automaattisia jälkitulkintamenetelmiä hierarkisen tietoaineiston tutkimiseen itseorganisoivan kartan avulla. Master's thesis, Helsinki University of Technology, 2000. In Finnish.
11. A. Ultsch. Knowledge extraction from self-organizing neural networks. In O. Opitz, B. Lausen, and R. Klar, editors, *Information and Classification*, pages 301–306. Springer Verlag, 1993.
12. Juha Vesanto and Esa Alhoniemi. Clustering of the self-organizing map. *IEEE Transactions on Neural Networks*, 11(3):586–600, May 2000.
13. Juha Vesanto, Petri Vasara, Riina-Riitta Helminen, and Olli Simula. Integrating environmental, technological and financial data in forest industry analysis. In Bert Kappen and Stan Gielen, editors, *Proceedings of 1997 Stichting Neurale Netwerke Conference*, pages 153–156, Amsterdam, the Netherlands, May 1997. World Scientific.

**Table 1.** Rules generated for cluster 1 ($n_i = 101$) of the mill map. The first 4 rows are conditions based on the 4 most significant components of the cluster, and the 3 last rows are their combinations such that all listed conditions must be true. The rules worked suprisingly well. The maximum significance was often achieved with 2-3 conditions and it was between 0.5–0.7. In this case, the most signicant rule contained 3 components and the significance was 0.91 (bolded).

| Rule | $n_{r\&i}$ | $n_r$ | $P(i\|r)$ | $P(r\|i)$ | $S_r(i,r)$ |
|---|---|---|---|---|---|
| (1) Price_vol > 0.75 | 101 | 274 | 0.37 | 1.00 | 0.37 |
| (2) Tot_chem > 0.74 | 93 | 400 | 0.23 | 0.92 | 0.21 |
| (3) Tot_sa > 0.67 | 93 | 372 | 0.25 | 0.92 | 0.23 |
| (4) Bl_sa > 0.64 | 88 | 292 | 0.30 | 0.87 | 0.26 |
| (1,2) | 93 | 101 | 0.92 | 0.92 | 0.85 |
| (1,2,3) | 93 | 94 | 0.99 | 0.92 | **0.91** |
| (1,2,3,4) | 81 | 82 | 0.99 | 0.80 | 0.79 |

**Figure 1.** The U-matrix (on left, white denotes large distance from neighbors), automatically produced clusters (middle) and most significant component according to $s_v$ (on right). On the whole lower half of the map, the most significant component is the industrial paper production capacity. For them (and other similar cases) differences between clusters only become apparent after investigation of the other significant components.

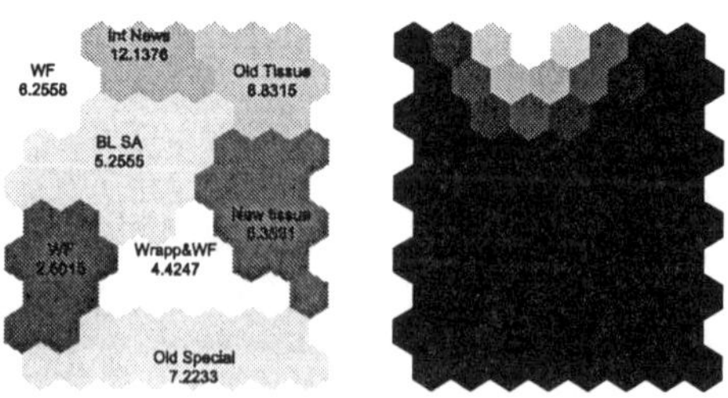


**Figure 2.** The clusters and most significant components (left) and the "Integrated News" component plane (right, white denotes high value), which clearly corresponds to the cluster on top middle of the SOM.

# VQ-Based Clustering Algorithm of Piecewise-Dependent-Data

Itshak Lapidot (Voitovetsky)[1] and Hugo Guterman[2]

[1]Department of Software Engineering
Negev Academic College of Engineering
P.O.B. 45 Beer-Sheva, 84100, Israel

[2]Department of Electrical and Computer Engineering
Ben-Gurion University of the Negev
P.O.B. 653, Beer-Sheva, 84105, Israel

itsik@nace.ac.il, hugo@ee.bgu.ac.il

Abstract

In this paper a piecewise-dependent-data (PDD) clustering algorithm is presented, and a proof of its convergence to a local minimum is given. A distortion measure-based model represents each cluster. The proposed algorithm is iterative. At the end of each iteration, a competition between the models is performed. Then the data is regrouped between the models. The "movement" of the data between the models and the retraining allows the minimization of the overall system distortion. The Kohonen Self-Organizing Map (SOM) was used as the VQ model for clustering. The clustering algorithm was tested using data generated from four generators of Continuous Density HMM (CDHMM). It was demonstrated that the overall distortion is a decreasing function.

# 1 Introduction

Many time signals can be viewed as time-dependent-data, e.g. speech signals, bio-signals, seismic signals, etc. In these, a signals dependence exists between consecutive samples or frames (e.g., the same speaker or the same sleep stage). Most of the clustering algorithms described in the literature are static [1]-[3]. This means that dependence between consecutive vectors is not taken into account. Although, several algorithms exist for clustering of piecewise-dependent-data [4]-[8], no theoretical proofs of convergence have been provided. The objective of this research is to present a piecewise-dependent-data algorithm. This paper is divided as follows: a multi-VQ-based iterative algorithm is presented in section 2. In section 3 a convergence of the iterative algorithm is proven for the minimal distortion sense. The results of an experiment using Kohonen SOM [9] as a VQ algorithm for the cluster models is presented in section 4. The overlapping synthesized database simulates four CDHMM [8] generators. Conclusions are given in section 5.

# 2 The VQ-Based Clustering Algorithm

For given piecewise-dependent-data, i.e., a database consisting of distinct segments while the vectors in each segment are dependent, the goal is to cluster the input data into $R$ clusters (Fig. 1). We assume that the switching points between non-dependent segments are known (the switching points refer to the boundary between two adjacent segments). The piecewise-dependent-data consists of $N$ vectors, $\mathbf{V}=\{v_n\}_{n=1,\dots,N}$. These vectors are partitioned into $M$ segments, $\mathbf{V}=\{\mathbf{V_m}\}_{m=1,\dots,M}$ (equation 1) according to the switching points. The segments have to be clustered into $R$ clusters ($R \le M$), i.e., two vectors that belong to the same segment must be clustered to the same cluster. In static algorithms each cluster is usually represented by one centroid. In the proposed algorithm each cluster is represented by a VQ-based model. For each model a CodeBook (CB) is created. Every $CB_r$ is of size $L_r$ and presents the $r$-th cluster (equation 2).

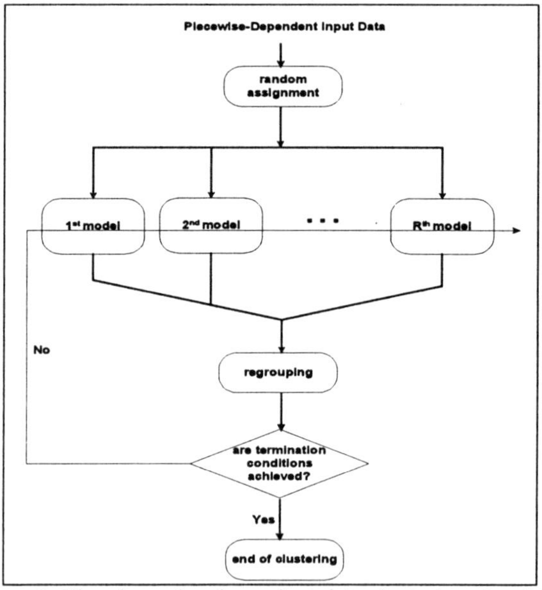


Figure 1. The piecewise-dependent-data clustering algorithm.

$$\begin{cases} \mathbf{V}=\left\{\underbrace{v_1^1,\dots,v_{n_1}^1}_{\mathbf{V_1}},\dots,\underbrace{v_1^m,\dots,v_{n_m}^m}_{\mathbf{V_m}},\dots,\underbrace{v_1^M,\dots,v_{n_M}^M}_{\mathbf{V_M}}\right\}= \\ \quad =\{\mathbf{V_m}\}_{m=1,\dots,M} \\ \sum_{m=1}^{M} n_m = N \end{cases} \qquad (1)$$

$$CB_r = \{c_r^l\}_{r=1,\dots,R\,;\,l=1,\dots,L_r}. \tag{2}$$

$\{c_r^l\}_{r=1,\dots,R\,;\,l=1,\dots,L_r}$ is the union of all the Code-Words (CW) that belongs to $CB_r$.

The initialization of the algorithm is performed by randomly assigning each of the $M$ segment to the $R$ codebooks; $\mathbf{V}^{\mathbf{r},\mathbf{0}}$ notates the segments that are partitioned to $CB_r$ at the beginning of the algorithm, and each model is trained with its segments. After the training, the regrouping process is applied and the models are retrained again. After $i$ iterations the partition will be:

$$\begin{cases} \mathbf{V}^{\mathbf{r,i}} = \{\mathbf{v}_{\mathbf{m}}^{\mathbf{r,i}}\}_{m=1,\dots M_{r,i}\,;\,r=1,\dots,R} \\ \mathbf{v}_{\mathbf{m}}^{\mathbf{r,i}} = \{v_{m,n}^{r,i}\}_{n=1,\dots,n_m\,;\,m=1,\dots M_{r,i}\,;\,r=1,\dots,R} \\ \sum_{r=1}^{R} M_{r,i} = M \end{cases} \tag{3}$$

and the code-books are:

$$CB_r^i = \{c_r^{l,i}\}_{r=1,\dots,R\,;\,l=1,\dots,L_r} \tag{4}$$

Different algorithms such as Kohonen SOM [9], LBG [10], fuzzy C-means [11], can be used for VQ training.

After the retraining the data is regrouped. The reordering of the data is attained by finding which $CB_k^i$ best fits every $\mathbf{v}_{\mathbf{m}} \in \mathbf{V}$ according to a given distance measure. Thus a new partition, $\mathbf{V}^{\mathbf{r},\mathbf{i+1}}$ is produced.

The system has to be retrained according to the new partition. The convergence condition is met when:

$$\mathbf{V}^{\mathbf{r,i}} = \mathbf{V}^{\mathbf{r,i+1}} \tag{5}$$

# 3 Proof of System Convergence

The distance measure can be any measure that meets the conditions that satisfy:

$$\begin{cases} d(x,y) \ge 0 \text{ where equality holds if and only if } x = y \\ d(x,y) = d(y,x) \\ d(x,y) \le d(x,z) + d(z,y) \end{cases}, \tag{6}$$

and the VQ algorithm must converge to at least a local minimum [9], [10].

After the $i$-th iteration the partition of the data between the models will be according to equation (3), where $\mathbf{V}^{\mathbf{r,i}}$ is the data set associated with the $r$-th model at the $i$-th iteration, and the $CB_r$ at the $i$-th iteration is $CB_r^i = \{c_r^{l,i}\}_{r=1,\dots,R\,;\,l=1,\dots,L_r}$.

Let the distance between the $m$-th vector of the $n$-th segment that belongs to $CB_q^{i-1}$ ($v_{m,n}^{q,i-1}$) and $c_r^{l,i}$ be $\mathrm{d}_{\mathrm{m,n}}^{l,\mathrm{i}}(r,q)$. Then the distance between $v_{m,n}^{q,i-1}$ and $CB_r^i$ is:

$$\mathrm{d}_{\mathrm{m,n}}^{\mathrm{i}}(r,q) = \min_{l=1,\ldots,L_r} \left\{\mathrm{d}_{\mathrm{m,n}}^{l,\mathrm{i}}(r,q)\right\}. \tag{7}$$

The distance between $\mathbf{v}_{\mathbf{m}}^{\mathbf{q,i-1}}$ and $CB_r^i$ is:

$$\mathrm{D}_{\mathrm{m}}^{\mathrm{i}}(r,q) = \sum_{n=1}^{n_m} \mathrm{d}_{\mathrm{m,n}}^{\mathrm{i}}(r,q), \tag{8}$$

and the minimal distance between the segment that belongs to $CB_q$ at iteration $(i-1)$ from all the $CB$s is:

$$\begin{aligned} &\mathrm{D}_{\mathrm{m}}^{\mathrm{i}}(j) = \min_{r=1,\ldots,R} \left\{\mathrm{D}_{\mathrm{m}}^{\mathrm{i}}(r,q)\right\} \\ &j = \arg\min_{r=0,\ldots,R} \left\{\mathrm{D}_{\mathrm{m}}^{\mathrm{i}}(r,q)\right\} \Rightarrow \mathbf{v}_{\mathbf{m}}^{\mathbf{q,i-1}} = \mathbf{v}_{\mathbf{m}}^{\mathbf{j,i}} \end{aligned}. \tag{9}$$

If after the $i$-th iteration the overall distance is calculated, the partition before regrouping is:

$$\mathrm{D}^{\mathrm{i}} = \sum_{r=1}^{R} \sum_{m=1}^{M_r} \mathrm{D}_{\mathrm{m}}^{\mathrm{i}}(r,r) \tag{10}$$

and the distance according to a new partition (after regrouping) is:

$$\tilde{\mathrm{D}}^{\mathrm{i}} = \sum_{r=1}^{R} \sum_{m=1}^{M_r} \mathrm{D}_{\mathrm{m}}^{\mathrm{i}}(r). \tag{11}$$

Because $\tilde{\mathrm{D}}^{\mathrm{i-1}}$ is the distance before the $i$-th retraining and $\mathrm{D}^{\mathrm{i}}$ is the minimum according to the previous partition after retraining, the next inequality holds:

$$\tilde{\mathrm{D}}^{\mathrm{i}} \le \mathrm{D}^{\mathrm{i}} \le \tilde{\mathrm{D}}^{\mathrm{i-1}} \tag{12}$$

if

$$\tilde{\mathrm{D}}^{\mathrm{i}} < \mathrm{D}^{\mathrm{i}}. \tag{13}$$

This means that there exists at least one segment, $\mathbf{v}_{\mathbf{m}}^{\mathbf{r,i}} = \mathbf{v}_{\mathbf{m}}^{\mathbf{q,i-1}}$, whose distances are:

$$\mathrm{D}_{\mathrm{m}}^{\mathrm{i}}(r) < \mathrm{D}_{\mathrm{m}}^{\mathrm{i}}(q,q) \tag{14}$$

and there exists a $CB_r$ so that:

$$\mathrm{D}_{\mathrm{m}}^{\mathrm{i}}(r,q) < \mathrm{D}_{\mathrm{m}}^{\mathrm{i}}(q,q). \tag{15}$$

In other words, there exists a better partition of $\mathbf{V}$ that gives a lower distance $\tilde{\mathrm{D}}^{\mathrm{i}}$. If the new partition is chosen, then the previous VQ is not optimal because it was designed with an other partition. It can be seen that from the $i$-th to $(i+1)$-th iteration the overall distance did not increase. The iterative process will stop when

$$\begin{cases} \widetilde{D}^{i+1} = D^{i+1} \\ \mathbf{V}^{r,i+1} = \mathbf{V}^{r,i} \end{cases}. \tag{16}$$

In this case there is no change in the partition between the two consecutive iterations.

# 4 Experiments and Results

In order to test the proposed algorithm a dataset consisting of four clusters was created. Each cluster was produced by a CDHMM generator, and consisted of three states with two two-dimensional Gausian's per state. The dataset consisted of 100 segments, the length of each segment being a random variable that was uniformly distributed, $U(5,100)$. Each model was a Kohonen SOM [9] of size $5\times5$. Clustering results and the distorsion are shown in Fig. 2 and 3, respectively. Although the data overlaps, the models fit the cluster properly. The error was only one segment out of a hundred. The segment was short (seven vectors out of 4726). An error percentage of only about 0.15% was found. Only five iterations were needed for the system to converge.

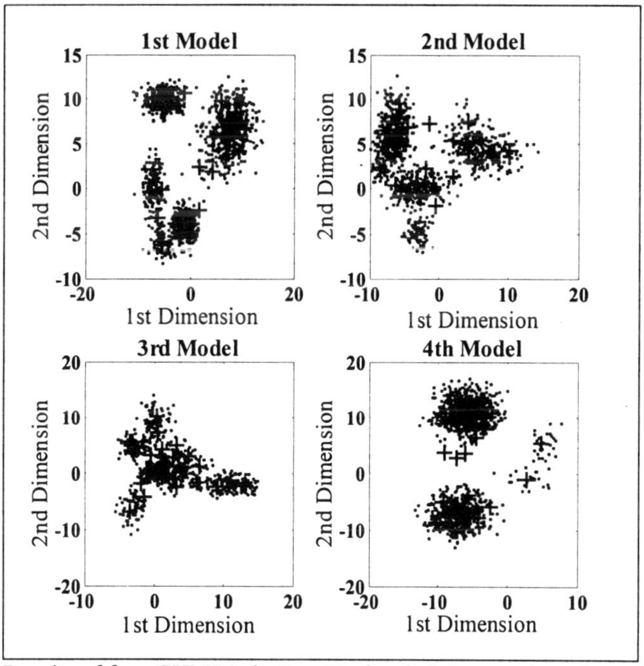


Figure 2. Results of four CHMM clusters. Each subplot presents each model's data (dots) and SOM convergence (+).

# 5 Conclusions

This paper presented an iterative algorithm for VQ-based PDD clustering. A general proof of the convergence of the algorithm was given as well. The effectiveness of the proposed algorithm was demonstrated on data generated from four CDHMMs. The models converged correctly to the clusters and the overall distortion function decreased.

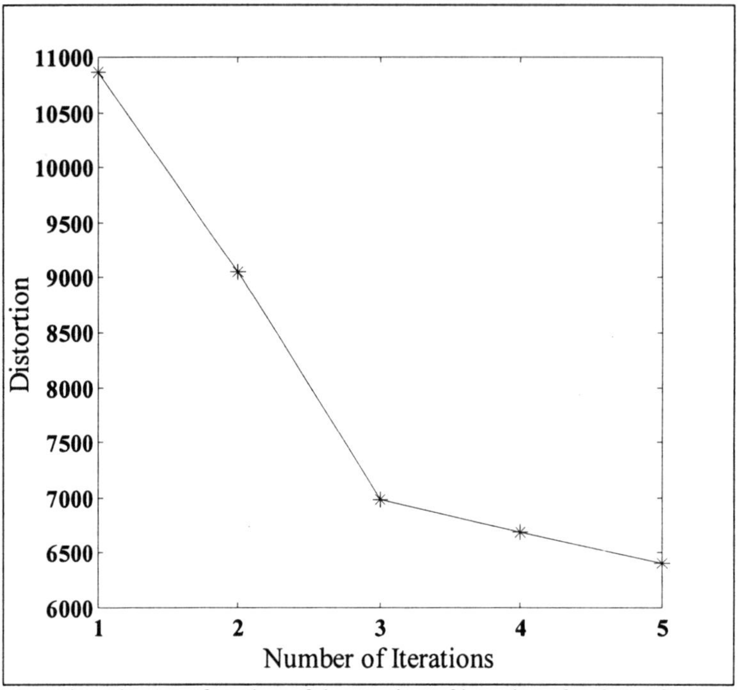


Figure 3. Distortion as a function of the number of iterations for data of 4 CHMMs.

**Reference**

1. Man Y. and Gath I.Detection and separation of ring-shaped clusters using fuzzy clustering. IEEE Trans. Patt. Anal. Machine Intell. 1994; 16:855-861
2. Rose K., Gurewitz E., and Fox G. C. Statistical mechanics and phase transitions in clustering. Physical Review Letters 1990; 65:945-948
3. Gath I. and Geva A. B. Unsupervised optimal fuzzy clustering. IEEE Trans. Patt. Anal. Machine Intell. 1989; 11:773-781
4. Voitovetsky I., Guterman H., and Cohen A. Unsupervised speaker classification using self-organizing maps (SOM). Proc. Neural Networks for Signal Processing VII IEEE Workshop 1997; 578-587

5. Cohen A. and Lapidus V. Unsupervised, text independent, speaker classification. Proc. of the Int. Conf. on Signal Processing Application and Technology 1996; 1745-1749
6. Kohlmorgen J., Muller K.-R., and Pawelzik K. Segmentation and identification of drifting dynamical systems. Proc. Neural Networks for Signal Processing VII IEEE Workshop 1997; 326-335
7. Moon T. K. Temporal pattern recognition using fuzzy clustering. Proc. of Third IEEE International Conference on Fuzzy Systems 1994; 1:432-435
8. Rabiner L. and Juang B.-H. Fundamentals of Speech Recognition. Prentice Hall, 1993
9. Kohonen T. The self-organizing map. Proc. IEEE 1990; 78:1464-1480
10. Linde Y., Buzo A., and Gray R. M. An algorithm for vector quantizer design. IEEE Trans. Commun. 1980; 28:84-95
11. Pedrycz W. Fuzzy sets in pattern recognition: methodology and methods. Pattern Recognition 1990; 23121-146

# Adaptive Subspace Encoders Using Stochastic Vector Quantisers

S P Luttrell
Defence Evaluation and Research Agency
Malvern, U.K.

Abstract

The theory of stochastic vector quantisers (SVQ) has been extended to allow the quantiser to develop invariances, so that only "large" degrees of freedom in the input vector are represented in the code. In particular, this may be applied to the problem of encoding data that predominantly lies on a manifold whose dimensionality is much less than the full dimensionality of the data space. In effect, the approach learns a subspace that approximates the local manifold subspace at each point on the manifold, so it is an example of an adaptive sub-space approach. To illustrate this approach, the technique has been applied to the problem of encoding data vectors that are a superposition of a "large" jammer and a "small" signal, so that only the jammer is represented in the code. This allows the jammer to be subtracted from the total input vector (i.e. the jammer is nulled), leaving a residual that contains only the underlying signal.

## 1 Introduction

In vector quantisation a code book is used to encode each input vector as a corresponding code index, which is then decoded (again, using the codebook) to produce an approximate reconstruction of the original input vector [1, 2]. The standard approach to vector quantiser (VQ) design [3] may be generalised [4] so that each input vector is encoded as a *vector* of code indices that are stochastically sampled from a probability distribution that depends on the input vector, rather than as a *single* code index that is the deterministic outcome of finding which entry in a code book is closest to the input vector. This will be called a stochastic vector quantiser (SVQ), and it includes the standard VQ as a special case.

One advantage of using the stochastic approach is that it automates the process of splitting high-dimensional input vectors into low-dimensional blocks before encoding them, because minimising the mean Euclidean reconstruction error can encourage different stochastically sampled code indices to become associated with different input subspaces [5, 6]. Another advantage is that it is very easy to connect SVQs together, by using the vector of code index probabilities computed by one SVQ as the input vector to another SVQ [7].

SVQ theory will be extended to the case of encoding noisy (or distorted) data, with the intention of subsequently reconstructing an approximation to the noiseless data. In particular, this may be applied to the problem of encoding data that predominantly lie on a manifold whose dimensionality is much less than the full dimensionality of the data space. In effect, the approach learns a subspace that approximates the local manifold subspace at each point on the manifold, so it is an example of an adaptive subspace approach [8, 9, 10]. This theory is then applied to the problem of encoding data vectors which are a superposition of a "large" jammer and a "small" signal, where the signal is regarded as a distortion superimposed on the jammer, rather than the other way around. The reconstruction is then an approximation to the jammer, which can thus be subtracted from the original data to reveal the underlying signal of interest.

In section 2 the underlying theory of SVQs is developed together with its extension to the encoding of noisy data, and in section 3 some simulations illustrating the application of SVQs to the nulling of jammers are presented.

# 2 Stochastic Vector Quantiser Theory

In section 2.1 the basic theory of folded Markov chains (FMC) is given [11], in section 2.2 FMC theory is extended to the case of encoding noisy or distorted data with the intention of eventually recovering the undistorted data, in section 2.3 this extended theory is applied to the problem of encoding data that contain unwanted "nuisance degrees of freedom", and in section 2.4 some constraints on the optimisation of the encoder are introduced to encourage the encoder to disregard the nuisance degrees of freedom (i.e. discover invariances).

## 2.1 Folded Markov Chains

The basic building block of the SVQ used in this paper is the folded Markov chain (FMC) [11]. An input vector $\mathbf{x}$ is encoded as a code index vector $\mathbf{y}$, which is then subsequently decoded as a reconstruction $\mathbf{x}'$ of the input vector. Both the encoding and decoding operations are allowed to be probabilistic, in the sense that $\mathbf{y}$ is a sample drawn from $\Pr(\mathbf{y} \mid \mathbf{x})$, and $\mathbf{x}'$ is a sample drawn from $\Pr(\mathbf{x}' \mid \mathbf{y})$, where $\Pr(\mathbf{y} \mid \mathbf{x})$ and $\Pr(\mathbf{x}' \mid \mathbf{y})$ are Bayes' inverses of each other, as given by $\Pr(\mathbf{x}' \mid \mathbf{y}) = \frac{\Pr(\mathbf{y}|\mathbf{x}) \Pr(\mathbf{x})}{\int dz \Pr(\mathbf{y}|\mathbf{z}) \Pr(\mathbf{z})}$, and $\Pr(\mathbf{x})$ is the prior probability from which $\mathbf{x}$ is sampled. In order to ensure that the FMC encodes the input vector optimally, a measure of the reconstruction error must be minimised. There are many possible ways to define this measure, but one that is consistent with many previous results, and which also leads to many new results, is the mean Euclidean reconstruction error measure $D$, which is defined as [11]

$$D \equiv \int d\mathbf{x} \Pr(\mathbf{x}) \sum_{\mathbf{y}} \Pr(\mathbf{y} \mid \mathbf{x}) \int d\mathbf{x}' \Pr(\mathbf{x}' \mid \mathbf{y}) \|\mathbf{x} - \mathbf{x}'\|^2 \tag{1}$$

where $\mathbf{y} = (y_1, y_2, \cdots, y_n)$, $1 \leq y_i \leq M$ is assumed, $\Pr(\mathbf{x}) \Pr(\mathbf{y} \mid \mathbf{x}) \Pr(\mathbf{x}' \mid \mathbf{y})$ is the joint probability that the FMC has state $(\mathbf{x}, \mathbf{y}, \mathbf{x}')$, $\|\mathbf{x} - \mathbf{x}'\|^2$ is the Euclidean reconstruction error, and $\int dx \sum_{y_1=1}^{M} \sum_{y_2=1}^{M} \cdots \sum_{y_n=1}^{M} \int dx'(\cdots)$ sums over all possible states of the FMC (weighted by the joint probability). The Bayes' inverse probability $\Pr(x' \mid y)$ may be integrated out of this expression for $D$ to yield [11]

$$D = 2 \int dx \Pr(x) \sum_{y} \Pr(y \mid x) \, \|x - x'(y)\|^2 \tag{2}$$

where the reconstruction vector $x'(y)$ is defined as $x'(y) \equiv \int dx \Pr(x \mid y)\, x$. Because of the quadratic form of the objective function, it turns out that $x'(y)$ may be treated as a free parameter whose optimum value (i.e. the solution of $\frac{\partial D}{\partial x'(y)} = 0$) is $\int dx \Pr(x \mid y)\, x$, as required.

## 2.2 Noisy Data

The FMC approach can be generalised to the problem of encoding noisy or distorted data, with the intention of eventually recovering the undistorted data. This generalisation is based on the results reported in [12]. The input vector is $x_0$, which is converted into the distorted input vector $x$ by a distortion process $\Pr(x \mid x_0)$, which is then encoded as a code index vector $y$, which is then subsequently decoded as a reconstruction $x_0'$ of the original input vector. This is described by the directed graph $x_0 \longrightarrow x \longrightarrow y \longrightarrow x_0'$. The mean Euclidean reconstruction error measure $D$ becomes (compare equation 1)

$$D = \int dx_0 \Pr(x_0) \int dx \Pr(x \mid x_0) \sum_{y} \Pr(y \mid x) \int dx' \Pr(x_0' \mid y) \, \|x_0 - x_0'\|^2 \tag{3}$$

The Bayes' inverse probability $\Pr(x_0' \mid y)$ may be integrated out of this expression for $D$ to yield (compare equation 2)

$$D = 2 \int dx_0 \Pr(x_0) \int dx \Pr(x \mid x_0) \sum_{y} \Pr(y \mid x) \, \|x_0 - x_0'(y)\|^2 \tag{4}$$

where the reconstruction vector $x_0'(y)$ is defined as $x_0'(y) \equiv \int dx_0 \Pr(x_0 \mid y)\, x_0$, which may be treated as a free parameter. Bayes' theorem $\Pr(x_0) \Pr(x \mid x_0) = \Pr(x) \Pr(x_0 \mid x)$ may be used to integrate out $x_0$ to yield

$$D = 2 \int dx \Pr(x) \sum_{y} \Pr(y \mid x) \, \|x_0(x) - x_0'(y)\|^2 + \text{constant} \tag{5}$$

where $x_0(x)$ is defined as $x_0(x) \equiv \int dx_0 \Pr(x_0 \mid x)\, x_0$.

## 2.3 Nuisance Degrees of Freedom

For convenience, split up the input space into (possibly non-orthogonal) subspaces as $(x_0, x_\perp)$, where all of the distortion is contained in $x_\perp$, which requires that any distortion that lies in the $x_0$ subspace is regarded as part of the undistorted input. The expression for $D$ becomes (compare equation 4).

$$D = 2 \int dx_0 \Pr(x_0) \int dx_\perp \Pr(x_\perp \mid x_0) \sum_{y} \Pr(y \mid x_0, x_\perp) \, \|x_0 - x_0'(y)\|^2 \tag{6}$$

Consider the related optimisation problem in which an attempt to to reconstruct $(\boldsymbol{x}_0, \boldsymbol{x}_\perp)$ is made. The corresponding objective function may be obtained by modifying equation 6, where the cross-term arising from non-orthogonal $(\boldsymbol{x}_0, \boldsymbol{x}_\perp)$ is omitted.

$$D = 2\int d\boldsymbol{x}_0 \Pr(\boldsymbol{x}_0) \int d\boldsymbol{x}_\perp \Pr(\boldsymbol{x}_\perp \mid \boldsymbol{x}_0) \sum_{\boldsymbol{y}} \Pr(\boldsymbol{y} \mid \boldsymbol{x}_0, \boldsymbol{x}_\perp) \begin{pmatrix} \|\boldsymbol{x}_0 - \boldsymbol{x}_0'(\boldsymbol{y})\|^2 \\ + \|\boldsymbol{x}_\perp - \boldsymbol{x}_\perp'(\boldsymbol{y})\|^2 \end{pmatrix} \quad (7)$$

Now introduce the following assumptions

$$\Pr(\boldsymbol{y} \mid \boldsymbol{x}_0, \boldsymbol{x}_\perp) = \Pr(\boldsymbol{y} \mid \boldsymbol{x}_0)$$
$$\Pr(\boldsymbol{x}_\perp \mid \boldsymbol{x}_0) = \Pr(\boldsymbol{x}_\perp)$$

which allow the objective function $D$ (see equation 7) to be replaced by the equivalent objective function

$$D = 2\int d\boldsymbol{x}_0 \Pr(\boldsymbol{x}_0) \sum_{\boldsymbol{y}} \Pr(\boldsymbol{y} \mid \boldsymbol{x}_0)\, \|\boldsymbol{x}_0 - \boldsymbol{x}_0'(\boldsymbol{y})\|^2 + \text{constant} \quad (9)$$

This is the standard FMC objective function (compare equation 2) for encoding and reconstructing the undistorted input, for which the directed graph is $\boldsymbol{x} \longrightarrow \boldsymbol{y} \longrightarrow \boldsymbol{x}_0'$. Note that, under the stated assumptions, the simplification in equation 9 occurs even if the two subspaces are not orthogonal to each other, the potential cross-term $\int d\boldsymbol{x}_\perp \Pr(\boldsymbol{x}_\perp \mid \boldsymbol{x}_0)\, (\boldsymbol{x}_0 - \boldsymbol{x}_0'(\boldsymbol{y})).(\boldsymbol{x}_\perp - \boldsymbol{x}_\perp'(\boldsymbol{y}))$ in equation 7 is zero.

## 2.4 Optimisation Constraints

Henceforth, only the scalar case will be considered, so the vector $\boldsymbol{y}$ is now replaced by the scalar $y$ $(1 \le y \le M)$. In order to implement a practical optimisation procedure for minimising $D$ it is necessary to introduce a variety of assumptions and constraints. Because $\Pr(y \mid \boldsymbol{x})$ is a probability it satisfies $\Pr(y \mid \boldsymbol{x}) \ge 0$ and $\sum_{y=1}^{M} \Pr(y \mid \boldsymbol{x}) = 1$, which is guaranteed if $\Pr(y \mid \boldsymbol{x})$ is written as $\Pr(y \mid \boldsymbol{x}) = \frac{Q(y|\boldsymbol{x})}{\sum_{y'=1}^{M} Q(y'|\boldsymbol{x})}$, where $Q(y \mid \boldsymbol{x}) \ge 0$. This removes the need to explicitly impose the constraint $\sum_{y=1}^{M} \Pr(y \mid \boldsymbol{x})$ during optimisation. The $Q(y \mid \boldsymbol{x})$ are the unnormalised likelihoods of sampling code index $y$ from the code book. However, $Q(y \mid \boldsymbol{x})$ itself needs to be described by a finite number of parameters in order that the values that minimise $D_1 + D_2$ may be derived from a finite amount of training data. It can be shown that the optimal form of $\Pr(y \mid \boldsymbol{x})$ is piecewise linear in $\boldsymbol{x}$ [7], and that for training data that lie on smooth curved manifolds the form of this solution is well approximated by a piecewise linear $Q(y \mid \boldsymbol{x})$ of the form [6]

$$Q(y \mid \boldsymbol{x}) = \begin{cases} \boldsymbol{w}(y).\boldsymbol{x} - a(y) & \boldsymbol{w}(y).\boldsymbol{x} \ge a(y) \\ 0 & \boldsymbol{w}(y).\boldsymbol{x} \le a(y) \end{cases}$$

However, the precise functional form of $Q(y \mid \boldsymbol{x})$ needs to exhibit this behaviour only in the vicinity of the data manifold, so in particular it can be allowed to saturate (i.e. $Q(y \mid \boldsymbol{x}) \longrightarrow 1$) as $\boldsymbol{w}(y).\boldsymbol{x} \longrightarrow \infty$. A convenient functional form that achieves this is the sigmoid, which is defined as $Q(y \mid \boldsymbol{x}) = \frac{1}{1+\exp(-\boldsymbol{w}(y).\boldsymbol{x}-b(y))}$. This reduces the problem of minimising $D$ to one of finding the optimal values of the $\boldsymbol{w}(y)$, $b(y)$ and $\boldsymbol{x}'(y)$. This may be done by using the gradient descent procedure described in [4].

If the input is an *undistorted* signal (i.e. $\boldsymbol{x} = (\boldsymbol{x}_0, \mathbf{0})$) which lies on a smooth curved manifold, then the sigmoids can cooperate in encoding this input as illustrated in figure 1, where the sigmoid threshold planes $\boldsymbol{w}(y).\boldsymbol{x} + b(y) = 0$ are shown slicing pieces off the curved manifold [6].

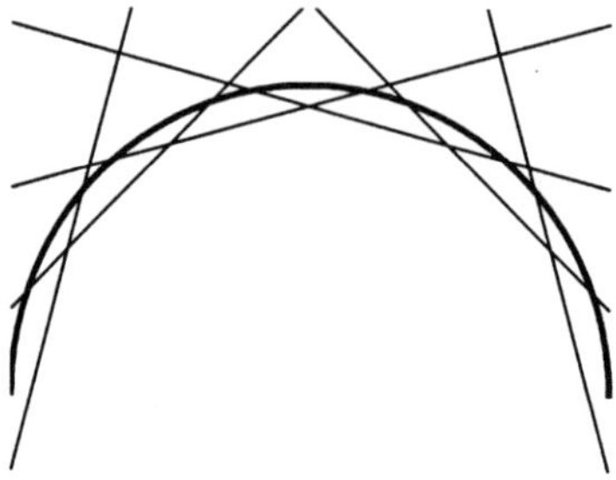

Figure 1. Illustration of how a number of sigmoids can cooperate to slice pieces off a signal manifold.

The additional constraints that are required in order to implement the behaviour described in section 2.3 will now be described. Thus the constraints must be such that the encoder disregards (or is invariant with respect to) the nuisance degrees of freedom $\boldsymbol{x}_\perp$ in the full input vector $(\boldsymbol{x}_0, \boldsymbol{x}_\perp)$. However, without knowing $\Pr(\boldsymbol{x}_0, \boldsymbol{x}_\perp)$ in advance (which would allow $\boldsymbol{x}_0(\boldsymbol{x})$ in equation 5 to be calculated), it is not possible to give a general approach that works in all cases. At best, an empirical approach must be used. A very simple and useful constraint is to impose a threshold constraint on the sigmoid function, which forces the value of the sigmoid to lie exactly halfway up its slope when the norm of its input vector is $\theta$. This is achieved by choosing $b(y) = -\theta \| \boldsymbol{w}(y) \|$, so that

$$Q(y \mid \boldsymbol{x}) = \frac{1}{1+\exp(-(\hat{\boldsymbol{w}}(y).\boldsymbol{x} - \theta) \| \boldsymbol{w}(y) \|)} \tag{11}$$

where $\| \boldsymbol{w}(y) \| \equiv \sqrt{\boldsymbol{w}(y).\boldsymbol{w}(y)}$ and $\hat{\boldsymbol{w}}(y) \equiv \frac{\boldsymbol{w}(y)}{\|\boldsymbol{w}(y)\|}$.

If the input is a *distorted* signal (i.e. $\boldsymbol{x} = (\boldsymbol{x}_0, \boldsymbol{x}_\perp)$) which lies on a "thickened" version of the smooth curved manifold of figure 1 (the thickness represents the nuisance degrees of freedom), then the sigmoids can cooperate in encoding this input as illustrated in figure 2, where the sigmoid threshold planes $\hat{\boldsymbol{w}}(y).\boldsymbol{x} = \theta$ are shown slicing pieces off the curved manifold in a way that disregards the nuisance degrees of freedom.

Note that in this figure 2 the representation of thickening is not complete, because it can actually occur in any direction orthogonal to the manifold, including directions orthogonal to the space in which the manifold is embedded; the radial direction in figure 2 does not include this latter possibility.

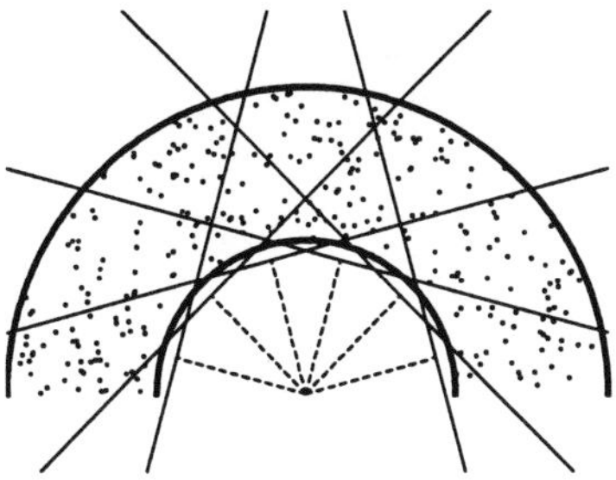

Figure 2. Illustration of how a number of sigmoids can cooperate to slice pieces off a signal manifold thickened by nuisance degrees of freedom.

# 3 Jammer Nulling Simulations

Consider the case of a weak signal that is masked by a strong jammer. If it is assumed that the jammer is the "large" degree of freedom and the signal is the "small" degree of freedom, the signal and jammer subspaces may be separated by adjusting the threshold parameter $\theta$, in which case the $\Pr(y \mid \boldsymbol{x})$ for $y = 1, 2, \cdots, M$ will all become invariant over the signal subspace. The jammer subspace is then spanned by the set of gradient vectors $\boldsymbol{\nabla} \Pr(y \mid \boldsymbol{x})$ for $y = 1, 2, \cdots, M$, which can thus be used to construct a projection operator $\boldsymbol{J}$ onto the jammer subspace, and a projection operator $\boldsymbol{1} - \boldsymbol{J}$ onto the signal subspace. This definition of the projection operator may also be used in cases where the jammer and signal subspaces are curved, so that the directions of their axes are functions of $\boldsymbol{x}$ defining a curvilinear coordinate system and its coordinate surfaces. Note that curved subspaces are the norm rather than the exception.

In these simulations the input vector $\boldsymbol{x}$ is 100-dimensional so that $\boldsymbol{x} = (x_1, x_2, \cdots, x_{100})$, and each vector in the training set is independently generated as a superposition of a pair of response functions $x_i = a_s \frac{\sin(\frac{i-i_s}{\sigma})}{\frac{i-i_s}{\sigma}} + a_j \frac{\sin(\frac{i-i_j}{\sigma})}{\frac{i-i_j}{\sigma}}$, where $a_s$ is the signal amplitude that is uniformly distributed in the interval $[-\sqrt{10^{-3}}, \sqrt{10^{-3}}]$ (this correponds to a signal level of -30dB), $a_j$ is the jammer amplitude that is uniformly distributed in the interval $[-1, 1]$ (this correponds to a jammer level of 0dB), $i_s$ is the signal location that is chosen to be 50, $i_j$ is the jammer location that is uniformly distributed in the interval $[38 - \Delta, 38 + \Delta]$ ($\Delta = 0, 2, 4$ is used in the simulations), and $\sigma$ is the width of the response function that is chosen to be 2. The peak and the first zero of the sinc function are separated by $\pi\sigma$, which defines the resolution cell size. The mean jammer position and the signal position satisfy $i_s - \langle i_j \rangle = 12$, which corresponds to a separation of $\frac{12}{\pi\sigma} \approx 2$ resolution cells. Random noise uniformly distributed in the interval $\left[-\sqrt{10^{-5}}, \sqrt{10^{-5}}\right]$ (this correponds to a noise level of -50dB) is also added to each component of the training vector.

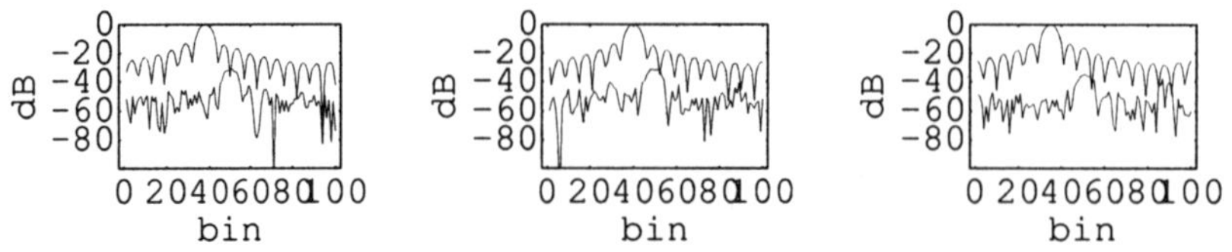


Figure 3. Plot of a typical input vector before and after jammer nulling for each of the scenarios $\Delta = 0, 2, 4$.

In figure 3 typical examples of an input vector together with how it appears after jammer nulling are shown for each of the jammer scenarios $\Delta = 0, 2, 4$. In every case the signal is clearly revealed at its correct location after nulling the jammer.

# 4 Conclusions

The theory of stochastic vector quantisers (SVQ) [4] has been extended to allow the quantiser to develop invariances, so that only "large" degrees of freedom in the input vector are represented in the code. In particular, this may be applied to the problem of encoding data that predominantly lies on a manifold whose dimensionality is much less than the full dimensionality of the data space. In effect, the approach learns a subspace that approximates the local manifold subspace at each point on the manifold, so it is an example of an adaptive sub-space approach. A nice property of the SVQ approach is that it is a natural generalisation of the standard approach to VQ design [3], in which the deterministic code book is replaced by a stochastic code book. This apparently innocuous change leads to wealth of new results.

To illustrate the SVQ approach, the technique has been applied to the problem of encoding data vectors that are a superposition of a "large" jammer and a "small" signal, so that only the jammer is represented in the code. This allows the jammer to be subtracted from the total input vector (i.e.the jammer is nulled), leaving a residual that contains only the underlying signal. Several numerical simulations have shown how that idea works in practice, even when the jammer location is uncertain so that the jammer subspace is curved.

**Acknowledgement**

This work was carried out as part of Technology Group 10 of the MoD Corporate Research Programme.

## References

1. Gray RM. Vector quantisation. IEEE Acoustics, Speech, and Signal Processing Magazine April 1984; 4-29

2. Gersho A and Gray RM. Vector quantisation and signal processing.. Kluwer, 1992.

3. Linde Y, Buzo A and Gray RM. An algorithm for vector quantiser design. IEEE Transactions on Communications 1980; 28:84-95

4. Luttrell SP. A theory of self-organising neural networks. In: Ellacott SW, Mason JC and Anderson IJ (eds) Mathematics of Neural Networks: Models, Algorithms and Applications. Kluwer, 1997, pp 240-244

5. Luttrell SP. Self-organisation of multiple winner-take-all neural networks. Connection Science (special issue on Combining Artificial Neural Nets: Modular Approaches) 1997; 9:11-30

6. Luttrell SP. Self-organised modular neural networks for encoding data. In: Sharkey AJC (ed) Combining Artificial Neural Nets: Ensemble and Modular Multi-Net Systems. Springer-Verlag, 1999, pp 235-263

7. Luttrell SP. An adaptive network for encoding data using piecewise linear functions. In: Proceedings of the 9th International Conference on Artificial Neural Networks (ICANN99), 1999, pp 198-203

8. de Ridder D, Lemmers O, Duin RPW and Kittler J. The adaptive subspace map for image description and image database retrieval. In: Ferri FJ, Inesta JM, Amin A, Pudil P (eds) Advances in Pattern Recognition, Proc. Joint IAPR International Workshops SSPR2000 and SPR2000. Springer Verlag, Berlin, 2000, pp 94-103 (Lecture Notes in Computer Science no. 1876)

9. de Ridder D, Kittler J, Lemmers O and Duin RPW. The adaptive subspace map for texture segmentation. In: Sanfeliu A, Villanueva JJ, Vanrell M, Alquezar R, Eklundh J-O and Aloimonos Y (eds) ICPR15, Proc. 15th Int. Conference on Pattern Recognition, Computer Vision and Image Analysis, vol 1. IEEE Computer Society Press, Los Alamitos, 2000, pp 216-220

10. Kohonen T, Kaski S and Lappalainen H. Self-organised formation of various invariant-feature filters in the adaptive-subspace SOM. Neural Computation 1997; 9:1321-1344

11. Luttrell SP. A Bayesian analysis of self-organising maps. Neural Computation 1994; 6:767-794

12. Ephraim Y and Gray RM. A unified approach for encoding clean and noisy sources by means of waveform and autoregressive model vector quantisation. IEEE Transactions on Information Theory 1988; 34:826-834

# SHAPESOM

P. N. Suganthan
School of Electrical and Electronic Engineering
Nanyang Technological University, Republic of Singapore, 639798
Email: epnsugan@ntu.edu.sg

**ABSTRACT:** In this paper, we propose a novel topology preserving mapping scheme called SHAPESOM to map structural shapes in a topology preserving manner. The structural information contained in geometrical structures is extracted using the pairwise geometric histograms. These histograms are quantised using a self-organising maps (SOM), as the SOMs offer a number of advantages over the standard equidistance histogram quantisation. Using this trained SOM, a global pairwise histogram is generated for every structural shape. These histograms are treated as input vectors to another SOM called the SHAPESOM. As these global histograms capture the shape properties of the objects, the SOM trained using these histograms naturally generates a topology conserving mapping for the structural shapes. The topological mapping can be made to invariant to some chosen transformations such as the translation, rotation, scale and affine. This scheme can be used to organise and index large structural databases for shape based image retrieval or hypotheses generation in object recognition systems.

## 1 Introduction

The SOM has been applied to solve several complex problems such as vector quantisation, speech recognition, combinatorial optimisation, control, pattern recognition, organisation of documents (WEBSOM), organisation of image databases and modeling the structure of the visual cortex [5,6,8,11,14]. In this paper, we propose a novel topology preserving mapping scheme called SHAPESOM to map structural/geometric shapes in a topological manner. The structural information contained in geometrical structures is extracted using the pairwise geometric histograms [4,13]. These histograms are quantised using an SOM. Using this trained SOM, a global pairwise histogram is generated for every structural shape. These histograms are treated as input vectors to another SOM called the SHAPESOM. As these global histograms capture the shape properties of the objects, the SHAPESOM trained using these histograms naturally generates a topology conserving mapping for the structural shapes. This structural topology preserving maps can be made invariant to chosen transformations such as the similarity or affine transformations by choosing attributes appropriately.

In this paper, we make use of the relational attributes developed by Huet et al. [4] to construct the relational histograms. However, instead of using a histogramming technique [4], we employ the SOM [5] to quantize five dimensional relational attribute vectors between a pair of line segments. The SOMs offer a number of advantages such as the ability to quantize adaptively depending on the dynamic ranges of the attributes and the ability deal with the curse of dimensionality in the histograms-based methods. For example, Huet et al. [4] define a number of relational attributes but use the best two attributes in their histogram based line patterns retrieval system. If they were to use all five or so attributes, then the line patterns retrieval will become a five dimensional histogram matching problem. Further, if the problem [4] does not require scale invariance, then quantisation of absolute lengths and distances may become a difficult problem. All these difficulties can be resolved satisfactorily by employing the SOM to perform the quantisation. The trained SOM will yield just one dimensional histogram for every line pattern object. As the SOM preserves the topological relationships in the data (i.e. similar relational attribute vectors are mapped to neighboring SOM units), we can combine neighboring elements in the histogram, if we desire to perform a coarse-to-fine search. In this paper, we make use of trademark images [1] approximated by line segments.

The most popular approach to indexing into image databases has been the color histogram indexing [12]. A natural extension to color histogram is to make use of texture and shape information available in the images [2,3,4,7]. Although the color histogram technique has been somewhat generalized to include texture and shape attributes, still there are a number of limitations associated with the histograms-based methods such as the need to quantize parameter spaces over varying ranges in color, texture, shape and relational domains and the curse of dimensionality. The SOM has also been proposed to organise and retrieve image using colour, texture and edge information [8,14].

Efficient organisation and indexing of model objects are important steps in the field of model-based structural object recognition [9,10]. Specifically, in the graph-based object recognition, it is critically important to effectively organize and index into the databases in order to confine the graph matching operation to some selected models, as the graph matching is a hard combinatorial optimization problem. For instance, handwritten Chinese character recognition problem may also be formulated as a line patterns matching problem. As there are over 10,000 distinct Chinese characters, organising them using the SHAPESOM would enable us to retrieve a limited number of potential model candidates for further rigorous matching. The purpose of this research is to demonstrate that the SHAPESOM can be used to effectively retrieve a number of potential model objects or images from databases for further rigorous comparison.

This paper is organized as follows. In Section 2, we present the relational attributes used to characterize a pair of line segments. In Section 3, the SOM algorithm as well as the applications of these algorithms for generating one-dimensional histograms and organization and indexation of databases using the SHAPESOM are described. Experimental results are presented in Section 4 and the paper is concluded in Section 5.

# 2 Relational Attribute Vectors

Depending on the requirements, the relational attributes may be chosen to possess various invariance properties such as invariance to translation, rotation, scale and more general affine transformations [4,13]. In this implementation, we have decided to include invariance to all similarity transformations.

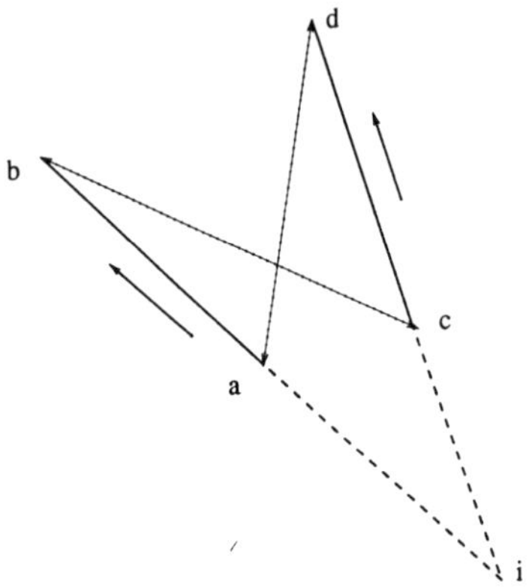


Figure 1: A figure illustrating the computation of directed relational attributes.

We observed that the attributes defined in [4] would be sufficient to characterise relational attributes between two line segments invariant of similarity transforms. Prior to computing the attributes, the intersection point between the two lines are computed as shown by "i" in Fig. 1. The end point of the first line closer to the intersection point is labeled as "a". The other end point of the first line is labeled as "b". Likewise the end points of the second line are also labeled as "c" and "d" as shown in Fig. 1. The following relational attributes are used: (1) $\theta_{ab,cd} = arccos\{\frac{\vec{ab}.\vec{cd}}{|\vec{ab}|.|\vec{cd}|}\}$. The angle returned is between 0 and $\pi$. However, if we identify the rotation from $\vec{ab}$ to $\vec{cd}$ as clockwise or counter-clockwise by evaluating the vector product between vectors $\vec{ab}$ and $\vec{cd}$, then we can compute the angle attribute between -$\pi$ to $\pi$ thereby improving the discrimination quality of this attribute, (2) Relative position ratio: $\frac{1}{\frac{1}{2}+\frac{l_{ib}}{l_{ab}}}$, (3) Line length ratio: $\frac{min\{l_{ab},l_{cd}\}}{max\{l_{ab},l_{cd}\}}$, (4) End point ratio: $\frac{min\{l_{ac},l_{bd}\}}{max\{l_{ac},l_{bd}\}}$, (5) Cross end point ratio: $\frac{min\{l_{ad},l_{bc}\}}{max\{l_{ad},l_{bc}\}}$.

As there are thousands of lines in the logo images, there would be a huge number of line pairs. Further, if the line patterns used to retrieve the images or model objects, are corrupted by noise, then making use of every line pairs may not be beneficial to the performance of the system. In such a situation, local neighborhood graphs with the neighborhood degree of 6 are known to yield the near optimum performance [4]. In this study too, the pairwise relational vectors are computed up to 6 nearest neighbor line segments of every line segment in the logo model base. Every pair of line segments generates a five dimensional vector. The vectors obtained from all line patterns are used to train an SOM as explained in the next section to perform an adaptive quantisation.

# 3 Self-organizing Maps

In this section, first the SOM algorithm [5] is briefly explained. Then the procedure for generating a one-dimensional relational histogram for every line pattern object using the trained SOM and developing the SHAPESOM are described.

Table 1: The SOM Algorithm

**SOM1:** Randomly initialize the weights for the given size map. Initialize the learning rate parameter, neighborhood size and set the number of unsupervised learning iterations.
**SOM2:** Present the input feature vector $\boldsymbol{x}= [x_1, x_2, ..., x_n, ..., x_N]$ in the training data set to the network.
**SOM3:** Determine the winner node $c$ such that $||\boldsymbol{x} - \boldsymbol{w}_c|| = \min_i\{||\boldsymbol{x} - \boldsymbol{w}_i||\}$
**SOM4:** Update the weights within the neighborhood of node $c$, $N_c(t)$ using the standard update rule [5]: $\boldsymbol{w}_i(t+1) = \boldsymbol{w}_i(t) + \alpha(t)[\boldsymbol{x} - \boldsymbol{w}_i(t)]$ where $i \in N_c(t)$. The neighborhood wraps around at edges, i.e. column and row indices are in modulo representation.
**SOM5:** Update learning rate and neighborhood size. $\alpha(t+1) = \alpha(0)\{1 - \frac{t}{K}\}$; $|N_c(t+1)| = |N_c(0)|\{1 - \frac{t}{K}\}$, where $K$ is a constant and is usually set to be equal to the total number of iterations in the self-organising phase.
**SOM6:** Repeat SOM2-5 for the specified number of unsupervised learning iterations.

The SOM algorithm is summarised in Table 1. In this application, the SOM is used to quantize the relational vectors obtained for every 6 nearest neighbour graphs in every line pattern model. The dimension of input vector is five. The number of output neurons will be identical to the number of bins that we wish to have in the one-dimensional histogram. As the relational attributes have known dynamic ranges, it is not necessary to make use of all relational attribute vectors to train the SOM. After completing the training of the SOM, we can extract the one-dimensional histogram by performing the following steps: (1) For every line pattern, initialize to zero a one-dimensional array of length identical to the number of output neurons in the SOM, (2) Present every five dimensional pairwise relational vector to the SOM and identify the winning neuron, (3) Increment the element in the one-dimensional array corresponding to the winning neuron, (4) Repeat steps 2 and 3 for every relational vector of the line pattern model. (5) Normalise the histogram, (6) Repeat steps 1-5 for every line pattern model in the database.

At the completion of the steps above, we have a one-dimensional normalized histogram for every line pattern (i.e. trade mark) model. These normalized histograms are the input vectors to construct the SHAPESOM using the same SOM algorithm in Table 1. Having trained the SHAPESOM, we associate the line pattern model identity with the neurons. Every line pattern is associated with three best matching neurons. With this, the organization phase is complete. During the application phase, we perform the following steps to extract potential model candidates: (1) Extract all pairwise relational attribute vectors from the query line pattern. (2) Present every vector to the SOM and obtain the one-dimensional normalized histogram. (3) Present the one-dimensional normalized histogram to the SHAPESOM and identify the best matching unit at of the SHAPESOM. (4) Extract all model line pattern numbers associated with the best matching unit and eliminate any duplication in the line pattern numbers to obtain the potential model candidates.

An advantage of using the SHAPESOM for this application is that there have been several adaptive and hierarchical/tree-structured SOM models [6,11] developed in the past. These adaptive models can be easily fine tuned to be used for the SHAPESOM. These adaptive methods will enable us to include additional line pattern models or remove line pattern models with minimal efforts.

# 4 Experimental Results

Experiments were conducted using about 1000 line pattern trademarks [1]. The number of line segments in the trademark objects varies between about 50 to slightly over 2000. In our experiments, we restricted the number of neurons in the SOM to 900 for the quantization of the relational vectors. The size of the SHAPESOM is 18x18. As naturally expected, the system was able to always retrieve the original line objects when noiseless original objects were presented. When noisy line pattern trade marks are presented, on average the SHAPESOM extracted about 45 line pattern models for further rigorous matching. This is about 5% of the total models in the database. The final retrieval can be performed using a more accurate matching method on the 45 or so potential candidates.

In order to show the mappings generated by the SHAPESOM, we present some of the line pattern trademarks represented by the neurons in Fig. 2. The first 4 images in row one are represented by the neuron at (1,1) in (row, column) order in the SHAPESOM. Images in 2nd row, 3rd row and 4th row are represented by neurons at (5,15), (15,5) and (16,16) respectively. We could observe the topology preservation in every row. Images in row 1 have a mixture of circular elements and straight lines. Pictures in row 2 have more longer straight line segments. Pictures in row 3 have a huge number of little line segments and a few long line segments. In the last row all pictures have multiple circles and characters written in a circularly oriented manner. Therefore, clearly the shapes are being mapped topologically in the SHAPESOM. Currently, we are investigating the applications of the SOM in generating hypotheses for the existence of model objects in a cluttered test scene. This is an important application when there are a large number of structural objects in the model-base. By generating reliable hypotheses using the SOM, we can pick the models most likely to be present in the cluttered scene for further matching, instead matching all the models in the model base sequentially against the cluttered test scene.

## 5 Conclusion

In this paper, we proposed a novel topology preserving mapping scheme for geometric structural objects by the SOM. We made use of five dimensional relational vectors. These relational vectors are quantized using an SOM and subsequently converted into a one dimensional histogram for every structural object. These one dimensional histograms are then used to train the SHAPESOM. The SHAPESOM can be used to retrieve structural objects from a database or organise structural objects in a model-based object recognition system. The proposed approach offers a number of advantages such as the ability to make use of several relational attributes (as opposed to the limited number of attributes that can be used in the histogram-based relational indexing methods [3,4,13]), the ability to perform a dynamic quantization, the flexibility in including and removing model objects and database images, possibility for efficient implementation such as the tree-structured SOM [5,8] and the ability to handle other attributes like color and texture in a homogeneous manner by the SOM [8,14]. The SHAPESOM is also capable of incorporating relational attributes invariant to selected transformations.

**Acknowledgement:** The author wishes to thank the authors of references [3,4] for kindly allowing me to use their datasets in this study.

# References


1. J. P. Eakins, J. M. Boardman and M. E. Graham, "Similarity retrieval of trademark images", *IEEE Trans. on Multimedia* Vol. 5, No. 2 , pp. 53-63, 1998.

2. T. Gevers and A. W. M. Smeulders, "PicToSeek: Combining color and shape invariant features for image retrieval", *IEEE Trans. Image Processing,* Vol.'9, No. 1, pp. 102-119.

3. B. Huet and E. R. Hancock, "Relational histograms for shape indexing", *In Proc. of ICCV'98,* pp. 563-569.

4. B. Huet and E. R. Hancock, "Line pattern retrieval using relational histograms", *IEEE Trans. PAMI,* Vol. 21, No. 12, pp. 1363-1369, 1999.

5. T. Kohonen, *Self-organising Maps,* $2^{nd}$ edition, Springer Verlag, 1997.

6. P. Koikkalainen and E. Oja, "Self-organizing hierarchical feature maps", *In Proc. IJCNN-90, International Joint Conference on Neural Networks,* 1990, vol. II, pp. 279–285,

7. W. Niblack, et al., "The QBIC project: Querying images by content using color, texture and shape", *Image and Visual Storage and Retrieval,* pp. 173-187, 1993.

8. E. Oja and S. Kaski (Editors), *Kohonen Maps, Proc. of WSOM'99,* Elsevier, 1999.

9. K. Sengupta and K. L. Boyer, "Organizing large structural modelbases", *IEEE Trans. PAMI,* Vol. 17, No. 4, pp. 321-332, 1995.

10. L. G. Shapiro and R. M. Haralick, "Organization of relational models for scene analysis", *IEEE Trans. PAMI,* Vol. 4, No. 11, pp. 595-602, 1982.

11. P. N. Suganthan, "Hierarchical overlapped SOM's for pattern classification", *IEEE Transactions on Neural Networks,* Vol. 10, No. 1, Jan. 1999.

12. M. Swain and D. Ballard, "Indexing via color histograms", *In Proc. of ICCV'98,* pp. 390-393.

13. N. A. Thacker, P. A. Riocreux, and R. B. Yates, "Assessing the completeness of properties of pairwise geometric histograms", *Image and Vision Computing,* Vol. 13, No. 5, pp. 423-429, 1995

14. H. Zhang and D. Zhong, "A scheme for visual feature based image indexing", *In Proc. of SPIE Conf. on Storage and Retrieval for Image and Video Databases,* San Jose, CA., Feb. 1995.

Figure 2: An illustration of topographical mapping generated for the trademark images.

# A New Interpolation Algorithm Employing a Self-Organizing Map

T. Yamakawa†* , K. Horio† , Y. Oosako‡ and T. Miki†*

† Department of Brain Science and Engineering
Graduate School of Life Science and Systems Engineering
Kyushu Institute of Technology
680-4 Kawazu, Iizuka, Fukuoka 820-8502 Japan
yamakawa@ces.kyutech.ac.jp

‡ Department of Computational Intelligence and Systems Engineering
Interdisciplinary Graduate School of Science and Engineering
Tokyo Institute of Technology
4259 Hagatsuta, Midori-ku, Yokohama, Kanagawa 226-6803 Japan

Abstract

In early learning stage of a self-organizing map (SOM), the weight vectors are arranged to interpolate among the input vectors. A new interpolation method employing the transient characteristics of the SOM is proposed in this paper. In order to realize a smooth interpolation, a modified updating equation based on compensation vector is used in the proposed method. Furthermore, we define the criterion of smoothness, and the reasonable number of learning iterations for the desired arrangement of the weight vectors can be decided. The performance of the proposed method is confirmed by the interpolation results in 2-dimensional and 3-dimensional spaces.

# 1 Introduction

It is called interpolation that arbitrary $x$ is evaluated between sampled data $x_k$ $(k = 1, \cdots, K)$. The well-known interpolation methods include a linear interpolation, a spline interpolation, Lagrange's interpolation and so on[1]. The linear interpolation method is very easy to use, however, the interpolated curve is not smooth. The spline and the Lagrange's interpolation methods can offer a smooth curve indeed but have a disadvantage of over fitting. Of course, the interpolation of multi-valued function can not be achieved by these conventional interpolation methods.

On the other hand, it is known that weight vectors of a self-organizing map (SOM) represent the distribution of the input vectors. In early learning stage, the weight vectors are arranged to interpolate among the input vectors.

*The author is also with the Fuzzy Logic Systems Institute (FLSI), Kawazu, Iizuka, Fukuoka 820-0067 Japan.

‡This work was achieved when the author was a Bachelor course student of Kyushu Institute of Technology.

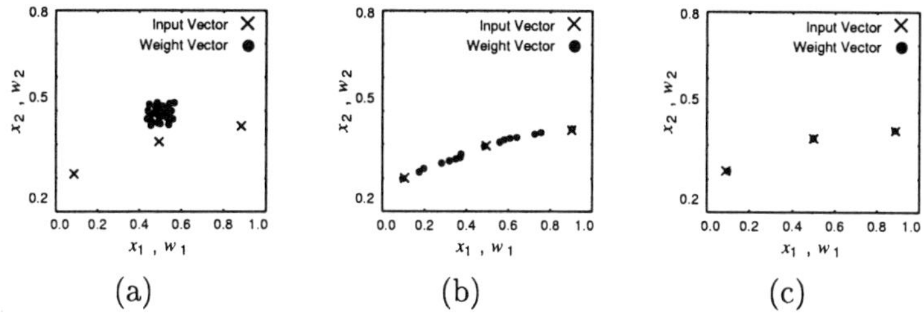


Figure 1: Distribution of weights in a SOM. (a) Initial state, (b) early stage of learning and (c) final stage.

In this paper, a new interpolation algorithm that the weight vectors are arranged to interpolate between the input vectors by using the special feature of the SOM is proposed. In the proposed method we define the smoothness measure, the reasonable number of iteration can be decided. Furthermore a new updating equation of the weight vector is proposed to realize the smooth interpolation. The interpolation results accord with human sensibility, furthermore the interpolation of multi-valued function can be successfully realized. The effectiveness of the proposed method is verified by applying it to interpolation problems in 2-dimensional and 3-dimensional spaces.

# 2 Interpolation Method by Using SOM

## 2.1 New Interpolation Method

Fig.1 shows a change of weights distribution during learning of the SOM including 20 units in the competitive layer. In Fig.1 the crosses and dots are input vectors and weight vectors of the units in the competitive layer, respectively. The initial values of the weight vectors are assigned by using random values (Fig.1(a)). Fig.1(b) and (c) show the weight vectors after 10 and 1000 learning iterations, respectively. It is obvious that the weight vectors are located between input vectors in early learning stage. The proposed interpolation method is based on this feature. However, in the algorithm of the conventional SOM, the weight vectors are linearly located between the input vectors. In this study, a new learning algorithm of the SOM which realizes smooth interpolation.

Consider that interpolation between 3 sampled data shown in Fig.2(a) is achieved. In the proposed method, the coordinates of the sampled data are used as the input vectors of the SOM, thus the number of the units in the input layer is 2. In the competitive layer, the units corresponding to the sampled data are prepared as shown in Fig.2(b). These units are represented as $U_{k-1}$, $U_k$ and $U_{k+1}$, where $k$ is the sampled data number. The weight vectors of these units are not updated in learning, so that these units are referred to as "fixed units". The weight vectors of the fixed units are initialized to the coordinates of the sampled data. A reasonable number of units are allocated

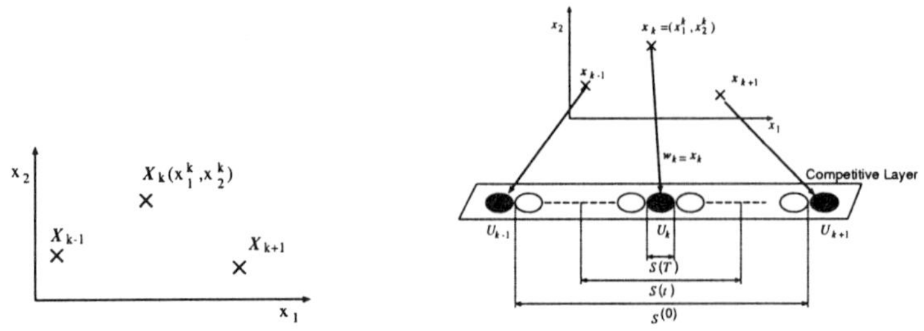


(a) (b)

Figure 2: The sampled data in 2-dimensional space allocated to units in one-dimensional competitive layer.

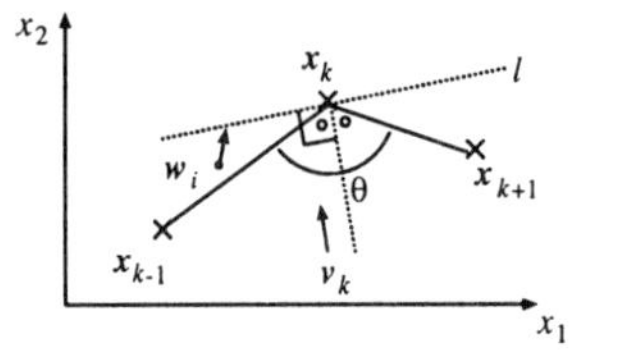

Figure 3: The compensation vector $\boldsymbol{v}_k$.

between the fixed units. The initial values of the weight vectors of the these units are 0, which is quite different from the conventional SOM. When order of the fixed units and number of the units between the fixed units are changed, various interpolations can be realized. The learning of the proposed method is achieved as follows.

When input vector $\boldsymbol{x}_k$ is applied to the input layer, the unit $U_k$ is defined as the winner unit. The units which are located in the neighborhood $S(t)$ of the winner unit shown in Fig.2(b) are defined as neighboring units. Here $t$ means the iteration number. At the start of learning, the area of $S(t)$ is assigned over inside $U_{k-1}$ and $U_{k+1}$, and it decreases with time. The weight vectors of the winner unit and neighboring units are updated by:

$$\boldsymbol{w}_j^{new} = \boldsymbol{w}_j^{old} + \alpha(t)N_c(t)(\boldsymbol{x}_k - \boldsymbol{w}_j^{old} + \beta((\boldsymbol{x}_k - \boldsymbol{w}_j^{old})\boldsymbol{v}_k)\boldsymbol{v}_k) \tag{1}$$

where $\beta$ and $\boldsymbol{v}_k$ are compensation coefficient and compensation vector, respectively. $N_c(t)$ is a function about neighborhood. $\boldsymbol{v}_k$ is the unit vector whose direction is equal to bisector of the angle defined by the three sampled data, and it is represented by:

$$\boldsymbol{v}_k = \frac{\boldsymbol{p}_k}{\| \boldsymbol{p}_k \|} \tag{2}$$

$$\boldsymbol{p}_k = \frac{\boldsymbol{x}_k - \boldsymbol{x}_{k-1}}{\| \boldsymbol{x}_k - \boldsymbol{x}_{k-1} \|} + \frac{\boldsymbol{x}_k - \boldsymbol{x}_{k+1}}{\| \boldsymbol{x}_k - \boldsymbol{x}_{k+1} \|} \tag{3}$$

Eq.(1) means that the weight vectors of the units between $U_{k-1}$ and $U_k$ are updated to be located between $\boldsymbol{x}_{k-1}$ and $\boldsymbol{x}_k$ and close to line $l$ shown in

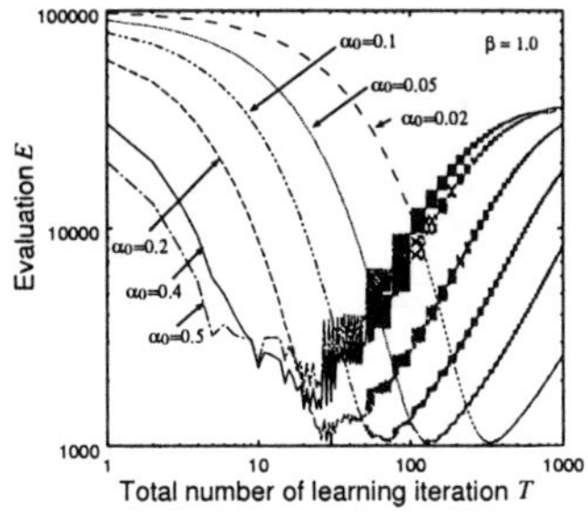


Figure 4: The relationship between evaluation function $E$ and total number of learning iteration $T$.

Fig.3. Thus the weight vectors are arranged to interpolate between $\boldsymbol{x}_{k-1}$ and $\boldsymbol{x}_k$ smoothly. After learning, the interpolated curve can be drawn by making straight lines between weight vectors.

## 2.2 Definition of Smoothness

In the proposed method, the performance of the interpolation is based on the number of learning iterations. In order to decide the reasonable number we define the evaluation function $E$ representing "smoothness" by:

$$E = \sum_{j=2}^{m-1} \| d_j \|^2 = \sum_{j=2}^{m-1} \| \boldsymbol{w}_{j-1} + \boldsymbol{w}_{j+1} - 2 \cdot \boldsymbol{w}_j \|^2 \tag{4}$$

$E$ means the declination of the angle defined by three weight vectors $\boldsymbol{w}_{j-1}$, $\boldsymbol{w}_j$ and $\boldsymbol{w}_{j+1}$ from $\pi$. In this study the minimum value of $E$ means that the weight vectors are arranged to interpolate between the sampled data most smoothly. The learning, in which the 5 sampled data $\boldsymbol{x}_1 = (10, 15)$, $\boldsymbol{x}_2 = (25, 40)$, $\boldsymbol{x}_3 = (50, -5)$, $\boldsymbol{x}_4 = (75, 55)$, $\boldsymbol{x}_5 = (100, 200)$ is used as the input vectors, is achieved. Fig.4 shows the relationship between $E$ and the total number of learning iterations $T$. It indicates that evaluation $E$ is large when total number of learning is large. In the proposed method, the random values are not used to initialize the weight vectors, so that the relationship between $E$ and $T$ is decided by only parameters ($\alpha(0)$, $\beta$). It is easy to determine the reasonable number of learning iteration.

Although it should be considered that it is reasonable to derive the updating equation from the evaluation function, the updating equation does not guarantee the global minimum.

# 3 Experimental Results

In order to verify effectiveness of the proposed interpolation method, we show some simulation results in which the sampled data in 2-dimensional and 3-dimensional spaces are interpolated.

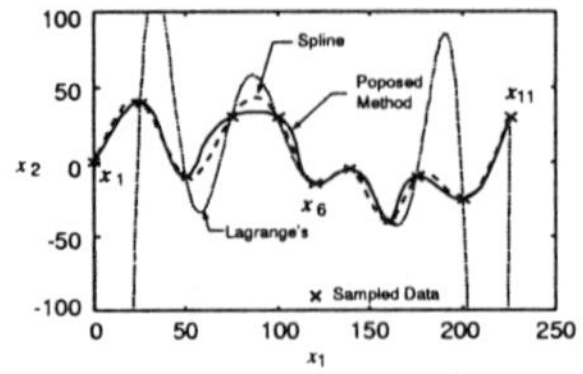


Figure 5: 11 sampled data and interpolation results in 2-D space by spline interpolation, Lagrange's interpolation and the proposed interpolation method.

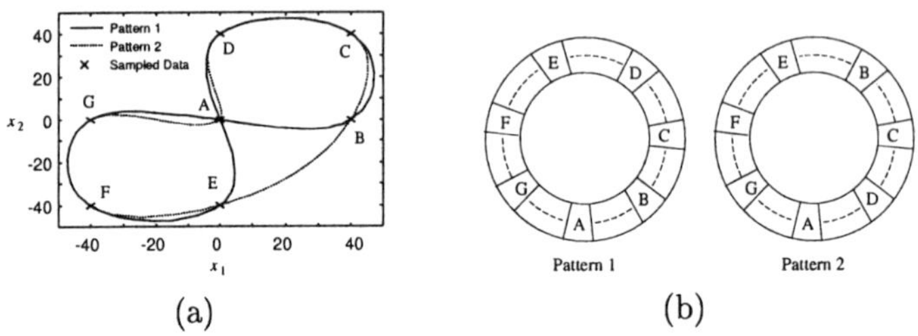


Figure 6: Interpolation results in 2-D space. (a) Interpolation results. (b) Arrangements of the fixed unit $U_k$.

## 3.1 Interpolation in 2-D Space

Interpolation between 11 sampled data shown in Fig.5 is achieved. 11 fixed units and 24 units between fixed units are prepared in the 1-dimensional competitive layer. Fig.5 shows interpolation results by spline, Lagrange's and the proposed interpolation method. In the proposed method, the initial value of learning rate $\alpha(0)$, the compensation coefficient $\beta$ and the total number of learning iteration $T$ are 0.4, 1.0 and 30, respectively. It is known that interpolation result by the proposed method accords with human sensibility in comparison with the conventional methods.

The interpolation of multi-valued function that can not be realized by the conventional methods is also achieved. Fig.6(a) shows 7 sampled data to be interpolated. The competitive layer is circular, and two kinds of arrangement of the fixed units are employed as shown in Fig.6(b). Number of the units between the fixed units, $\alpha(0)$, $\beta$ and $T$ are 24, 0.4, 1.0 and 30, respectively. Fig.6(a) shows the interpolation results by the proposed method. It is shown that the sampled data are interpolated smoothly and that some kind of interpolation results can be obtained by change of order of the fixed units.

## 3.2 Interpolation in 3-D Space

Consider that the interpolation among 16 sampled data in 3-dimensional space shown in Fig.7(a) is achieved. The competitive layer is 1-dimensional and number of the units between the fixed units is 30. $\alpha(0)$, $\beta$ and $T$ are 0.4, 1.0 and 30, respectively. Smooth interpolation between the sampled data is

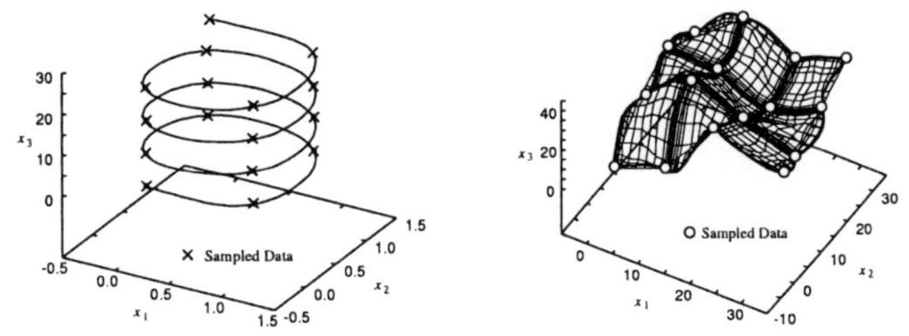


(a) (b)

Figure 7: Interpolation results in 3-D space by the proposed interpolation method.

realized (Fig.7(a)).

Interpolation between 16 sampled data shown in Fig.7(b) with curved surface is achieved. 1-dimensional competitive layer can be expanded to 2-dimensional one. In this case vector $\boldsymbol{p}_k$ in Eq.(3) is represented by:

$$\boldsymbol{p}_k = \frac{\boldsymbol{x}_k - \boldsymbol{x}_{k-4}}{\| \boldsymbol{x}_k - \boldsymbol{x}_{k-4} \|} + \frac{\boldsymbol{x}_k - \boldsymbol{x}_{k-1}}{\| \boldsymbol{x}_k - \boldsymbol{x}_{k-1} \|} + \frac{\boldsymbol{x}_k - \boldsymbol{x}_{k+1}}{\| \boldsymbol{x}_k - \boldsymbol{x}_{k+1} \|} + \frac{\boldsymbol{x}_k - \boldsymbol{x}_{k+4}}{\| \boldsymbol{x}_k - \boldsymbol{x}_{k+4} \|} \quad (5)$$

The number of the unit between the fixed units, $\alpha(0)$, $\beta$ and $T$ are 15, 0.4, 1.0 and 30, respectively. It is obvious that reasonable interpolation result with curved surface is obtained (Fig.7(b)).

## 4 Conclusions

In this paper, a new interpolation algorithm by using the special feature of the SOM in early learning stage is proposed. In the proposed method, weight vectors are updated by the modified learning algorithm with compensation vectors, and the smooth interpolations between sampled data are realized. The effectiveness of the proposed method is verified by applying it to interpolation problems in 2-dimensional and 3-dimensional spaces.

## References

[1] Y. Ohno and K. Isoda, *Numerical Calculation Handbook*, Ohm, 1990.

[2] T. Kohonen, "Self-Organized of Topologicaly Correct Feature Maps," *Biol. Cybern.*, Vol.43, pp.59-69, 1982.

[3] T. Kohonen, *Self-organization and associative memory*, Springer-Verlag, 1988.

# SOM-Based Exploratory Analysis of Gene Expression Data

Samuel Kaski

Helsinki University of Technology
Neural Networks Research Centre
P.O. Box 5400, FIN-02015 HUT, Finland
samuel.kaski@hut.fi

**Abstract.** Applications of new SOM-based exploratory data analysis methods to bioinformatics are described. Cluster structures are revealed in data describing the expression of a set of yeast genes in several experimental treatments. The structures are visualized in an intuitive manner with colors: The similarity of hue corresponds to the similarity of the multivariate data. The clusters can be interpreted by visualizing changes of the data variables (expression in different treatments) at the cluster borders. The relationship between the organization of the SOM and the functional classes of the proteins encoded by the genes may additionally reveal interesting relationships between the functional classes, and substructures within them.

## 1 Introduction

Recent methodological breakthroughs are transforming the cellular-level research of biological organisms. Several large-scale projects to decipher the genetic sequences have been carried out, and new methods for genome-wide measurement of gene expression have been introduced. The three-dimensional structures of the proteins are being modeled. All these approaches generate massive amounts of data, into which important knowledge about the function of the genes and proteins is hidden. It has been said that biology is changing from a hypotheses-driven science to a data-driven science. The currently very active field of bioinformatics provides the necessary new data-analysis methods. It is the glue between biology, computer science, and various modeling approaches.

Currently the challenges in genetics are shifting from analyzing the genetic sequences to analyzing the function of the genes. The Self-Organizing Map (SOM) has potential as a tool in the required large-scale analyses. SOMs have already been applied to clustering of yeast gene expression data [8,9].

In this paper SOM-based methods for exploratory data analysis will be described and applied to yeast gene expression analysis. We have developed methods for uncovering and visualizing cluster structures in an easily understandable manner, and for interpreting them in terms of the original data variables.

## 2 Gene Expression Data

The expression of even hundreds of thousands of genes can be measured simultaneously with DNA microarrays or "DNA chips." They are essentially arrays of DNA from each gene, placed as a matrix on a "chip." The array can be used to measure the amount of messenger RNA of each gene, an indicator of the activity of the gene, at a certain time. The effects of a drug treatment, for example, can be monitored by comparing the expression with a control condition. When this monitoring is done in a genome-wide scale and in many treatments, a huge amount of data about gene function is generated. One aim of our bioinformatics project is to analyze this data with SOMs.

The data set used in the experiments consists of measurements of the expression level of 1551 genes of the budding yeast *Saccharomyces cerevisiae.* The measurements containing 79 time points in total were made in 8 treatments, resulting in an "expression profile" for each gene. The data is available at `http://genome-www.stanford.edu/clustering`. For more details on the data and its analyses see [3].

## 3 SOM-based Data Exploration

The Self-Organizing Map [6,7] has been a very useful tool in several kinds of applications. Over 4000 scientific articles and several books have already been written on the SOM (`http://www.cis.hut.fi/research/som-bibl/`; cf. [1]).

One of the main applications of the SOM is data analysis. Similarity relationships within a data set can be visualized on a graphical SOM display. Also other aspects of the data, clusteredness or density structures, and distribution of the values of the data variables, can be visualized on the same display. Combined use of these visualizations helps in gaining an overview of the data set, and in interactive exploration of its properties. The SOM-based methods are perhaps at their best in the exploratory or data-driven phase of data analysis and research, in which hypotheses concerning properties or contents of the data set are formed. The exploratory phase is often followed by a hypothesis-driven phase in which the formed hypotheses are then confirmed.

There are many subtle differences between the SOM and alternative methods. For example, experimental results indicate that visualizations formed by the SOM are more trustworthy in the sense that if two data samples are close-by on the SOM display then they are close-by in the original space as well, at least more often than for alternative methods. This result was obtained empirically by comparing the results of the SOM and traditional multidimensional scaling-based non-linear projection methods [11]. Such trustworthiness is of course important in data analysis.

### 3.1 Visualization of Cluster Structures

Each data sample, here a gene expression profile, is mapped onto a certain point on the SOM grid. As a result of the SOM algorithm the data becomes organized on the grid so that close-by points contain genes having similar expression profiles.

The locations of genes coding proteins of three sample functional classes have been shown in Figure 1. Based on the displays it is evident that one of the classes (**a**) contains mostly genes that are expressed similarly in the set of treatments. The other classes are more diverse, containing subgroups that can be identified based on the SOM visualizations. We are currently investigating the biological significance of these results.

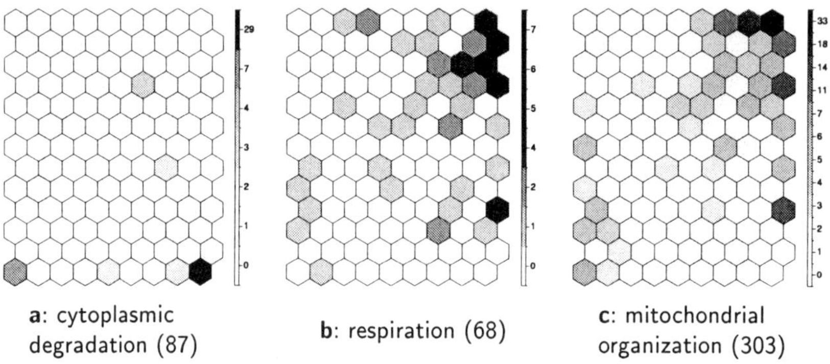


**Fig. 1.** Distribution of genes of three functional classes on the map. The number in parentheses denotes the total number of genes in the class. The scale shows the number of genes mapped onto the map units.

**U-matrix.** The SOM is defined by an ordered set of model vectors, one vector attached to each map unit or grid point. The model vectors represent the data in an ordered fashion, and the point density of the model vectors in the data space reflects the probability density of the data. Hence density structures of the data, i.e., cluster structures, can be visualized by visualizing the density of the model vectors. The so-called U-matrix display [10] visualizes the relative distances between neighboring model vectors by shades of gray.

The U-matrix of the gene expression data is shown in Figure 2**a**. Some clustered areas are visible, but the noisiness of the visualization is problematic. In fact, for this map all pairwise distances between neighboring model vectors are relatively similar as can be seen in Figure 3**a**.

**A new method for detecting cluster borders.** We have developed [2] a method that uses both the model vectors and the local distribution of the

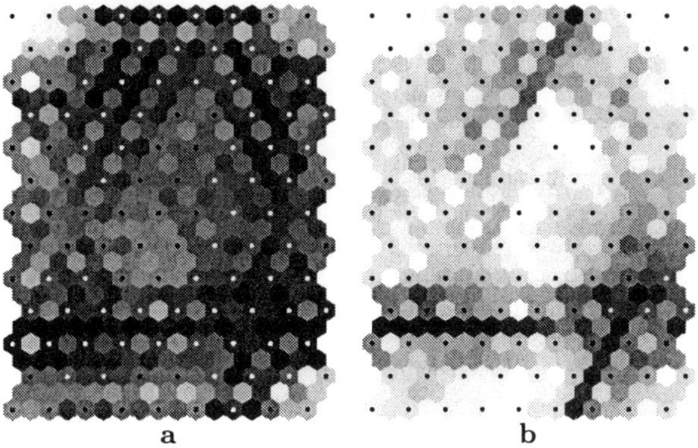

**Fig. 2.** Visualizations of the cluster structure in the gene expression data. **a** U-matrix. **b** Our new method for detecting cluster borders. Light shades: clusters; dark shades: sparser areas or gaps in between clusters. The dots denote the map units.

data to detect and visualize salient cluster borders not necessarily visible in the U-matrix type displays. It first computes, for each location of the SOM grid, a (heuristic) approximation of the local gradient of the data density. It is given by the difference between the model vector and the centroid of a set of closest data points. Then the gradient approximations at neighboring map units are compared by computing the norm of their difference, and the result of the comparison is visualized as a gray level on the map. If the difference in the gradients is large (dark shade in the visualization), the two map units are close to different clusters and there is a cluster border between them.

The cluster structures of the gene data have been visualized in Figure 2**b**. Two cluster borders are clearly visible as dark stripes.

**Illustrative Visualization with Colors.** The cluster structures are visible in the displays of Figure 2. However, the interpretation of the gray shades still requires background knowledge on the SOM methodology. One of the main goals in data exploration is to automatically form overviews that are immediately understandable. An analyst could use such visualizations as reports for executives or laymen, and the visualizations could even form the basis of a decision support system.

Color is a good candidate for grouping data according to its properties; color-based visualizations of data groups have already been used so pervasively that their interpretation is common knowledge. We have introduced a method for assigning colors to SOM displays such that the perceptual differences of the colors reflect the distances in the original data space as faithfully as possible. Similar data are colored similarly.

This seemingly simple task is actually far from trivial if the data space is high-dimensional, like the gene expression space is. The coloring corresponds to a (non-linear) projection of the model vectors of the SOM to the color space. We have used a constant-lightness "slice" of the CIELab color space, and discouraged very non-saturated colors. Then the resulting colors differ predominantly only in one perceptual dimension, the hue, which makes the interpretation easier. The available color space is further bounded by the limits of the display device, and hence the projection must be able to flexibly conform to a very irregular-shaped two-dimensional area while still remaining ordered and retaining color resolution. A suitable yet computationally relatively demanding projection method has been described in more detail in [5].

When the SOM of Figure **2a** is projected to the color space it can be seen that all distances between neighboring model vectors are relatively uniform (Fig. **3a**). The color will then change in a uniform fashion on the map. The new cluster display in Figure **2b** can be projected similarly, resulting in a display where the found cluster borders are reflected as rapid changes in colors (Fig. **3b**). Different clusters will then get different hues. The hue will change more slowly within the clusters, while still reflecting their inner substructures. The resulting colored maps can be viewed at `http://www.cis.hut.fi/sami/wsom01/`.

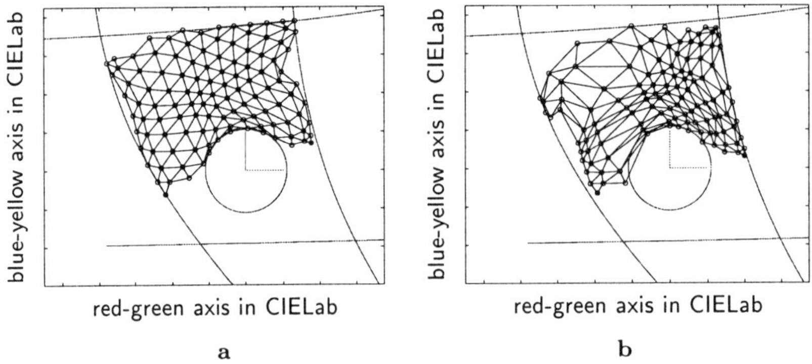


**Fig. 3.** Projection of the SOM grid onto the constant-lightness cross-section of the color space by preserving **a** the pairwise distances between neighboring model vectors, and **b** the indicators of cluster borders (shades of gray in Fig. **2b**). The outer borders denoted by the long curves limit the colors displayable by a typical CRT tube, and a set of non-saturated colors in the middle has been encircled.

### 3.2 Interpretation of the Structures

The SOM displays are commonly interpreted utilizing three kinds of visualizations. Well-known samples, or distributions of classes of samples as in Figure 1, can be used as "landmarks" on the map. Another option is to plot the model vectors as curves or histograms at each SOM grid point. In the present case the model vectors are typical gene expression profiles that could be depicted within the map units. The third possibility is to plot the distribution of the values of the original data variables, or even some new variables, with gray shades on the map. Visualizations of the values of a variable in the model vectors are often called "component planes."

The problem with all these interpretation methods is that for large maps and large, high-dimensional data sets it may be difficult to uncover the essentials. Summaries of relevant information are therefore needed. Our approach to interpretation is to interactively choose an area of the map, usually a cluster, and characterize it by showing what distinguishes it from its surroundings.

In earlier works we have computed contributions of the original variables to the "local factors" that describe the local change in the SOM [2]. A cluster is then characterized by the set of variables having the largest contributions on the cluster borders. In this paper a computationally simpler and perhaps more easily interpretable method is demonstrated.

A sample cluster is characterized in Figure 4 in terms of how the data variables change on its borders. The average difference between neighboring model vectors on different sides of the border is computed and shown as a bar graph. From the visualization it can be seen that, for instance, the genes in the cluster are consistently more up-regulated in the DTT shock treatment than the genes in the surroundings. This is biologically understandable since the shock affects protein folding, and misfolded proteins must be degraded. It is known from Figure 1 that the cluster contains proteins related to cytoplasmic degradation.

Note that to characterize *any* area of the map, not only a clear-cut cluster, it can be interpreted in the same fashion. The result then tells in what sense the data within an area is similar, and in what sense it is different from the data of the surrounding areas.

## 4 Conclusions

The main conclusion of this paper is that SOM-based exploratory data analysis methods are promising tools for bioinformatics applications, in particular for the analysis of gene expression and function. The investigation of the biological meaning of the discoveries is still underway, but already now it is safe to say that the methods are applicable.

The cluster structures in gene expression data were visualized in new ways which emphasize cluster borders and make the clusters easily interpretable by

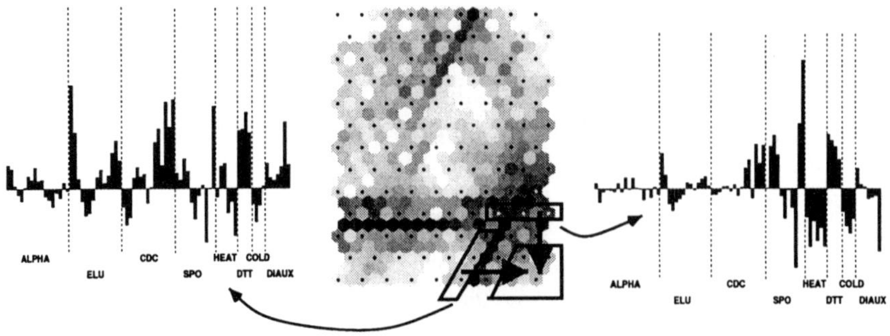


**Fig. 4.** A characterization of a sample cluster in terms of the original variables. The profiles show how the expression of the genes changes (on the average) when moving on the SOM according to the arrows. The changes can be used to characterize what is special in the expression of the genes. Abbreviations: ALPHA=cell-cycle: alpha factor arrest and release, ELU=cell-cycle: elutriation, CDC=cell-cycle: cdc15 arrest and release, SPO=sporulation, HEAT=heat shock, DTT=dtt shock 1mm, COLD=cold shock, DIAUX=diauxic shift.

coloring the data according to their similarity. The clusters were interpreted by computing characteristic changes occurring in the data at the cluster borders.

The present study was based on the usual inner-product similarity measure between gene expression profiles. It is evident, however, that not all time points and treatments are equally important and they should be weighted according to their importance. More generally, the metric of the data space should concentrate on measuring the important dissimilarities. In subsequent studies we will apply new methods [4] that *learn* a suitable distance measure. The metric is learned from data in which the expression profiles have been paired with auxiliary data that implicitly defines what is important. For gene expression data the auxiliary data could denote for instance the functional class of the proteins the genes encode.

### Acknowledgement

The research was supported by the Academy of Finland, in part by the grant 50061. I would like to thank our bioinformatics group, the bioinformatics group of the A.I. Virtanen Institute of the University of Kuopio, and Mr Jarkko Venna for the contributions.

## References

1. S. Kaski, J. Kangas, and T. Kohonen. Bibliography of self-organizing map (SOM) papers: 1981–1997. *Neural Computing Surveys*, 1(3&4):1–176, 1998.

Available in electronic form at http://www.icsi.berkeley.edu/~jagota/NCS/: Vol 1, pp. 102–350.
2. S. Kaski, J. Nikkilä, and T. Kohonen. Methods for exploratory cluster analysis. In *Proceedings of SSGRR 2000, International Conference on Advances in Infrastructure for Electronic Business, Science, and Education on the Internet.* Scuola Superiore G. Reiss Romoli, 2000. (Proceedings on CD-ROM, ISBN 88-85280-52-8).
3. S. Kaski, J. Nikkilä, P. Törönen, E. Castren, and G. Wong. Analysis and visualization of gene expression data using self-organizing maps. In *Proceedings of NSIP-01, IEEE-EURASIP Workshop on Nonlinear Signal and Image Processing.* 2001. Accepted for publication.
4. S. Kaski, J. Sinkkonen, and J. Peltonen. Bankruptcy analysis with self-organizing maps in learning metrics. *IEEE Transactions on Neural Networks,* 2001. Accepted for publication.
5. S. Kaski, J. Venna, and T. Kohonen. Coloring that reveals cluster structures in multivariate data. *Australian Journal of Intelligent Information Processing Systems,* 6:82–88, 2000.
6. T. Kohonen. Self-organized formation of topologically correct feature maps. *Biological Cybernetics,* 43:59–69, 1982.
7. T. Kohonen. *Self-Organizing Maps.* Springer, Berlin, 1995. (Third, extended edition 2001).
8. P. Tamayo, D. Slonim, J. Mesirov, Q. Zhu, S. Kitareewan, E. Dmitrowsky, E. S. Lander, and T. R. Golub. Interpreting patterns of gene expression with self-organizing maps: Methods and application to hematopoietic differentiation. *Proceedings of the National Academy of Sciences, USA,* 96:2907–2912, 1999.
9. P. Törönen, M. Kolehmainen, G. Wong, and E. Castrén. Analysis of gene expression data using self-organizing maps. *FEBS Letters,* 451:142–146, 1999.
10. A. Ultsch. Self-organizing neural networks for visualization and classification. In O. Opitz, B. Lausen, and R. Klar, editors, *Information and Classification,* pages 307–313. Springer-Verlag, Berlin, 1993.
11. J. Venna and S. Kaski. Neighborhood preservation in nonlinear projection methods: An experimental study. In *Proceedings of ICANN'01, International Conference on Artificial Neural Networks.* 2001 Submitted.

# Exploring power transformer database using Self-Organising Maps (SOM) and Minimal Spanning Tree (MST)

K. Obu-Cann, K. Fujimura, H. Tokutaka, M. Ohkita, M. Inui and S. Yamada
Tottori University
Koyama-Minami 4-101, Tottori, 680-8552, Japan

Abstract

Data mining or exploration is part of a larger area of recent research in Artificial Intelligence and Information Processing and Management otherwise known as Knowledge Discovery in Database (KDD). The main aim here is to identify new information or knowledge from database in which the dimensionality or amount of data is so large that it is beyond human comprehension. Self-Organising Map and Minimal Spanning Tree are used to analyse power transformer database from one of the electric energy providers in Japan. Evaluation of the clusters generated by SOM is usually done by human eye. Due to its qualitative nature, the evaluator may either overestimate or underestimate the number of clusters formed on the map. With this approach, the exact number of clusters generated by the map cannot be confirmed because of the misinterpretation of the grey level expression. This paper looks at clustering with Minimal Spanning Tree (MST).

## 1 Introduction

Data is useless to mankind if no meaningful information can be derived from it. The SOM [1], [2] is a powerful tool for data mining, knowledge discovery and visualisation of high dimensional data. SOM and MST are applied to power transformer database from one of the electric energy providers in Japan. The aim is to apply a data-mining tool based on SOM to learn more about database. The data was analysed to identify the energy consumption pattern of the consumers, season type pattern classification and day type pattern classification. In day type pattern classification, the MST algorithm [3], [4] was applied to augment the efforts of SOM in clustering. Prediction of the oil temperature changes in a power distribution transformer has been practiced using conventional explicit numerical calculations. This paper looks at the application of SOM to the prediction of the oil temperature changes.

# 2 Methodology for SOM Clustering

## 2.1 Season Type Pattern Classification

SOM is applied in the classification of the input patterns into the various seasons of the year namely winter, spring, summer and autumn. The data set for the SOM comprises of 4 input measurements (percentage load on the transformer, the transformer oil temperature, atmospheric temperature and transformer-surrounding temperature) for the months of January, March, July and September 1996. Each month is represented by two different day type patterns. Figure 1 illustrates the seasonal feature map obtained for the year 1996.

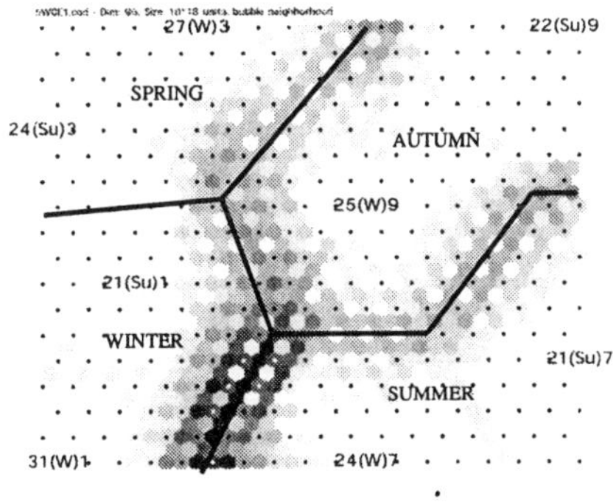


Figure 1: Seasonal SOM for 1996.

The eight input patterns, which represent the various seasons of the year, were classified into four separate regions on the map. Between winter and summer, the darkest grey separation was identified on the map. This is because during winter, though it is very cold, heating is mostly done with paraffin other than electric energy. In summer a lot of electric energy goes into cooling thereby resulting in a drastic increase in energy consumption. The next in the degree of grey is the difference between winter and autumn followed by winter and spring. The pattern for autumn is closer to spring than summer. This is because in autumn, the heating fuel is also paraffin. Looking at the SOM of figure 1 a lot can be said about the electric energy consumption patterns of the inhabitants of the area.

## 2.2 Day Type Pattern Classification

SOM is utilised for the classification of daily transformer oil temperature patterns. In SOM, the output units of the network are not specified. It is proposed that the SOM will generate the output based on the different patterns available in the input data space [5]. An input pattern comprises of the 24 hourly transformer oil temperature readings for a day. The 31 input patterns for the month of January 1996 were mapped onto a two-dimensional output network shown in figure 2(a). Figure 2(b) is a summary of the results obtained from figure 2(a). The holidays in January 1996 were 1/1(M) located at node 320, 3/1(W) at node 28, 4/1(TH) at node 285 and 2/1(T), 5/1(F) and 15/1(M) at node 198. The patterns for 2/1(T) and 15/1(M) were identified as weekend patterns. Comparing these with results obtained when the other months were analysed, new output nodes, each corresponding to a group of input patterns with a particular feature emerged after the training process. Moreover, not all the days appeared on the map, this is because days with similar transformer oil temperature patterns are mapped to the same output node as the same day type. The following day types were identified. Weekdays, Holidays and Weekends.

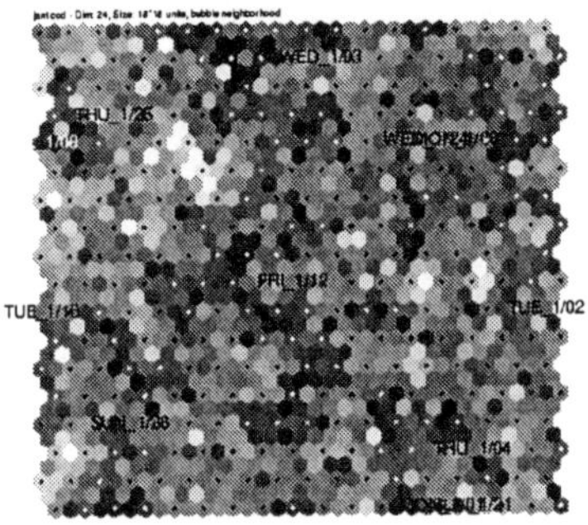


| Output node | Date (day) |
|---|---|
| 28 | 3(W) |
| 57 | 25(TH) |
| 73 | 9(T), 10(W), 11(TH), 17(W), 18(TH), 19(F), 22(M), 23(T), 26(F), 31(W) |
| 86 | 24(W) |
| 87 | 8(M), 30(T) |
| 171 | 12(F) |
| 181 | 16(T), 29(M) |
| 198 | 2(T), 5(F), 6(S), 7(SU), 13(S), 14(SU), 15(M), 20(S), 27(S) |
| 256 | 28(SU) |
| 285 | 4(TH) |
| 320 | 1(M) |
| 321 | 21(SU) |

Figure 2: (a) SOM for transformer oil temperature in January 1996. (b) Summary of the SOM for January 1996.

## 3 Methodology for MST Clustering

The MST method is a kind of tree like structure that links all the units by the shortest path [3], [4]. This structure defines the mutual distance and also describes the similarity relationship of any two points on the map [4], [6]. This algorithm assigns arcs between the nodes in such a manner that all nodes are connected together by single linkages. The lengths of the arcs are defined as the non-weighted norms of the vectorial differences between corresponding reference vectors as follows [4]:

$$\|x - m_c\| = \min_i \{\|x - m_i\|\} \qquad (1)$$

Where $m_i$ is the reference vector of node $i$ and $m_c$ is the reference vector of the winner node or unit (BMU). The neighbourhood of the cell in the MST topology is defined along the arcs that emanates from that cell. Learning begins with a wide neighbourhood, which reduces as learning progresses. This wide neighbourhood means traversing more MST arcs of the selected cell in order to make up the neighbourhood [4].

## 3.1 MST 1 (First Winner)

Looking at figure 2(a), it is difficult to tell the numbers of cluster off-hand from the map. In order to enhance the visual characteristics of the SOM map, the MST 1 was applied. This algorithm is based on figure 3a, the nodes selected to form the neighbourhood in this section comprised of only the first winner nodes.

## 3.2 MST 2 (Second Winner)

Depending on the degree of similarity among the data, the MST is extended to embrace other winners or nodes in close proximity to the node selected as the winner by the first algorithm i.e. MST 1. The neighbourhood in MST 1 is extended to include the second winner and its neighbours (see figure 3b). The neighbourhood this time contains far more nodes than that of the former algorithm. The procedure of extending the neighbourhood to include other winners is continued until there is no appreciable change in the clusters formed by connecting the MST arcs together.

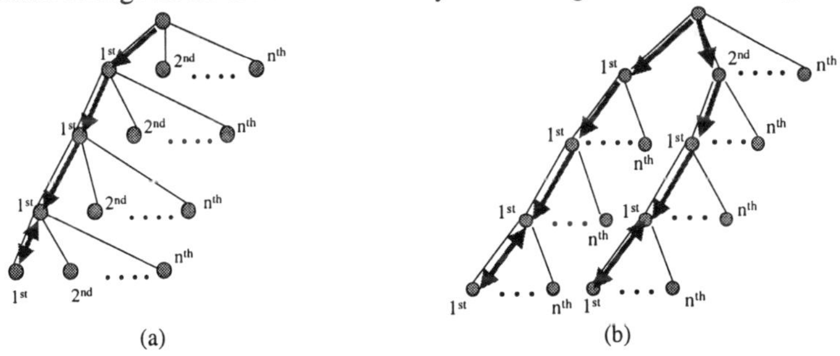


Figure 3: (a) Examples of nodes selected in the MST Topology using only the first winners. The selection process continues until a node, which has already been selected is selected again. (b) Examples of nodes selected using first and second winners.

Figure 4(a) illustrates the application of MST to the SOM map for the transformer oil temperature in January 1996. In each of the months analysed, three day-type patterns emerged after the training process. These are Weekdays, Holidays and

Weekends. The MST algorithm provided a physical link among the data within a particular cluster thereby providing a more visual map than the previous method.

(a)

| Output node | Date (day) |
|---|---|
| 1 | 9(T), 10(W ), 26(F), 31(W) |
| 8 | 25(TH) |
| 18 | 6(S), 11(TH), 17(W), 18(TH), 19(F), 22(M) |
| 72 | 13(S) |
| 111 | 5(F) |
| 124 | 27(S) |
| 149 | 7(SU), 14(SU) |
| 160 | 15(M), 16(T), 23(T) |
| 178 | 20(S) |
| 240 | 12(F) |
| 284 | 28(SU) |
| 299 | 29(M) |
| 310 | 21(SU) |
| 321 | 1(M), 2(T), 3(W), 4(TH) |
| 379 | 8(M) |
| 392 | 30(T) |
| 396 | 24(W) |

(b)

Figure 4: (a) MST map for transformer oil temperature in January 1996. Algorithm for this map was extended to include the 2nd winner. (b) Summary of the MST map for January 1996

# 4 Comparison of results for SOM and MST

The day type pattern classification explained in section 2.2 was repeated for the other months (February – October). In each of the months analysed, three day-type patterns emerged after the training process. These are Weekdays, Holidays and Weekends. The MST algorithm maintained the same day types that were obtained in section 2.2. The advantage of the MST algorithm is that it provides a physical link among the data within a cluster thereby making it easier to visualize the clusters on the map. Figure 4(b) is a summary of the results obtained from figure 4(a). This procedure was repeated for the other months in the year and the results compared. In section 2.2, since the learning process is to identify new day types and combine similar day types together, only a rough learning was conducted. For the case of MST, a more detailed learning was performed. The learning parameters for the SOM were as follows: Learning cycles of 2000, learning coefficient factor of 0.008 and a radius of 20. Here the similarity in the data was based on the Euclidean distance between the nodes after the SOM learning.

# 5 Transformer oil temperature forecast with SOM

During summer, the power system is put under heavy load. The temperature of the insulation oil for cooling the transformer windings rises. The continuous rise in the oil temperature causes the dielectric strength of the oil to deteriorate. The oil then loses its insulating properties thus creating a short circuit within the windings of the transformer. Under such conditions, if the temperature change of insulation oil of the transformer can be predicted, necessary countermeasures can be implemented to forestall any damage to the transformer. Hence, efficient operation of the transformer can be achieved and also the reliability may be improved.

## 5.1 Data set

In forecasting the transformer oil temperature, the atmospheric temperatures from meteorological data were used. The input data vector for the SOM comprised of the highest value of the atmospheric temperature for the day, the lowest value of the atmospheric temperature for the day, the percentage loading on the transformer and the transformer oil temperature. Since these are hourly data, the input vector was 50 dimensional.

## 5.2 Forecast with 3-month data set

The input data comprised of data for 3 months. The month of the day to be forecasted is selected as the centre month and the month before and after are selected to form an input data set. I.e., the input data set for predicting 7th August will be July, August and September. The input data for the day to be predicted was made up of the highest value of the atmospheric temperature for the day, the lowest value of the atmospheric temperature for the day and the percentage loading on the transformer. The original oil temperature data for the day to be predicted was presumed unknown and used as evaluation data. Figure 5(a) illustrates the comparison of the original transformer oil temperature readings for 7th August, the SOM forecast values and the Conventional forecast values. The SOM predicted results were mostly proportional to the measured data and this confirms the effectiveness of the SOM in prediction of the temperature changes of the oil in the transformer. The results obtained were compared to the recorded values on 7th of August using equation (4). This approach resulted in a Mean Squared Errors (MSE) of 0.585 (see figure 5(b)).

## 5.3 Forecast with 10-months data

In order to improve the prediction accuracy for the SOM, the input data set was increased from a 3-month data set to a 10-month data set. The input data comprised of data for 10 months, from January to October 1996. The MSE was reduced from

0.585 to 0.072. Figure 5(b) compares the MSE between the predicted data and the recorded data for 7th August.

## 5.4 Forecast with Conventional Method

Forecasting of the oil temperature of the transformer according to the operational guideline of oil filled transformers [7] entails explicit numerical calculation using various constants, such as time constant necessary for the calculation of the corresponding optimal cooling conditions of the installation environment of the transformer. This method resulted in a MSE of 16.38 for the forecast of the 7th August temperature values (see figure 5(b)).

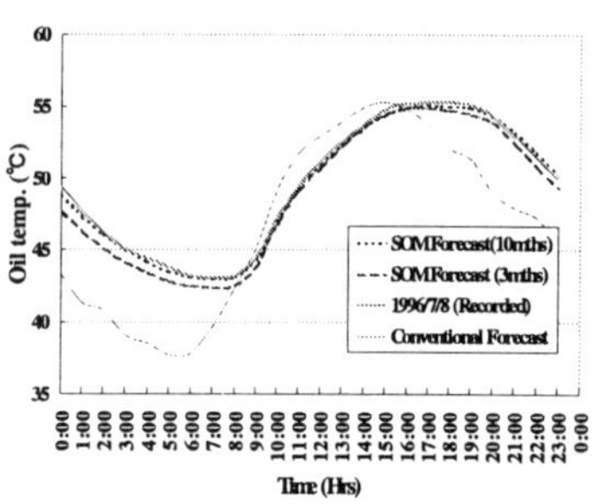


| Date (Day) | Mean Squared Error 10-month | Mean Squared Error 3-month | Mean Squared Error Conventional |
|---|---|---|---|
| 12/2/96 (Monday) | 0.665 | 1.052 | 71.724 |
| 5/3/96 (Tuesday) | 1.659 | 3.816 | 63.357 |
| 10/4/96 (Wednesday) | 0.532 | 0.956 | 29.970 |
| 8/5/96 (Wednesday) | 1.885 | 1.954 | 17.855 |
| 6/6/96 (Thursday) | 18.451 | 12.475 | 12.150 |
| 17/7/96 (Wednesday) | 2.221 | 1.068 | 12.958 |
| 7/8/96 (Wednesday) | 0.072 | 0.585 | 16.377 |
| 12/9/96 (Thursday) | 0.677 | 2.639 | 16.844 |

(a) (b)

Figure. 5a: Recorded transformer oil temperature readings for 7th August, prediction by SOM (SOM prediction was for 10-month and 3-month data sets) and prediction by conventional method. (b) Mean squared error of the predicted results of oil temperature using 10-month data set, 3-month data set and conventional method.

The hourly temperature changes are calculated using the following equations:

$$\theta_{OK_1} = \theta_{ON} \left(\frac{K_1^2 R + 1}{R + 1}\right)^m \quad (2)$$

Where $\theta_{ON}$ highest rise in temperature at normal load, $K_1$ is the ratio of the load $P_1$ to the normal load $P_N$, $R$ is the ratio of loss at normal load to loss at no load and m is a constant determined by the mode of cooling. $\theta_{ON}$ and $R$ can be obtained from data from factory test conducted on the transformer at the time of manufacture. In transformer operation guidelines it is recommended that, in the case where transformer cooling depends on the natural circulation of the oil, the constant $m$ be taken as 0.8. Furthermore, after more than a unit time has elapsed after a change of transformer load from $K_1P_N$ to $K_2P_N$, the rise in oil temperature is as shown in equation (3). $\tau$ is the time constant of the oil temperature change.

$$\theta_{O(K_1 - K_2)} = (\theta_{OK_2} - \theta_{OK_1})(1 - \exp(\frac{-t}{\tau})) + \theta_{OK_1} \quad (3)$$

## 5.5 Comparison of forecast by SOM and Conventional method

The predicted results obtained from SOM and numerically calculated values, which were based on the conventional method, were compared to the recorded data. The comparison was done based on the mean squared error shown in equation (4).

$$Err = \frac{1}{N}\sum_{j=1}^{N}(r_j - p_j)^2 \qquad (4)$$

Where, $r_j$ is the recorded oil temperature, $p_j$ is the predicted oil temperature and $N$ is the number of components per data vector. The SOM predictions resulted in a better approximation of the recorded data as compared to the conventional method.

# 6 Conclusions

In its application to power transformer database, the SOM provided in-depth knowledge about the consumption pattern of the consumers in a particular day or season. This provides the energy producer with a guide as to the consumption pattern of its consumer in any particular day or season. This could be very vital information for planning engineers in their load forecasting and other prediction and planning activities. Superimposing the MST onto the SOM also provided a more visual approach in identifying the days that belong to a particular cluster. Forecasting of the oil temperature changes in a power distribution transformer has been done using conventional methods based on explicit numerical calculation. With such techniques of forecasting, the forecasting accuracy is affected due to assumptions made in the characteristics of the transformer. By the use of SOM, this problem is eliminated because the transformer characteristics are not included in the input data set for the SOM.

## References

1. Kohonen T. Biological Cybernetics, 43(1): 59-69,1982, Springer-Verlag, 1997
2. Kohonen T. The Self-Organizing Map, proceedings of IEEE, 78:1464-1480, 1990
3. Obu-Cann K., Tokutaka H., Fujimura K. and Yoshihara K. Technical report of IEICE, NC99-133, pp 121-128, March 2000
4. Kohonen T. Self-Organizing Maps, Springer-Verlag, 1997
5. Murray A. Applications of Neural Network, Kluwer Academic Publishers, pp 157-189.
6. Obu-Cann K., Iwamoto K., Tokutaka H. and Fujimura K. ICONIP'99, pp 986-991
7. Technical report of the Institute of Electrical Engineers of Japan, Part 1, No. 143: "The operation guideline of the oil filled transformers (in Japanese)", 1978.

# Recent Advances with the Growing Hierarchical Self-Organizing Map

Michael Dittenbach[1], Andreas Rauber[2], and Dieter Merkl[2]

[1] E-Commerce Competence Center – EC3,
Siebensterngasse 21/3, A–1070 Wien, Austria
[2] Institut für Softwaretechnik, Technische Universität Wien,
Favoritenstraße 9–11/188, A–1040 Wien, Austria
www.ifs.tuwien.ac.at/{~mbach, ~andi, ~dieter}

**Abstract.** We present our recent work on the *Growing Hierarchical Self-Organizing Map*, a dynamically growing neural network model which evolves into a hierarchical structure according to the necessities of the input data during an unsupervised training process. The benefits of this novel architecture are shown by organizing a real-world document collection according to semantic similarities.

## 1 Introduction

The *Self-Organizing Map* (*SOM*) [3,4] has shown to be exceptionally successful in arranging high-dimensional input data along its two-dimensional output space such that similar inputs are mapped onto neighboring regions of the map. In other words, the similarity of the input data is preserved as faithfully as possible within the representation space of the *SOM*.

Despite the large number of research reports on applications of the *SOM* some deficiencies remained largely untouched. First, the *SOM* uses a static network architecture both in terms of number and arrangement of neural processing elements, which have to be defined prior to the start of training. Second, hierarchical relations between the input data are rather difficult to detect in the map display. So far, both issues are have been addressed separately by means of adaptive architectures, e.g. the *Growing Grid* [2], or hierarchies of independent *SOMs*, e.g. the *Hierarchical Feature Map* [8] or the *Tree Structured SOM* [5,6].

With the *Growing Hierarchical Self-Organizing Map* (*GHSOM*) [1] we proposed a novel neural network model that addresses both deficiencies as outlined above within one framework. Basically this neural network model is composed of independent *SOMs*, each of which is allowed to grow in size during the training process until a quality criterion regarding data representation is met. This growth process is further continued to form a layered architecture such that hierarchical relations between input data are further detailed at deeper layers of the neural network.

In this paper we report on our recent work on the *GHSOM* where particular emphasis is directed to the orientation of the various layers during training. In particular, we discuss the positive effects on input data representation when the initial orientation of deeper layers is chosen according to their respective higher layer maps. We can thus maintain the already learned similarities between input data during the establishment of the hierarchical structure of the *GHSOM*. As a consequence, the negative effects of generating strictly disjoint clusters are alleviated because neighboring maps in deeper layers of the hierarchy show similar characteristics at their respective borders.

# 2 Growing Hierarchical Self-Organizing Map

## 2.1 Architecture and training

The key idea of the *GHSOM* is to use a hierarchical structure of multiple layers where each layer consists of a number of independent *SOMs*. One *SOM* is used at the first layer of the hierarchy. For every unit in this map a *SOM* might be added to the next layer of the hierarchy. This principle is repeated with the third and any further layers of the *GHSOM*.

Since one of the shortcomings of *SOM* usage is its fixed network architecture we rather use an incrementally growing version of the *SOM*. This relieves us from the burden of predefining the network's size which is rather determined during the unsupervised training process. We start with a layer 0, which consists of only one single unit. The weight vector of this unit is initialized as the average of all input data. The training process basically starts with a small map of, say, $2 \times 2$ units in layer 1, which is self-organized according to the standard *SOM* training algorithm.

This training process is repeated for a fixed number $\lambda$ of training iterations. Ever after $\lambda$ training iterations the unit with the largest deviation between its weight vector and the input vectors represented by this very unit is selected as the error unit. In between the error unit and its most dissimilar neighbor in terms of the input space either a new row or a new column of units is inserted. The weight vectors of these new units are initialized as the average of their neighbors.

An obvious criterion to guide the training process is the quantization error $q_i$, calculated as the sum of the distances between the weight vector of a unit $i$ and the input vectors mapped onto this unit. It is used to evaluate the mapping quality of a *SOM* based on the mean quantization error (*MQE*) of all units in the map. A map grows until its *MQE* is reduced to a certain fraction $\tau_1$ of the $q_i$ of the unit $i$ in the preceding layer of the hierarchy. Thus, the map now represents the data mapped onto the higher layer unit $i$ in more detail.

As outlined above the initial architecture of the *GHSOM* consists of one *SOM*. This architecture is expanded by another layer in case of dissimilar input data being mapped on a particular unit. These units are identified by a rather high quantization error $q_i$ which is above a threshold $\tau_2$. This threshold basically indicates the desired granularity level of data representation as a

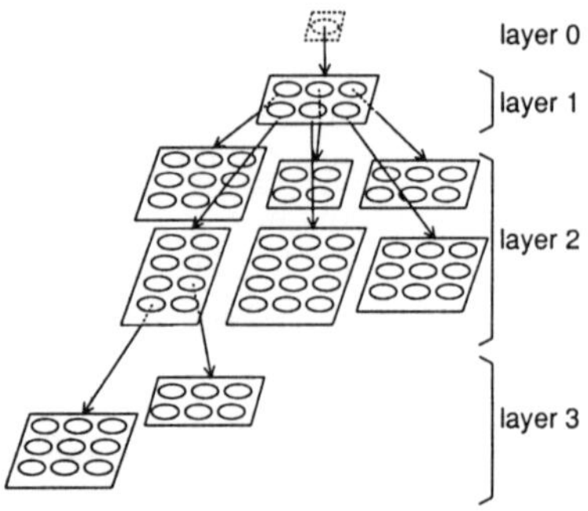


**Fig. 1.** *GHSOM* reflecting the hierarchical structure of the input data.

fraction of the initial quantization error at layer 0. In such a case, a new map will be added to the hierarchy and the input data mapped on the respective higher layer unit are self-organized in this new map, which again grows until its *MQE* is reduced to a fraction $\tau_1$ of the respective higher layer unit's quantization error $q_i$. Note that this may not necessarily lead to a balanced hierarchy. The depth of the hierarchy will rather reflect the diversity in input data distribution which should be expected in real-world data collections.

Depending on the desired fraction $\tau_1$ of *MQE* reduction we may end up with either a very deep hierarchy with small maps, a flat structure with large maps, or – in the most extreme case – only one large map, which is similar to the *Growing Grid.* The growth of the hierarchy is terminated when no further units are available for expansion.

A graphical representation of a *GHSOM* is given in Figure 1. The map in layer 1 consists of $3 \times 2$ units and provides a rough organization of the main clusters in the input data. The six independent maps in the second layer offer a more detailed view on the data. Two units from one of the second layer maps have further been expanded into third-layer maps to provide sufficiently granular input data representation.

### 2.2 Orientation

The hierarchical structuring imposed on the data results in a separation of clusters mapped onto different branches. While this, in principle, is a desirable characteristic helping to understand the cluster structure of the data, it may lead to misinterpretations when large clusters are mapped and expanded on two neighboring, yet different units. Similar input data are thus rather arbitrarily separated in different branches of the hierarchy.

In order to provide a global orientation of the individual maps in the various layers of the hierarchy, their orientation must conform to the orientation of the data distribution on their parents' maps. This can be achieved by creating a coherent initialization of the units of a newly created map.

Let unit $p$ be expanded to form a new $2 \times 2$ map in the subsequent layer of the hierarchy. This map's four weight vectors $s_1$ to $s_4$ are initialized to

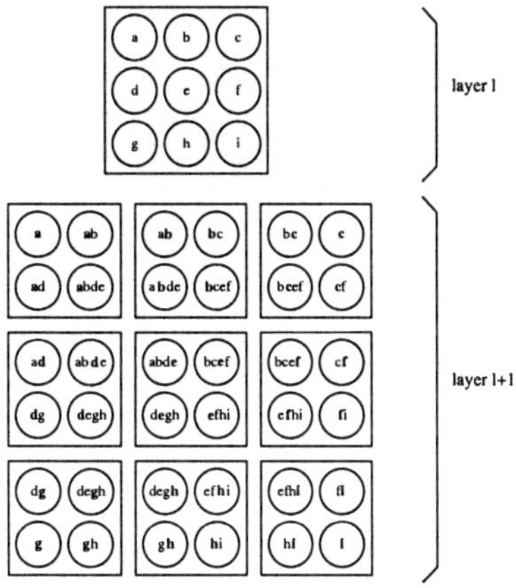


**Fig. 2.** Initialization of units for expanded maps

mirror the orientation of neighboring units of $p$. This is achieved by adding a fraction of the weight vectors in the neighborhood of $p$. Figure 2 provides an illustration of the initialization of new maps. Given a $3 \times 3$ map with weight vectors $a$ to $i$, the weight vectors of the four units representing the maps expanded from units $a$ and $e$ are initialized as provided in Expression 1.

$$\begin{array}{lll|lll} a_1 & = & a & e_1 & = & e + \frac{(a-e)+(b-e)+(d-e)}{3} \\ a_2 & = & a + \frac{(b-a)}{2} & e_2 & = & e + \frac{(b-e)+(c-e)+(f-e)}{3} \\ a_3 & = & a + \frac{(d-a)}{2} & e_3 & = & e + \frac{(d-e)+(g-e)+(h-e)}{3} \\ a_4 & = & a + \frac{(b-a)+(d-a)+(e-a)}{3} & e_4 & = & e + \frac{(f-e)+(h-e)+(i-e)}{3} \end{array} \quad (1)$$

This initial orientation of the map is preserved during the training process. While new units may be inserted in between, the 4 corner units will still be the most similar to the respective corner units of the maps in neighboring branches.

# 3 Experiments

For the experiments presented hereafter we use a collection of 11,627 articles from the Austrian daily newspaper *Der Standard* covering the second quarter of 1999. To be used for map training, a vector-space representation of the single documents is created by full-text indexing. Instead of defining language or content specific stop word lists, we rather discard terms that appear in more than 813 (7%) or in less than 65 articles (0.56%). We end up with a vector dimensionality of 3.799 unique terms. The individual documents are then represented by feature vectors using a $tf \times idf$, i.e. term frequency times inverse document frequency, weighting scheme [9]. This weighting scheme assigns high values to terms that are important as to describe and discriminate between the documents. These feature vectors are used to train the *GHSOM*.

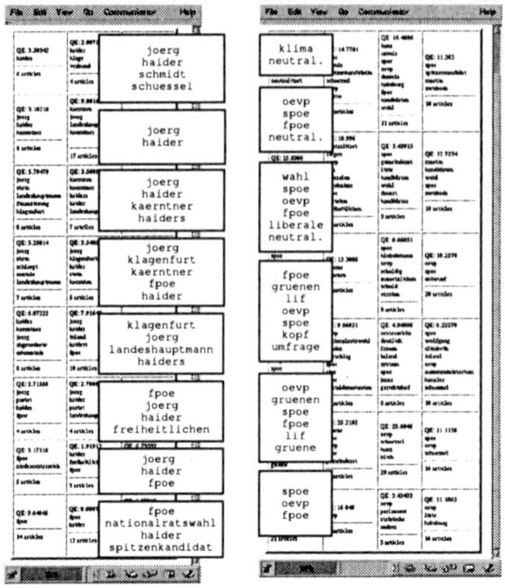


**Fig. 3.** Two neighboring second-layer maps on Internal Affairs

Training the *GHSOM* with parameters $\tau_1 = 0.035$ and $\tau_2 = 0.0035$ results in a shallow hierarchical structure of up to 7 layers. The layer 1 map grows to a size of 7×4 units, all of which are expanded at subsequent layers. We find the most dominant branches to be, for example, *Sports*, located in the upper right corner of the map, *Internal Affairs* in the lower right corner, *Internet*-related articles on the left hand side of the map, to name but a few. However, due to the large size of the resulting first layer map, a fine-grained representation of the data is already provided at this layer. This results in some larger clusters to be represented by two neighboring units already at the first layer, rather than being split up in a lower layer of the hierarchy. For example, we find the cluster on *Internal Affairs* to be represented by two neighboring units. One of these, on position (6/4)[1], covers solely articles related to the *Freedom Party* and its political leader *Jörg Haider*, representing one of the most dominant political topics in Austria for some time now, resulting in an accordingly large number of news articles covering this topic. The neighboring unit to the right, i.e. located in the lower right corner on position (7/4), covers other *Internal Affairs*, with one of the main topics being the elections to the *European Parliament.* Figure 3 shows these two second-layer maps.

We also find, for example, articles related to the *Freedom Party* on the branch covering the more general *Internal Affairs*, reporting on their role and campaigns for the elections to the *European Parliament.* As might be expected these are closely related to the other articles on the *Freedom Party* which are located in the neighboring branch. Obviously, we would like them

[1] We refer to a unit located in column $x$ and row $y$ as $(x/y)$ starting with (0/0) in the upper left corner.

to be presented on the left hand side of this map, so as to allow the transition from one map to the next, with a continuous orientation of topics. Due to the initialization of the added maps during the training process, this continuous orientation is preserved, as can easily be seen from the automatically extracted labels [7] provided in Figure 3. Continuing from the second layer map of unit (6/4) to the right we reach the according second layer map of unit (7/4) where we first find articles focusing on the *Freedom Party*, before moving on to the *Social Democrats*, the *People's Party*, the *Green Party* and the *Liberal Party*.

## 4 Conclusions

In this paper we have presented our recent work on the *Growing Hierarchical Self-Organizing Map*. The major features of this neural network are its hierarchical architecture, where the depth of the hierarchy is determined during the unsupervised training process. Each layer in the hierarchy consists of a number of independent *SOMs* which determine their size and arrangement of units during training. Thus, this model is especially well suited for applications which require hierarchical clustering of the input data. We have shown that significant improvement in data representation can be achieved by directing particular emphasis at the orientation of the various *SOMs* constituting the different branches of the hierarchy. Maps of neighboring branches now show the same orientation as the map they are derived from. Therefore the similarity of different clusters is presented in an intuitive fashion.

## References

1. M. Dittenbach, D. Merkl, and A. Rauber. The growing hierarchical self-organizing map. In *Proc Int'l Joint Conf Neural Networks*, Como, Italy, 2000.
2. B. Fritzke. Growing Grid – A self-organizing network with constant neighborhood range and adaption strength. *Neural Processing Letters*, 2(5), 1995.
3. T. Kohonen. Self-organized formation of topologically correct feature maps. *Biological Cybernetics*, 43, 1982.
4. T. Kohonen. *Self-organizing maps*. Springer-Verlag, Berlin, 1995.
5. P. Koikkalainen. Fast deterministic self-organizing maps. In *Proc Int'l Conf Neural Networks*, Paris, France, 1995.
6. P. Koikkalainen and E. Oja. Self-organizing hierarchical feature maps. In *Proc Int'l Joint Conf Neural Networks*, San Diego, CA 1990.
7. D. Merkl and A. Rauber. Automatic labeling of self-organizing maps for information retrieval. In *Proc Int'l Conf Neural Information Processing*, Perth, Australia, 1999.
8. R. Miikkulainen. Script recognition with hierarchical feature maps. *Connection Science*, 2, 1990.
9. G. Salton. *Automatic Text Processing: The Transformation, Analysis, and Retrieval of Information by Computer*. Addison-Wesley, Reading, MA, 1989.

# Self-Organizing Maps of Web Link Information

Sami Laakso, Jorma Laaksonen, Markus Koskela, and Erkki Oja

Laboratory of Computer and Information Science
Helsinki University of Technology
P.O. Box 5400, Fin-02015 HUT, Finland

**Summary.** We have developed a method that utilizes hypertext link information in image retrieval from the World Wide Web. The basis of the method consists of a set of basic relations that can take place between two images in the Web. Our method uses the SHA-1 message digest algorithm for dimension reduction by random mapping. The Web link features have then been used in creating a Self-Organizing Map of images in the Web. The method has been effectively tested with our PicSOM content-based image retrieval system using a Web image database containing over a million images. The method can as such be used also in other Web applications not related to content-based image retrieval.

## 1 Introduction

There is a growing interest in the ability to search the World Wide Web for various data. Automated methods for information retrieval from the Web have in recent years attracted significant research interest and several search engines have been developed and made available for public use. These search engines allow the user to search for multimedia resources, including images, inside Web documents.

A recent approach to the problem of locating relevant images in huge databases such as the Web is content-based image retrieval (CBIR). It is based on visual features that can automatically be extracted from images without human intervention or interpretation. Such features can, for example, be the colors and textures found in the image as well as shape, structure and composition of the image scene. Besides visual features, the hyperlink structure of the Web can also be a rich source of information about the content of the environment, and some promising methods have already been proposed.

According to [1], useful documents in the Web can be categorized as either authorities or hubs. An authority is a source of specific information about a certain topic and a hub is a collection of links pointing to authorities. A good authority is recognized from the fact that it is pointed at by many hubs and a good hub is therefore a document that contains links to many good authorities. Another page ranking method, which is now used by the popular text-based search engine Google (*http://www.google.com*), was proposed in [2]. It uses a hyperlink graph constructed on the entire set of documents retrieved from the Web. Given a query, Google retrieves the documents containing the query string in the order specified by the probability of a random walker to visit the page.

The problem with these methods is that they have been designed for text-based page search and cannot therefore be directly generalized into image search. One solution would be to make a normal query and then show images from those pages which were ranked best. However, this approach assumes that the images in a certain page are generally related to the surrounding text, which certainly is not always true.

## 2 Web Relation Feature Extraction

We have developed a mechanism which utilizes the information about the location of images in the Web and the hypertext link structure between them. The basis of the method consists of a set of basic Web relations that can take place between two images. For example, if one image acts as a hypertext link to another image (e.g. thumbnails) it can be assumed that the two images are closely related. Lesser but still highly informative image relation occurs if the images are located in the same Web page, or in the same directory, or at least in the same domain. The same relations can also be applied to Web pages that contain images. Although it would be possible to perform also a deeper study of links between image pages, we consider that images beyond two link steps are unlikely to be related. Another benefit of using only one level deep linkage information is that now only the links of image pages need to be saved. A deeper study would require the information about the entire indexed hypertext structure, which can be very space consuming.

Every object in the Web has its own unique URL. However, in order to efficiently exploit this location information and information concerning inter-location relations, we need to convert them into mathematical form. A trivial solution would be to form a relation vector whose dimensionality equals to the number of images in the database. Then a certain weight value would be set to each vector component according to the relation between the corresponding images. However, this would require storing $N^2$ relation values which is not feasible with large databases. In our case $N$ was of the order of one million and therefore the dimensionality of the relation data had to be reduced.

*Random mapping* [4] provides a computationally feasible method for reducing the dimensionality of data so that mutual similarities between the data vectors are approximately preserved. The dimensionality of the representations is reduced by replacing the original orthogonal base with a lower-dimensional almost orthogonal base.

*Secure Hash Algorithm* (SHA-1) [3] is a powerful method for computing a condensed representation of a message.When a message of length $< 2^{64}$ bits is input, the SHA-1 produces a 160-bit output called a *message digest.* The

**Table 1.** Used Web relations and the corresponding weights.

| Web relation | Weight |
|---|---|
| URL of an activation link | 1.5 |
| URL of the image | 1.4 |
| URL of the image's Web directory | 1.3 |
| URL of the image page | 1.2 |
| URL of the image page's Web directory | 1.15 |
| URL of the image page's domain | 1.1 |
| URL of a link to other image or page | 1.0 |

SHA-1 has been designed so that it is computationally infeasible to find a message which corresponds to a given message digest, or to find two different messages which produce the same message digest. The latter property was the main inspirer for us to use this algorithm as an indexation tool.

First, all URLs were extracted from the collected images pages. These URLs contained all external links, the locations of the images and image pages themselves, and the directories and domains of the image pages. Then, SHA-1 message digests were calculated for each URL. The first 8 hex characters (32 bits) of the digests were used to determine a sparse random vector by interpreting them as four 8-bit values. The first value was used as an index in the range $[0, 255]$, the second in $[256, 511]$, the third in $[512, 767]$, and the fourth in $[768, 1023]$. These four indices were used in setting four components of otherwise zero 1024-dimensional vector to value one. These 1024-dimensional vectors were then considered as being random base vectors of an almost orthogonal base whose dimensionality was 2 622 472, the count of unique URLs extracted from the image pages.

For each image in the database, the random base vectors corresponding to the extracted URLs in the image's Web page were multiplied with a weight value that depended on the type of the URL in question, as shown in Table 1. All the resulting random vectors for one particular image were combined into a relation feature vector so that for each component the maximum value was chosen. As a result of the formation of the link feature vectors, images which have close Web relations will also have small Euclidean distances between their feature vectors.

An example of forming one random base vector follows. Consider an imaginary URL of *http://www.cis.hut.fi/images/image1.jpg*. From that URL the SHA-1 algorithm results a 160-bit message digest which has the hex form of 4CDF3EEB45A12D6044A96A911E7559084B3F037F. The first 32 bits of the message digest are then used to determine the index values and the result is a random projection vector which has weight values in indices 76 ($= 4C_{16}$), 479 ($= 256 + DF_{16}$), 574 ($= 512 + 3E_{16}$) and 1003 ($= 768 + EB_{16}$), as shown in Table 2. The final relation feature vector would then be the combination of this vector multiplied by 1.4, the random vector of the image directory URL (that would be *http://www.cis.hut.fi/images/*) multiplied by 1.3, and the random vectors of other related URLs weighted accordingly.

**Table 2.** An example of using SHA-1 message digest for random mapping.

| Bit positions | 0...7 | 8...15 | 16...23 | 24...31 | 32...159 |
|---|---|---|---|---|---|
| SHA-1 hex value | 4C | DF | 3E | EB | 45A12D6... |
| Decimal value | 76 | 223 | 62 | 235 | |
| Random indices | 76 | 479 | 574 | 1003 | |
| Random base vector | $0\cdots010\cdots010\cdots010\cdots010\cdots0$ | | | | |

## 3 PicSOM

The PicSOM system [5] is a framework for research on algorithms and methods for content-based image retrieval. PicSOM implements *relevance feedback* by using Tree Structured Self-Organizing Map (TS-SOM) [6] in storing the user responses and in selecting the images. The TS-SOM differs from the original SOM [7] in that TS-SOM consists of a stack of SOM layers. The BMUs are first searched for on the topmost layer and the search is then continued on the next layer in a restricted area centered below the BMU on the above map. This makes the BMU search much faster, otherwise the properties of SOM and TS-SOM are similar.

Images that are similar to each other with respect to a particular feature extraction method are clustered together on the corresponding TS-SOM map. When the user's relevance feedback is marked on the maps and this spatial relevance function low-pass filtered, what results is automatic adaptation to the user's conception of image similarity and relevance. The mutual weighting of different feature types is performed simultaneously, as features that map relevant images in tight clusters are automatically given more weight than the others. A genuine characteristic of PicSOM is thus its ability to automatically adapt to the user's perception of similarity of images based on their low-level visual content even though humans perceive image similarity on abstract semantic level. An on-line demonstration of PicSOM and comprehensive documentation of it can be found at *http://www.cis.hut.fi/picsom/*.

## 4 Web-fi Image Database

Our experimental Web image database was retrieved in summer and autumn 2000. The original plan was to collect all images existing in the registered domains in Finland (i.e. all the domains whose name end in **.fi**). It was assumed that there exists a path from the host's main page to a majority of pages in that domain, and therefore the search was started from every domain's root URL. To avoid retrieving too many thumbnails and icons, our Web robots retrieved only images whose width and height were both more than 50 pixels and the image also had to contain at least five distinct colors. Accepted image formats were JPEG, GIF, TIFF, and PNG. The collection process was stopped when the count of retrieved images exceeded one million. Overall, the

Web robots examined 2 176 261 pages in 12 991 different domains. The total count of unique images was 1 008 844. Included in the examined domains are all domains owned by Finnish cities, municipalities, universities, and polytechnics. During the collection, link structure information corresponding to each retrieved image was also collected. After the collection process was completed, the Web relation feature vectors for all images were calculated. Then a five-layer TS-SOM, with layers of sizes $4 \times 4$, $16 \times 16$, $64 \times 64$, $256 \times 256$, and $1024 \times 1024$, was constructed.

## 5 Results

Figure 1 shows a partial surface of the $1024 \times 1024$-sized SOM formed from Web-fi database. The benefits of the link information can be seen, as there are certain clearly visible clusters of images, for example a group of trucks in the bottom left corner of the map and a group of images from technical books in the bottom right corner. Although, for example, some of the truck images are visually quite dissimilar, they are still mapped close to each other. This can be regarded as a promising achievement, as the system uses no visual nor textual data, but only information about the Web structure.

**Fig. 1.** Partial surface of $1024 \times 1024$-sized SOM formed from Web structure data.

Figure 2 shows another example of the discrimination power of our method. The figure shows the mappings of all images from four distinct domains. The selected domains were *hut.fi* (Helsinki University of Technology), *oulu.fi* (University of Oulu), *ouka.fi* (City of Oulu) and *utu.fi* (University of Turku), and the corresponding image counts were 103 613, 47 845, 21 786 and 21 495 images, respectively. The resulting image densities show that some of the domains are very tightly clustered, whereas others are spread more widely.

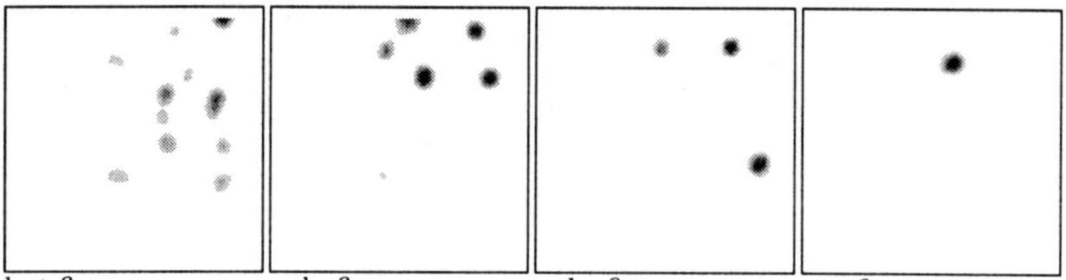


**Fig. 2.** Mappings of different domains on the lowest-level SOM ($1024 \times 1024$). The distributions have been low-pass filtered to ease inspection.

## 6 Conclusions

We have presented a method for utilizing the Web link structure in information retrieval. Our method requires no information about the entire hypertext structure, but only URLs included in Web pages that contain images. The preliminary results are very promising and indicate that the method can as such be used also in other applications not related to content-based image retrieval. However, the Web link feature is not intended to be a stand-alone feature, but rather to be used in combination with other features. Especially in CBIR, the link structure information may greatly enhance the search.

The huge size of the resulting TS-SOM is also worth noticing: The five layers contain 1 118 480 map units altogether, which results in the total of 1 145 323 520 real-valued weights in the map.

## References


1. Kleinberg, J. (1997) Authoritative sources in a hyperlinked environment, *IBM Technical Report RJ 10076*, May 1997.
2. Brin, S. and Page, L. (1998) The Anatomy of a Large-Scale Hypertextual Web Search Engine, *Proceedings of Seventh International World Wide Web Conference, April 1998, Brisbane, Australia.*
3. FIPS PUB 180-1 Secure Hash Standard (1995) *http://www.itl.nist.gov/fipspubs/fip180-1.htm*
4. Kaski, S. (1998) Dimensionality Reduction by Random Mapping: Fast Similarity Method for Clustering, *Proceedings of IEEE International Joint Conference on Neural Networks, May 1998, Anchorage, Alaska.*
5. Laaksonen, J., Koskela, M., Laakso, S. and Oja, E. (2000) PicSOM - Content-based image retrieval with self-organizing maps, *Pattern Recognition Letters* **21**(13-14): 1199–1207.
6. Koikkalainen, P. and Oja, E. (1990) Self-organizing hierarchical feature maps, *Proceedings of 1990 International Joint Conference on Neural Networks, January 1990, Washington, USA*, Vol. 2, pp. 279–284.
7. Kohonen, T. (2001) *Self-Organizing Maps*, Vol. 30 of *Springer Series in Information Sciences*, Springer-Verlag. Third Edition.

# A Design Method of DNA Chips using Self Organizing Maps

Hiroshi Douzono  Shigeomi Hara  Yoshio Noguchi

Faculty of Science and Engineering, Saga University
1-Honjyo Saga 840-8502 JAPAN

Abstract

In this paper, we introduce a design method of DNA chips using Self-Organizing Maps(SOM). DNA chips are powerful tools for sequencings and SNP(Single Nucleotide Polymorphism) analyses of DNA sequences. A DNA chip is an array of DNA probes which are hybridized with the compliment sub-sequences in the target sequence. However, conventional DNA chips are showing tendency to be comprised of longer probes and get larger in size to achieve a higher resolution. To shrink the size of DNA chips, the design is considered to be important. To solve this problem, we applied SOM to obtain common features of DNA sequences. Further, we improved the SOM algorithm to generate small number of probes which have different length and efficiently cover the target sequence with sufficient resolution for finding the correct position of SNPs. We evaluated the DNA chips designed by SOM with computer simulations of SNP analyses.

## 1 Introduction

Recently almost all sequences of human genome were determined. DNA sequencers were used to read the actual sequences. But the DNA sequencer can read the sequence of about 1000 bp at most, so longer sequences had to be divided into many sub-sequences which were shorter than the limits and each of them were read by sequencer. Iterations of biological experiments with DNA sequencers bring high cost and long time. On the other hand, the SNP[1] analysis receives much attention recently. SNP is variation of single oligo-nucleotide which is found in every 100-500 oligo-nucleotides. SNPs have some effects and the side effects of medicines for some diseases, so it is important to find SNPs. SNPs can be detected by cDNA micro array as the lacks of cDNA sequences in target sequences. The exact positions of SNPs can be known with sequencings.

As one of the rapid sequencing method, sequencing by hybridization [2] is under development. This method uses a sequencing chip, which is the array of short DNA probes (4-10bp), and the target sequences are determined by only one biological experiment of hybridization of the target sequence with DNA chip. A DNA chip comprised of longer probes can determine longer target sequences, but the number of probes increases at an exponential rate. To determine the DNA sequences longer than 1000bp, the DNA chip comprised of probes which are longer than 8bp is desirable[3], but the size of DNA chip may be huge without any designs, because it needs $4^8$=65536 probes. Therefore, the design of DNA chip is considered to be important.

For this problem, we applied Self Organizing Map (SOM) to select a set of probes which were adequate to represent actual DNA sequences. While the conventional clustering algorithm may be applicable by clustering the sub-sequences which are obtained from the reference DNA sequences, we improved our SOM algorithm as to organize the probes adaptively tuning the length of themselves. Some works applying SOM to the analysis of DNA sequences were reported [4] and they used SOMs to analyze the relations between sequences and functions. We applied SOMs to find representative probes which were sufficient to recombine the target sequences, detecting features of DNA sequences. Further, resulting map may indicate unknown function of the sub-sequence or unknown relations among the functions concerning the sub-sequences.

We made some experiments on feature detection of generic DNA sequences from the reference sequences obtained from genome database using our SOM algorithm. And we made the computer simulations of SNP analyses using the probes obtained from the map to evaluate adequacy of the sets of organized probes.

# 2 DNA Chip and Sequence Analyses

The sequencing by hybridization can identify the long DNA sequence with only one biochemical experiment. DNA chip is a biochemical chip on which probes are printed as an array of short(4bp-10bp) sub-sequences. DNA is represented by 4 symbols (A,G,T,C) and each DNA hybridizes with its complement DNA, such as A with T and G with C. In biochemical experiments, clones of target sequence are directly reacted with the DNA chip, then the probes which are complement to the sub-sequences in the target sequence are hybridized with the cloned sequences. For example, probe AGGTC will hybridize with the sequence GTTCCAGAT.

After hybridization, the DNA chip is examined to detect which of the probes are hybridized. The probes which are compliment to any parts of target sequence are hybridized, so we can get the information of target sequence as a set of hybridized probes. The target sequence can be reconstructed using the result of hybridization. This process is shown in Figure 1.

```
Original sequence      AGTCGAAT
Hybridization          TCAGC
                        CAGCT         Hybridized probes
                         AGCTT
                          GCTTA

     AGTCG, TCGAA, CGAAT, GTCGA       Compliment subsequences

Reconstruction   AGTCG
                  GTCGA
                   TCGAA        Reconstructed sequence AGTCGAAT
                    CGAAT
```

**Fig.ure 1 Sequencing by hybridization**

Although this process requires very large amount of computation compared with conventional sequencing method, recent processors have enough computation powers to do that. Unfortunately, the reconstructed sequence can not be determined a unique one because of the ambiguity reconstructing a long sequence from a large number of short probes, but features of the target sequences, such as the existence of motifs, can be analyzed using the resultant sequences. We have made simulations of sequencing by hybridization on computers. From these experiments, we found that DNA chips which are made from probes longer than 8bp are desirable to reconstruct sequences of 1000 bp, and for sequences of 10000bp, we will need probes of 9bp. If we don t have any designing strategies, all combinations of the DNAs of the specific length will be set up on the chip. Thus, if length of the probes is 8bp, $4^8$=65536 types of the probes are printed on DNA chip in 256x256 square array. For 9bp probes, $4^9$=262144 types of probes are printed on 512x512 square array.

DNA chips can also be used to detect SNPs (Single Nucleotide Polymorphism). SNPs are variations of DNAs which are found in every 100-500 nucleotides, and they correspond to the effectiveness and side-effect of the medicines. Thus, it considered to be very important to identify the SNPs of patients for some diseases. Generally, SNPs can be detected by CGH(Comparative Genomic Hybridization) on the DNA micro array or the DNA chip by hybridizing the target sequences and reference sequences stained with different colors respectively. The cDNA sequences, which are found as functional genes in genome analyses, are printed on DNA micro array, so DNA micro array can detect SNP as lacks of gene functions, not as changes of oligo-nucleotides. Using DNA chips, the position of the SNP is identified on probes and using combination of hybridized probes, the position may be focused on a oligo-nucleotide. If the reference sequence is known, SNPs can be detected by hybridizing the target sequence and each position of SNP can be determined exactly in a probe, Furthermore in a oligo-nucleotide. We also made the computer simulation of SNP analysis and found that SNPs can be detected by shorter probes than those for sequencing. The DNA chip made from 7bp probes can detect almost all SNPs in the sequences of 10000bp, but it still needs $4^7$=16384 type of probes.

# 3 Feature Mapping of DNA sequence by SOM

Some cDNA micro arrays are organized as arrays of known or required cDNA sequences for medical demands. But, they can only detect the existences of SNPs in cDNA sequences and their resolutions are limited in cDNA sequence level. The generic DNA chips mentioned above can detect the position of SNPs in less than the length of probes and can detect the SNPs located out of the cDNA sequences. Furthermore, the generic DNA chips will costs much lower than DNA micro arrays because the probes to be organized are much shorter than cDNA sequences. But the number of the probes to be organized on the chip becomes very large without any design strategies. Thus, it is very important for the design of DNA chips to reduce the number of probes to be printed on the chips.

We used SOM for this problem because SOM can organize the generic feature of the DNA sequences by sufficient learning of known DNA sequences. We developed an algorithm to train the self organizing map of fixed length probes which are the vector of discrete values 'A','C','G', T and made some experiments changing the length of probes and map sizes. This algorithm could organize the map with up to 8bp probes well, but not so good for longer probes because it requires large amount of computation to make a map using long probes directly.

To overcome this difficulty, we improved the algorithm as to modify the length of each probe adaptively. This learning algorithm is shown as follows.

## Algorithm

**Step-1** Initialize the map of probes using random sequences of initial length.
**Step-2** Select a position of reference sequence(RS) randomly and find the closest probes on the map to the sub-sequence which starts from that position.
**Step-3** Update the closest probe Pr found in Step-2 as follows.

For each symbol in the probe
  If the symbol is A (G,T,C)
    Then modify the value Pr.A(G,T,C)=Pr.A(G,T,C)+1

If Pr.A(G,T,C) > Th-U

Then update the symbol to A(G,T,C) and set All Pr.A(G,T,C) to 0.

**Step-4** Update the probe Pr whose distance is closer than M-Dist from Pr geometrically using the same procedure in **Step-3**

**Step-5** For each updated probe

If more than Th-N of neighboring probes are identical,

then increase the length of the probes of neighbors and itself by 1bp.

Repeat **Step-2** to **Step-5**, changing the value Th-U and M-Dist

The expression A(G,T,C) represents iterations changing the character as A,G,T,C. The values Pr.A, Pr.G,Pr.T and Pr.C are intermediary value which are introduced to gradually update the discrete values A,G,T,C as continues values in conventional SOM. The value is updated if it reached the threshold Th-U. In **Step-5**, if sufficient number of neighboring probes become identical, the probes are extended by 1bp. Using this method, the length of the probe is adaptively tuned. For the training of a map using probes of fixed length, **Step-5** should be skipped.

# 4 Experimental results

We trained some 2-dimentional maps changing size (64x64, 128x128, 256x256) and changing the length of probes (fixed to 7, 8 and variable). The initial length and maximum length of the probes are set to 6 and 10 respectively for variable probe length. To train the map, we used a reference sequence whose length are sufficiently long to detect the common feature of DNA sequences. In this paper, we used a sequence of 1835600bp as reference sequence. After the training, we made SNP analysis using each set of the probes obtained from each map. We selected a relatively long (10000bp) sub-sequence taken from reference sequence and calculated the rates detecting SNPs of all single nucleotide changes

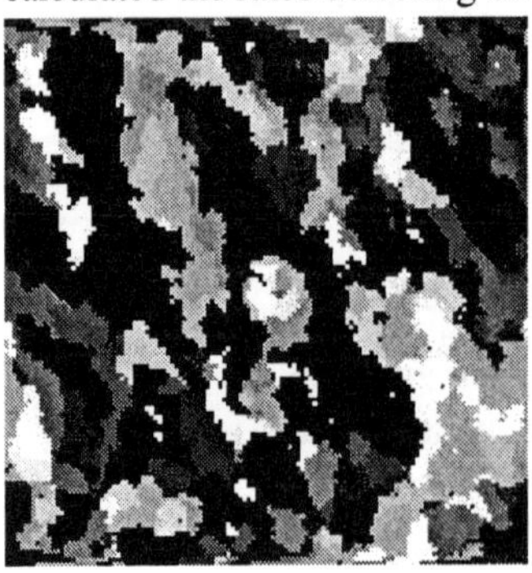

| | | |
|---|---|---|
| TTAAATTTT | TTAAATTTT | TTACATTT |
| TTAAATTTTC | TTAAATTTTC | TTAAATTTT |
| TTAAATTTTC | TTAAATTTTG | TTATATTTTT |
| TTAAATTTT | TTATATTTTT | TTATATTTTT |
| CTATATTTTG | TTATATTTTG | CTATATTTTT |
| CTATATTTTG | CTATATTTTG | CTATATTTTT |
| CTATAATTTG | CTATATTTTG | CTATATTTTT |

**Fig.ure 2 Map for 128x128 probes**

**Figure 3 Magnified map**

Figure 2 shows the trained map of 128x128 probes of 8bp. The probes are represented as grayscaled points. Figure 3 shows the magnified map for 256x256 probes of variable length. The similar probes are organized closely on the map and the probes of different length are organized on a map. Table.1 shows the number of different probes found in each map, number of probes for each length and mean length of the probes.

**Table 1 Number of organized probes and mean length**

| Map size<br>Probe size | 64x64<br>variable | 128x128<br>variable | 256x256<br>variable | 256x256<br>7bp | 256x256<br>8bp |
|---|---|---|---|---|---|
| Number of probes | 1794 | 5714 | 14520 | 5966 | 8864 |
| Mean length of probes | 9.79 | 9.52 | 8.4 | 7 | 8 |
| Number of 6bp probes | 0 | 19 | 4096 | | |
| Number of 7bp probes | 1 | 96 | 694 | 5966 | |
| Number of 8bp probes | 40 | 452 | 1107 | | 8864 |
| Number of 9bp probes | 275 | 1416 | 2473 | | |
| Number of 10bp probes | 1477 | 3730 | 6150 | | |

For each map, the number of the probes is less than its size because some probes are organized as groups. The longer the length of the probes becomes, the smaller maps are organized because this algorithm works only to extend the probes. Table 2 shows the covering rates of the probes on reference sequences.

**Table 2 Covering rates of reference sequence**

| Map size<br>Probe size | 64x64<br>variable | 128x128<br>variable | 256x256<br>variable | 256x256<br>7bp | 256x256<br>8bp |
|---|---|---|---|---|---|
| Covering rates | 0.0713 | 0.334 | 1.00 | 0.980 | 0.838 |

All nucleotides of the reference sequence are covered using the 14520 probes of variable length organized by 256x256 map, because all probes 6bp long are remained as the probes for this case. Table 3 shows the rates of successful detection of SNP.

**Table 3 Evaluation of SNP detection**

| Map size<br>Probe size | 64x64<br>variable | 128x128<br>variable | 256x256<br>variable | 256x256<br>7bp | 256x256<br>8bp |
|---|---|---|---|---|---|
| Complete detection rates | 0.0744 | 0.291 | 0.824 | 0.898 | 0.894 |
| Potential detection rates | 0.0779 | 0.328 | 1.00 | 0.998 | 0.944 |

We define two types of detections of SNPs for this evaluation. Complete detection is defined as the detection with complete lack of the hybridized probes for a SNP. Thus, this type of SNP can be easily detected as vivid change of the color observed on DNA chip. Potential detection is defined as the detection with partial lack of the hybridized probes for a SNP, which will change the number of hybridized sequences to probes on DNA chips. Thus this type of SNP may not be easily detected because it can slightly change the color observed on DNA chip. Using the probes of variable length organized by 256x256 map, 100% of the SNPs are potentially detected and about 80% of the SNPs are completely detected. Using all combinations (4096) of 6bp probes, only 50% of the SNPs are completely detected,. Therefore, the longer probes organized by SOM were considered to work to improve this rate. Using the probes of fixed length (7 or 8), the rate of completely detected SNPs becomes better. From these results, the DNA chip with probes of 7bp may be sufficient for SNP analyses. But, DNA chips with longer probes will help to reconstruct the sequences if they are applied to sequencing by hybridization. Anyway, the number of the probes selected by our method becomes much less compared with the number of all combination of the probes.

# 5 Conclusions

In this paper, we introduced a design method of DNA chips using SOM. Our SOM algorithm can select a set of small amount of probes that represent the feature of DNA sequence for SNP analyses. Furthermore, our SOM algorithm can tune the length of probes automatically. We made the simulation of SNP detection on computer and have shown that the organized probes are adequate for SNP analyses.

Looking towards the future, we should improve the algorithm to organize the probes which can detect more SNPs. And we must examine the performance of the DNA chip designed by this method if it is applied to sequencing by hybridization.

## References

1 Mark Stoneking, Single nucleotide polymorphisms, Nature : the human genome, pp.821-823,no.6822,(2001)

2. Pevzner,P.A. and Lipshutz, R.J.,Towords DNA sequencing chips, 19th Symp. Mathematical Foundation of Computational Science, Lecture Notes in Computer Science, 841:143-158,(1994)

3. Dozono,H. and Noguchi,Y., An Application of Genetic Algorithm to DNA Sequencing by Oligonucleotide Hybridization, Proc. of IEEE International Joint Symposia on Intelligence and Systems, pp.92-98, May(1998)

4. Giuliano,F and et.al. Potentially functional regions of nucleic acids recognized by a Kohonen's self organizing map, Comput. Appl. Biosci. 9, 687-93, (1993)

# Multi-Dimensional Self-Organizing Maps on Massively Parallel Hardware

Udo Seiffert and Bernd Michaelis
Institute for Electronics, Signal Processing and Communications
University of Magdeburg
Magdeburg, Germany

Abstract

Although available (sequential) computer hardware is very powerful nowadays, the implementation of artificial neural networks on massively parallel hardware is still undoubtedly of high interest, not only under an academic point of view. This paper presents an implementation of multi-dimensional Self-Organizing Maps on a scalable *SIMD* structure of a *CNAPS* computer with up to *512* parallel processors.

## 1 Introduction

In general there are two main reasons to implement artificial neural networks on parallel hardware. The first one comes from the internal data flow. Due to their parallel data processing, most artificial neural networks are predestined to be run on parallel processors. This was not enough to justify the expenditure of a parallel implementation. However, the second motive becomes evident when dealing with demanding real-world applications, when training times are increasing up to and above the pain threshold.

No matter how powerful a particular available workstation processor is, if it is possible to design suitable programs that scale from one processor to hundreds, there will be many neural network applications that can take advantage of the additional computational power due to the parallelism. After all also high-end workstation manufacturers acknowledge this by providing parallel computer hardware and support for workstation clusters as well.

The level of inherent parallel data flow and consequently the possible acceleration depends on the neural network type and the utilized hardware architecture. The given implementation details and results mainly focus on the distribution of the neurons among the available processors, because this is probably the most important task.

## 2 Multi-Dimensional Self-Organizing Maps

As already mentioned, the benefits of a parallel implementation depend on the particular network model. Self-organizing maps [1] have a relatively high level of parallel data flow, since all neurons belong to the same layer. There are a few

implementations of standard SOMs [2] (see [1] for some basic ideas and [3] for a comprehensive review of parallel SOM implementations) and with a slightly modified configuration to meet the requirements of a particular problem or the utilized parallel hardware ([4], [5], [6], [7], [8]). Almost all implementations use a *node parallel algorithm*, where all neurons are divided among the available processors. In contrast, a *training example parallel algorithm* divides all training vectors among the processors. This seems to be an interesting alternative especially for relatively small problems, as reported in [9]. Also a combination of both approaches has been described in [9]. A completely different way is *direct* hardware implementation using neuro-chips, which has been illustrated among others in [10].
Apart from the traditional two-dimensional neuron grid within the Kohonen layer, a multi-dimensional SOM (Figure 1) has been suggested to process image data against the background of motion detection in image sequences ([11], [12], [13]). Later on this model has been improved by a growing algorithm ([14], [15]) based on [16].

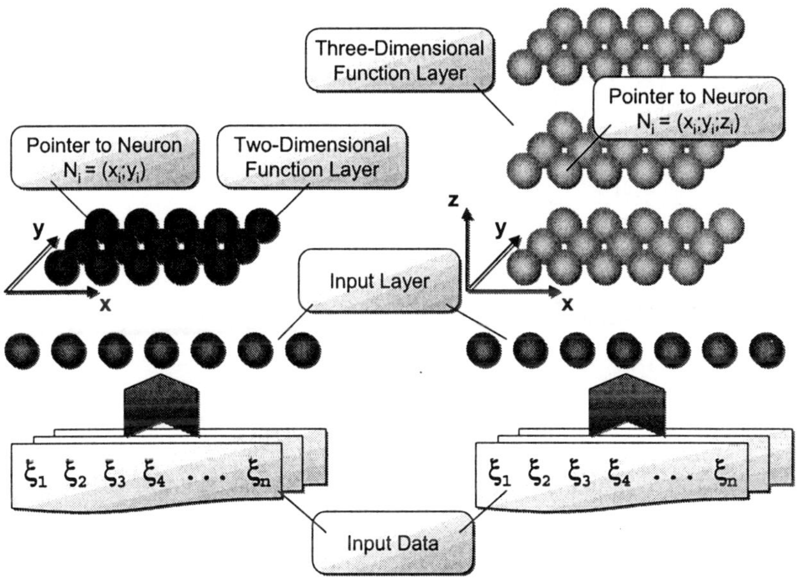


Figure 1: Standard topology with traditional two-dimensional function layer vs. multi-dimensional SOM (here: three-dimensional function layer).

This network model has been successfully applied to a number of real-world problems, especially where a shift independent classification of input patterns is desired and an extensive pre-processing is not applicable. Since these applications are not focus of this paper, please refer to [11] and [13] for further details.
In many image processing applications real-time handling is desired. Due to the huge number of neurons in multi-dimensional SOMs, typically several hundred up to a few thousand, a real-time work was impossible with a standard simulation on a single

processor. Hence, a parallel implementation is not only of academic interest or simply to accelerate the algorithm.

# 3 CNAPS Server 512 Development System

The **CNAPS** (Connected **N**etwork of **A**daptive **P**rocessors) system [17] has been developed since 1993 and has proved to be one of the most powerful **SIMD** (**S**ingle **In**struction **M**ultiple **D**ata) parallel processor architecture. The system is scalable from *16* up to *512* parallel processors with *40* Mbytes per second internal data transfer. In spite of the quite low clock rate of *20* MHz the system has a peak performance of *10.24* billion operations (additions or multiplications) per second.
Since each processor is rather simple and the entire machine is comparable to a very fast matrix multiplier, it is particularly suitable for algorithms with just little non-linear computation and a high part of parallelism, such as large-scale SOMs.

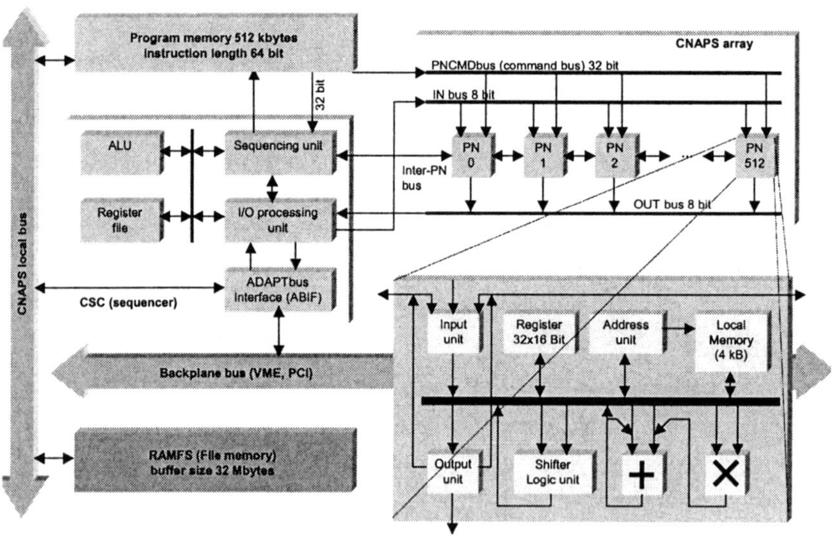


Figure 2: General CNAPS architecture and one single node (lower right part).

The sequencing unit (CSC) acts as controller decoding instructions (upper *32* bits) coming from the program memory and passing them to all nodes within the CNAPS array (single instruction). All nodes execute each instruction with data holding in their local memory (multiple data).

# 4 Distribution of Neurons Among Available Processors

The most important task is to find an efficient model to use the local memory. Since each node has only limited local memory, care must be taken to find an optimal mapping of neurons to the available nodes (node parallel mode). This depends on some conditions. In general each unnecessary I/O-operation should be avoided, because it takes time and can not be performed in parallel. That means, once a particular neuron has been assigned to a particular node, which holds from now on all the data (in the first place the weights), it should not be moved while the network runs. From that point of view the ideal case was, if the entire network had less or equal neurons as processors and the complete weight set could be stored in the local memory. Although the system has *512* processors, this is still more a theoretical case. Some common constellations are reflected in the following modes (**n** .. # **n**eurons in the net; **p** .. # available **p**rocessors; **w** .. required **w**eight space for one neuron; **m** .. available local **m**emory per node for weight storage, normally less than *4* kbytes):

- ❑ Mode 0: $n<p$ & $w<m$: ideal case; each processor represents just one neuron; no additional I/O-operations necessary; some processors idle; small networks.
- ❑ Mode 1: $n>p$: each processor represents $r$ ($r>1$) resp. $r-1$ neurons;
  Submode 1.1: $r \cdot w<m$: no additional I/O-operations necessary; bigger nets;
  Submode 1.2: $r \cdot w>m$: local memory too small to hold all network weights; additional I/O-operations to swap data with shared memory; bigger nets or high-dimensional input (long input vectors).
- ❑ Mode 2: $n<p$ & $w>m$: each processor represents just one neuron; $s$ processors idle; insufficient local memory; small nets with high-dimensional input;
  Submode 2.1: $n \cdot (w-m)<s \cdot m$: weights are also stored in local memory of idle processors, which allows faster access than shared memory;
  Submode 2.2: $n \cdot (w-m)>s \cdot m$: total amount of local memory (used and idle neurons) insufficient; additional I/O-operations to swap data with shared memory.

Most applications using the *basic* SOM algorithm run either in submode 0 or in submode 1.1. Since *multi-dimensional* SOMs are usually composed of many more neurons, they often run in submode 1.1 or submode 1.2. A single weight occupies usually *32* bits (single precision). Depending on further locally stored parameters, up to about *1000* weights can be stored in each processor.

# 5 Results

According to Amdahl's law ([18], [19]) and assuming that $p$ is again the number of available processors, $t_s$ is the estimated amount of time spent by one (serial) processor on serial parts and $t_p$ is the estimated amount of time spent by one processor on parts of the program that can be run in parallel, then the speed-up $S(p)$ is given by

$$S(p) = \frac{t_s + t_p}{t_s + \frac{t_p}{p}} \qquad (1)$$

However, this assumes all processors are equally working to capacity. Because this is just a special case in a neural network implementation, Eq. 1 has to be modified to reflect the number of actually used processors $p_a$ and their employment coefficient $c$. If, for example, a map consisting of *2025* neurons is run on *512* processors, then

*23* ($p_{a1}$) of them represent *3* and *489* ($p_{a2}$) represent *4* neurons each. In order to predict the speed-up, Eq. 2 can be used.
The coefficients are $c_1=0.75$ and $c_2=1$ for the described example. This has been used as benchmark to evaluate the properties of different runs using a different number of processors. All further parameters, especially those controlling the network behaviour, remained fixed during the test.

$$S(p_a) = \frac{t_s + t_p}{t_s + \frac{t_p}{c_1 \cdot p_{a1} + c_2 \cdot p_{a2}}} \tag{2}$$

In Figure 3 the actually achieved as well as theoretically estimated speed-up is plotted against the number of available processors. Both lines are starting at *(1,1)*. From now on the line showing the actual speed-up never meets the theoretical one again. The steps are caused by the above described neuron's distribution among the processors. More available neurons do not always lead to faster execution. The position and height of the steps depend in the first place on the number of neurons and the ratio of $t_p$ to $t_s$. In this example (main frame) the maximum speed-up of *129.3*, compared to serial execution, is achieved between *507* and *512* processors.

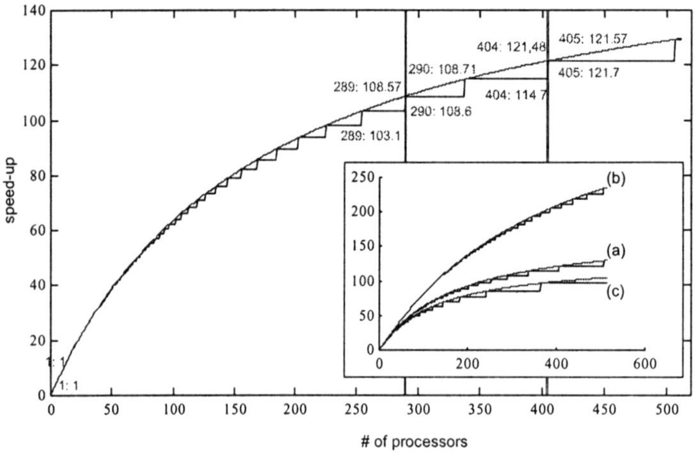


Figure 3: Actual speed-up (*2025* neurons, $t_p$:$t_s$=*172.3*, submode 1.1) operating up to *512* CNAPS processors in comparison to theoretical estimation using Eq. 1 (dotted line).
Inlet: Comparison of three multi-dimensional SOMs of different size; reference, same as big frame (a); *6561* neurons, $t_p$:$t_s$=*434*, submode 1.1 (b); *729* neurons, $t_p$:$t_s$=*140*, submode 1.1 (c).

The inlet shows a smaller net trained also on image blocks of *8x8* pixels (*64* inputs) with a maximum speed-up of *97*, achieved between *365* and *512* processors. The bigger net reaches with *419904* weights almost the limit of available local memory. The maximum speed-up is *234*.
If double precision of the weights (*64* bits) is required, several submodes are already entered with smaller networks. The curves shown in Figure 3 look slightly different.

# 6 Conclusion

This paper shows, by means of some typical examples, the properties and excellent results, achieved with an implementation of multi-dimensional SOMs on a CNAPS parallel computer.
Beyond it, these results are of general interest and transferable to other hardware. The classification of patterns derived from image blocks is possible in real-time now up to an image size of *512x512* pixels. Thus standard image sequences of *25* fps can be processed now using the fully equipped development system.
An improved numerical implementation and some further alterations to the algorithm are expected to save computation time once more. Almost *60%* acceleration can be achieved with just *128* processors, available on the much cheaper pci-board. Once an application can be run in real-time on the pci-board, these results should be very interesting for industrial realizations of the described motion detection system and for many further applications of multi-dimensional SOMs as well.

**References**


1. Kohonen T. Self-Organizing Maps. Springer-Verlag, Berlin, 1995
2. Hammerström D, Nguyen N. An implementation of Kohonen's self-organizing map on the Adaptive Solutions neurocomputer. In: Kohonen T et al. (eds) Artificial neural networks, North-Holland, Amsterdam, 1991, vol 1, pp 715-720
3. Hämäläinen T. Parallel implementations of Self-Organizing Maps. In: Seiffert U et al. (eds) Self-organizing neural networks. Recent advances and applications, Springer, Heidelberg, 2001, pp 245-278
4. Obermayer K, Ritter H, Schulten K. Large-scale simulation of a self-organizing neural network: Formation of a SOM totopic Map. In: Eckmiller R et al. (eds) Parallel processing in neural systems and computers, North-Holland, Amsterdam, 1990, pp 71-74
5. Kotilainen P, Saarinen J, Kaski K. Mapping of SOM neural network algorithms to a general purpose parallel neurocomputer. In: Gielen S, Kappen B (eds) Proc. of ICANN'93, Springer-Verlag, London, 1993, pp 1082-1087
6. Guan H, Li C, Cheung T, Yu S. Parallel design and implementation of SOM neural computing model in PVM environment of a distributed system. In: Proc. of advances in parallel and distributed computing, IEEE Comput. Soc. Press, Los Alamitos, 1997, pp 26-31
7. Hämäläinen T, Klapuri H, Saarinen J, Kaski K. Mapping of SOM and LVQ algorithms on a tree shape parallel computer system. Parallel Computing 1997; 23:271-289
8. Yasunaga M, Tominaga K, Kim JH. Parallel self-organization map using multiple stimuli. In: Proc. of IJCNN'99. IEEE Service Center, Piscataway, 1999, vol 2, pp 1127-1130
9. Myklebust G, Solheim JG. Parallel Self-Organizing Maps for actual applications. In: Proc. of ICNN'95. Causal Productions, Adelaide, 1995, vol 2, pp 1054-1059
10. Rueping S, Porrmann M, Rueckert U. SOM accelerator system. Neurocomputing 1998; 21:31-50
11. Seiffert U, Michaelis B. Three-dimensional Self-Organizing Maps for classification of image properties. In: Proc. of ANNES'95. IEEE Computer Society Press, Los Alamitos, 1995, pp 310-313
12. Schünemann S, Seiffert U, Michaelis B. Two more modifications of SOMs to handle signals with special properties. In: Proc. of WSOM'97, HUT, Helsinki, 1997, pp 292-297

13. Seiffert U, Michaelis B. Estimating motion parameters with three-dimensional Self-Organizing Maps. Information Sciences 1997; 101:187-201
14. Seiffert U, Michaelis B. Growing multi-dimensional Self-Organizing Maps. International Journal of Knowledge-Based Intelligent Engineering Systems 1998; 2:42-48
15. Fritzke B. Growing self-organizing networks - why? In: Proc. of ESANN'95. D-Facto Publishers, Brussels, 1995, pp 61-72
16. Fritzke B. Growing grid - a self-organizing network with constant neighborhood range and adaption strength. Neural Processing Letters 1995; 5:1-5
17. Documentation of the CNAPS Server 512 Development System. Adaptive Solutions Inc., 1996
18. Amdahl G. Validity of the single-processor approach to achieving large scale computing capabilities. In: Proc. of AFIPS'67, AFIPS Press, Reston, 1967, pp 483-485
19. Gustafson JL. Re-evaluating Amdahl's Law. Communications of the ACM 1988; 31(5)

# A New Method of Hough Transform by Using SOM with Input Vector Transformation

T. Yamakawa[†*], K. Horio[†], S. Izumi[‡] and T. Miki[†*]

[†] Department of Brain Science and Engineering
Graduate School of Life Science and Systems Engineering
Kyushu Institute of Technology
680-4 Kawazu, Iizuka, Fukuoka 820-8502 Japan
yamakawaa@ces.kyutech.ac.jp

[‡] Vehicle Control Group
The Third Department of Systems Research
Hitachi Research Laboratory, Hitachi,Ltd.

Abstract

In this paper, we propose a new method of the shape detection by using a self-organizing method with input vector transformation. It is difficult to detect a shape from a noisy image by the conventional Hough Transform. The proposed method enables to detect the shape automatically. Two SOMs are used in the proposed method, the first SOM determines the number of typical shapes included in the image and the second SOM determines the parameters of the shapes. The performance of the proposed method is confirmed by detection of lines and circles from noisy images.

## 1 Introduction

In the fields of image processing and computer vision it is very important to detect basic shapes, for example, lines or circles, from landmarks plotted in images. Hough proposed Hough Transform (HT) in 1962[1]. HT is powerful to detect the lines even if some landmarks are failed or are covered up. In HT the landmarks are mapped to a parameter space as trajectories. The trajectories vote for relative subspace in the parameter space, and the parameters of the lines to be detected are determined in consideration of the poll. However, if the size of the subspaces is small or large the votes straggle or the error of the detected parameters becomes large, so that it is difficult to decide reasonable size of the subspaces. Furthermore the problems become increasingly severe when the landmarks include noises.

---

*The author is also with the Fuzzy Logic Systems Institute (FLSI), Kawazu, Iizuka, Fukuoka 820-0067 Japan.

‡This work was achieved when the author was a Graduate course student of Kyushu Institute of Technology.

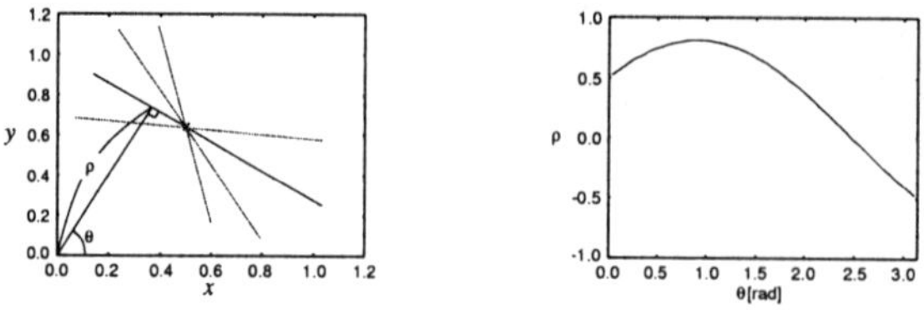


(a) (b)

Figure 1: Landmark in image space and relationship between $\theta$ and $\rho$ in parameter space. (a) Landmark. (b) Relationship between $\theta$ and $\rho$ for the landmark.

In this paper, we propose a new method of HT by using self-organizing maps (SOMs). Two SOMs are used in the proposed HT method. The number of the shapes to be detected is determined in the first SOM, and the parameters of the shapes are done in the second SOM. The coordinates of the landmarks as the input vectors are transformed into parameters before supplying them to first SOM, and the winner unit is defined in accordance with similarity of the parameters to the corresponding weights. The input vectors are applied to the first SOM after learning again, and the number of the shapes to be detected can be determined based on the numbers of times that the units are defined as the winner unit. In the second SOM, the number of the units of the competitive layer is equal to the number of the shapes determined in the first SOM, and the weight vectors of the first SOM after learning are used as the input vectors. The weight vectors of the unit in the second SOM after learning are employed as the parameters of the shapes to be detected.

The effectiveness and the validity of the proposed HT method are verified by detection of lines and circles from noisy images.

# 2 Hough Transform

HT is one of the methods which detects basic shapes from some landmarks. The cross in Fig.1(a) is a landmark in the image space. There are many lines including the landmark, and each line can be represented by two parameters, angle from $x$ axis $\theta$ and distance from origin $\rho$. The relationship between $\theta$ $(0 \leq \theta \leq \pi)$ and $\rho$ for the landmark is shown in Fig.1(b). Consider that one line is detected from the landmarks shown in Fig.2(a). Fig.2(b) shows the relationships between $\theta$ and $\rho$ for all the landmarks. The parameter space ($\theta$-$\rho$ space) is divided into many small subspaces, and the relationships vote for the corresponding subspace. The central values of the subspaces which have many votes are selected as the parameters of the lines to be detected. Fig.2(c) shows the detected line and the landmarks.

In the conventional HT, it is very difficult to decide reasonable size of the subspace and the threshold value of the votes. Large subspaces cause large

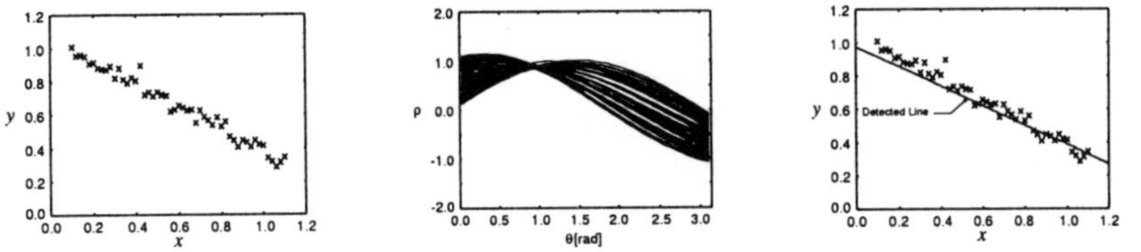


(a) (b) (c)

Figure 2: An example of detection of one line by HT. (a) 100 landmarks. (b) The relationships between $\theta$ and $\rho$ for the landmarks. (c) The results of the detection of one line.

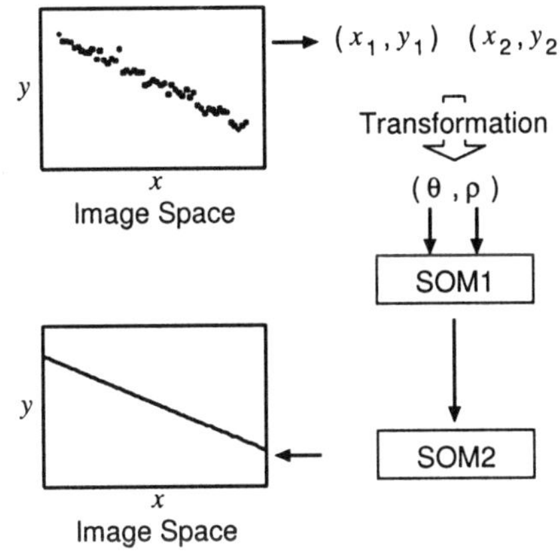


Figure 3: The conceptual diagram of the proposed method.

error, and small subspaces need much time. The threshold value of the vote decides the number of shapes to be detected, and it was usually decided by user with his/her experience.

# 3 SOM with Input Vector Transformation

In this paper we propose a new HT method which can determine the number of shapes to be detected automatically (SOM1) and can determine parameters of the shapes with high accuracy (SOM2) as shown in Fig.3. In this section, the proposed method for detection of lines is explained as an example. It is very easy to expand the proposed method to detection of circles.

The input vectors of the SOM1 are coordinates of two landmarks $(x_1, y_1)$ and $(x_2, y_2)$, and they are transformed into parameters $\theta$ and $\rho$ which define the line including two landmarks by:

$$\theta = \tan^{-1}\left(-\frac{x_2 - x_1}{y_2 - y_1}\right), \tag{1}$$

$$\rho = (-\frac{(y_2 - y_1)x_1}{x_2 - x_1} + y_1)\sin\theta. \tag{2}$$

The unit $j$ in the competitive layer has the weight vector $\boldsymbol{w}_j = [w_{j\theta}, w_{j\rho}]$. The winner unit $c$ is defined by:

$$c = \arg\min_{j=1}^{N} \sqrt{(\theta - w_{j\theta})^2 + (\rho - w_{j\rho})^2}, \tag{3}$$

where $N$ is the number of the units in the competitive layer. The weight vectors of the winner unit and their neighboring units are updated by:

$$w_{j\theta}(t+1) = w_{j\theta}(t) + \alpha(t)(\theta - w_{j\theta}(t)), \tag{4}$$

$$w_{j\rho}(t+1) = w_{j\rho}(t) + \alpha(t)(\rho - w_{j\rho}(t)), \tag{5}$$

where $t$ is learning iteration and $\alpha(t)$ is a learning rate at $t$. All combinations of the landmarks are applied to the SOM1 as the input vectors. After the learning all the input vectors are applied again, and the numbers of times that the units are defined as the winner units are counted. We focus the feature of the SOM that the units, which have the weight vectors located in the area where there are many input vectors, are defined as the winner units more than other units. The number of the shapes can be determined based on the number of times that the units are defined as the winner units.

In the SOM2, the number of the units in the competitive layer is same to the number of the shapes determined in the SOM1. The weight vectors of the units which won many times in SOM1 are employed as the input vectors of the SOM2. This means that the parameters of the shapes can be adjusted in the SOM2 and that detection error can be decreased.

# 4 Experimental Results

In this section, the experimental results, in which some shapes (lines and circles) are detected, are described.

## 4.1 Detection of Lines

Consider that lines are detected from 100 landmarks shown in Fig.4(a). The number of the input vectors of the SOM1 is 4950 (${}_{100}C_2$), and learning is achieved by using these input vectors. In the learning, the number of the learning iterations, the number of the units in the competitive layer, the initial value of the learning rate and the initial value of the range of the neighborhood are 10000, 100, 0.5 and 100, respectively. Fig.4(b) shows the arrangement of the weight vectors $(w_{j\theta}, w_{j\rho})$ after the learning. Fig.4(c) shows the number of times that each unit is defined as the winner unit when the input vectors are applied again after the learning (winner histogram). The number of the lines can be determined as 2, because the graphical form of Fig.4(c) is double-peaked pattern.

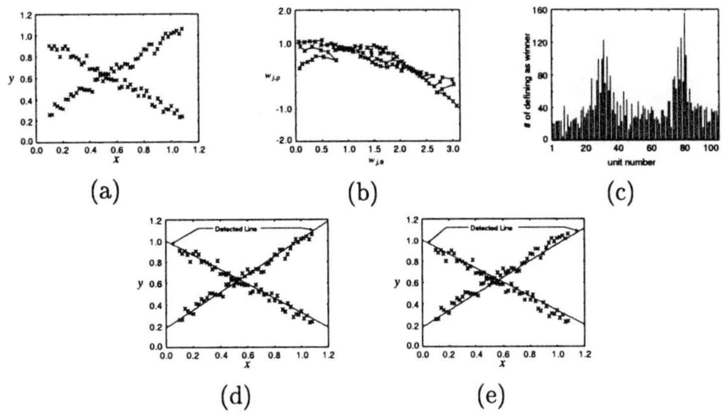


Figure 4: Experimental results of detection of the lines. (a) 100 landmarks. (b) Arrangement of the weight vectors of the SOM1 after learning. (c) Winner histogram. (d) Detected lines by the proposed HT. (e) Detected lines by the conventional method.

In the SOM2, the number of the units in the competitive layer is assigned to 2, and the number of the input vectors is 50 (weight vectors of the units 21-45 and units 66-90 of SOM1). In the learning, number of the learning iterations, the initial value of the learning rate and the initial value of the range of the neighborhood are 10000, 0.5 and 2, respectively. Fig.4(d) and (e) show the detected lines by the proposed HT method and by the conventional HT. Although the proposed method and the conventional HT give the same number of lines, the former exhibits higher accuracy of alignment of these lines than the later. Because the former processes the distribution of two peaks in the winner histogram to obtain the parameters with high accuracy in the second SOM.

## 4.2 Detection of Circles

Consider that a circle is detected from 100 landmarks shown in Fig.5(a). In the SOM1 the input vector is coordinates of three landmarks $(x_1, y_1)$, $(x_2, y_2)$ and $(x_3, y_3)$. Three landmarks are transformed into three parameters, *i.e.*, the center position $(a, b)$ and the radius of the circle including three landmarks by:

$$a = \frac{(x_2^2 - x_1^2 + y_2^2 - y_1^2)(y_3 - y_1) - (x_3^2 - x_1^2 + y_3^2 - y_1^2)(y_2 - y_1)}{2((x_2 - x_1)(y_3 - y_1) - (x_3 - x_1)(y_2 - y_1))} \tag{6}$$

$$b = \frac{(x_2^2 - x_1^2 + y_2^2 - y_1 - 2) - 2(x_2 - x_1)a}{2(y_2 - y_1)} \tag{7}$$

$$r = \sqrt{(x_1 - a)^2 + (y_1 - b)^2} \tag{8}$$

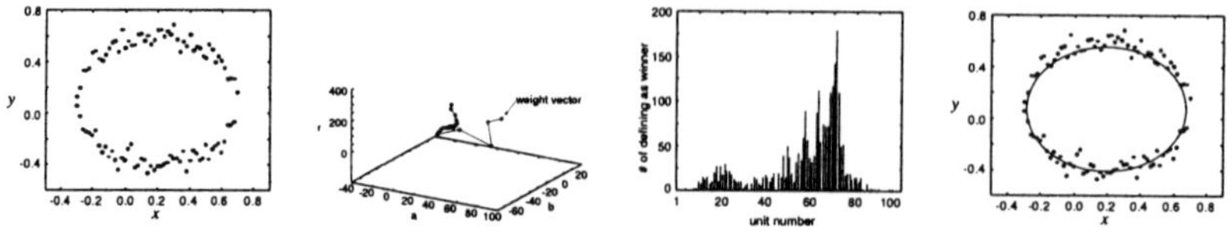


(a) (b) (c) (d)

Figure 5: Experimental results of the detection of circle. (a) 30 landmarks. (b) Arrangement of the weight vectors of the SOM1 after learning. (c) Winner histogram. (d) Detected circle by the proposed method.

The number of the input vectors of the SOM1 is 161700 (${}_{100}C_3$), and the learning is achieved by using these input vectors. In the learning, the number of the learning iterations, the number of the units in the competitive layer, the initial value of the learning rate and the initial value of the range of the neighborhood are assigned to 10000, 100, 0.5 and 100, respectively. Fig.5(b) shows the arrangement of the weight vectors $(w_{ja}, w_{jb}, w_{jr})$ after the learning. Fig.5(c) shows the winner histogram. Number of the circle can be determined as 1, because the graphical form of Fig 5(c) is single-peaked pattern.

In the SOM2, the number of the units in the competitive layer is 1, and the number of the input vectors are 50 (weight vectors of the units 41-90 of SOM1). In the learning, the number of the learning iteration, the initial value of the learning rate and the initial value of the range of the neighborhood are 10000, 0.5 and 1, respectively. Fig.5(d) shows that reasonable circle detected by proposed method.

# 5 Conclusions

In this paper, we proposed the new HT method in which two SOMs are used. The first SOM can determine the number of the shapes automatically and the second one can do the parameters of the shapes with high accuracy. The validity of the proposed method is verified by applying it to detection of lines and circles.

# References

[1] P. V. C. Hough, "Method and Means for Recognizing Complex Patterns," U. S. Patent, 3069 654, 1962.

[2] T. Kohonen, "Self-organized formation of topologically correct feature maps," *Biol. Cybern.*, Vol.43, pp.59-69, 1982.

[3] T.Kohonen, *Self-organization and associative memory*, Springer-Verlag, 1988.

# Estimating Relevant Input Dimensions for Self-organizing Algorithms

Barbara Hammer[1] and Thomas Villmann[2]

[1] University of Osnabrück, Department of Mathematics/Computer Science, Albrechtstraße 28, 49069 Osnabrück, Germany
[2] University of Leipzig, Clinic for Psychotherapy and Psychosomatic Medicine, Karl-Tauchnitz-Straße 25, 04107 Leipzig, Germany

**Summary.** We propose a new scheme for enlarging generalized learning vector quantization with weighting factors for the several input dimensions which are adapted according to the specific task. This leads to a more powerful classifier with little extra cost as well as the possibility of automatically pruning irrelevant input dimensions. The method is tested on real world satellite image data and compared to several well known algorithms which determine the intrinsic data dimension.

## 1 Introduction

Self-organizing methods such as the self-organizing map (SOM) or vector quantization (VQ) as introduced by Kohonen provide a successful and intuitive method of processing data for easy access [12]. Assumed data are labeled, an automatic clustering can be learned via attaching maps to the SOM or enlarging VQ with a supervised component to so-called learning vector quantization (LVQ) [13,17]. Various modifications of LVQ exist which ensure faster convergence, a better adaptation of the receptive fields to optimum Bayesian decision, or an adaptation for complex data structures, to name just a few [13,20,21]. Concerning SOM, one major problem consists in finding an appropriate topology of the initial lattice of codebooks such that the prior topology of the neural architecture mirrors the intrinsic topology of the data. Hence various approaches exist in order to measure the degree of topology preservation, to adapt the topology to the data, to define the lattice a posteriori, or to evolve structures which are appropriate for real world data [2,6,14,19,25]. In all tasks the intrinsic dimensionality of the data plays a crucial role since it determines large parts of the optimum neural network, i.e., the lattice for SOM. Moreover, superfluous data dimensions slow down the training for LVQ as well. They may even cause a decrease in accuracy since they add possibly noisy or misleading terms to the Euclidian metric where LVQ is based on. Hence a data dimension as small as possible is desirable for the above mentioned methods in general, for the sake of efficiency, accuracy, and simplicity of neural network processing. We will consider possibilities of pruning irrelevant data dimensions and computing the intrinsic data dimension in this paper. Approaches like [11] clearly indicate that often a considerable reduction of the data dimension is possible without loss of information.

We will focus on LVQ since it combines the elegancy of simple and intuitive updates in unsupervised algorithms with the accuracy of supervised methods. The main idea of our approach is to introduce weighting factors to the data dimensions which are adapted automatically such that the classification error becomes minimal. Small factors indicate that the respective data dimension is irrelevant. This idea can be applied to any generalized LVQ (GLVQ) scheme as introduced in [20] such that a robust and efficient method results which can push the classification borders near to the optimum Bayesian decision. This method, generalized relevance LVQ (GR-LVQ), generalizes relevance LVQ (RLVQ) [3] which is based on simple Hebbian learning and leads to worse and instable results in case of noisy real life data. However, like RLVQ, GRLVQ has the advantage of an intuitive update rule and allows efficient input pruning compared to other approaches which adapt the metric to the data involving additional transformations as proposed in [7,9,22] or depend on less intuitive differentiable approximations of the original dynamics [15]. We will apply GRLVQ to classify a real life satellite image with approx. 3 mio. data points. Apart from the above mentioned methods, dimensionality reduction is possible via standard methods like principal component analysis or independent component analysis [10,18]. Furthermore, a growing SOM (GSOM) automatically adapts the lattice of neurons to the data and hence gives hints about the intrinsic dimensionality as well. We compare our GRLVQ experiments to the results provided by GSOM and a Grassberger-Procaccia analysis respectively, and obtain comparable results concerning the intrinsic dimensionality of our data.

## 2 The GRLVQ Algorithm

Assume a finite training set $X = \{(x^i, y^i) \subset \mathbb{R}^n \times \{1, \ldots, C\} \mid i = 1, \ldots, m\}$ of training data is given and the clustering of the data into $C$ classes is to be learned. We denote the components of a vector $x \in \mathbb{R}^n$ by $(x_1, \ldots, x_n)$ in the following. GLVQ chooses a fixed number of vectors in $\mathbb{R}^n$ for each class, so called codebooks. Denote the set of codebooks by $\{w^1, \ldots, w^K\}$ and assign the label $c^i = c$ to $w^i$ iff $w^i$ belongs to the $c$th class, $c \in \{1, \ldots, C\}$. The receptive field of $w^i$ is defined by $R^i = \{x \in X \mid \forall w^j \, (j \neq i \rightarrow |x - w^i| < |x - w^j|)\}$. The training algorithm adapts the codebooks $w^i$ such that for each class $c \in \{1, \ldots, C\}$, the corresponding codebooks represent the class as accurately as possible. That means, the difference of the points belonging to the $c$th class, $\{x^i \in X \mid y^i = c\}$, and the receptive fields of the corresponding codebooks, $\bigcup_{c^i=c} R^i$, should be as small as possible for each class. For a given data point $(x, y) \in X$ denote by $\mu(x)$ some function which is negative if $x$ is classified correct, i.e., it belongs to a receptive field $R^i$ with $c^i = y$, and which is positive if $x$ is classified wrong, i.e., it belongs to a receptive field $R^i$ with $c^i \neq y$. Denote by $f : \mathbb{R} \rightarrow \mathbb{R}$ some monotonically increasing function. The general scheme of GLVQ consists in minimizing the error term

$$S = \sum_{i=1}^{m} f(\mu(x_i)) \tag{1}$$

via a stochastic gradient descent. The concrete choice of $f$ as the identity and $\mu(x) = \eta \cdot d$, $d$ being the squared Euclidian distance of $x$ to the nearest codebook and $\eta = 1$ if $x$ is classified correct, $\eta = -1$, if $x$ is classified wrong, yields the standard LVQ update

$$w^j := \begin{cases} w^j + \epsilon(x^i - w^j) & \text{if} \quad y^i = c^j \\ w^j - \epsilon(x^i - w^j) & \text{otherwise} \end{cases} \tag{2}$$

where $\epsilon \in (0,1)$ is the so called learning rate. LVQ2.1 can be obtained in a similar fashion. The choice of $f$ as the sigmoidal function and

$$\mu(x) = \frac{d_1 - d_2}{d_1 + d_2}, \tag{3}$$

where $d_1$ is the squared Euclidian distance to the next codebook labeled with $y$, say $W_1$, and $d_2$ is the squared Euclidian distance to the next codebook labeled with $c \neq y$, say $W_2$, yields a particular powerful and noise tolerant behavior since it combines adaptation near the optimum Bayesian borders like LVQ2.1, whereby prohibiting the possible divergence of LVQ2.1 as reported in [20]. We refer to the update as GLVQ:

$$\Delta W_1 := \epsilon \cdot \frac{f' \cdot d_2}{(d_1 + d_2)^2}(x - W_1), \; \Delta W_2 := -\epsilon \cdot \frac{f' \cdot d_1}{(d_1 + d_2)^2}(x - W_2) \tag{4}$$

Obviously, the success of GLVQ crucially depends on the fact that the Euclidian metric is appropriate for the data and the several input dimensions are approximately equally scaled and equally important. Here, we introduce input weights $\lambda = (\lambda_1, \ldots, \lambda_n)$, $\lambda_i \geq 0$ in order to allow a different scaling of the several input dimensions hence making possibly time consuming preprocessing of the data superfluous. Substituting the Euclidian metric $||x - y||$ by

$$||x - y||_\lambda^2 = \sum_{i=1}^{n} \lambda_i (x_i - y_i)^2, \tag{5}$$

the receptive field of codebook $w^i$ becomes $R^i_\lambda = \{x \in X \mid \forall w^j \, (j \neq i \rightarrow \|x - w^i\|_\lambda < \|x - w^j\|_\lambda)\}$. Substituting $R^i$ by $R^i_\lambda$ in the error function $S$ yields to a different weighting of the several input dimensions. Appropriate weighting factors $\lambda$ can be determined via a stochastic gradient descent as well. Hence (2) is accompanied by the update

$$\lambda_l := \begin{cases} \lambda_l - \epsilon_1 (x^i_l - w^j_l)^2 & \text{if} \quad y^i = c^j \\ \lambda_l + \epsilon_1 (x^i_l - w^j_l)^2 & \text{otherwise} \end{cases}, \tag{6}$$

$\epsilon_1 \in (0,1)$, followed by a normalization to obtain $||\lambda|| = 1$ in order to avoid numerical instabilities for the weighting factors. This constitutes RLVQ as proposed in [3]. (4) is accompanied by

$$\lambda_l := \lambda_l - \epsilon_1 f' \cdot \left( \frac{d_2}{(d_1 + d_2)^2}(x^i_l - W^1_l)^2 - \frac{d_1}{(d_1 + d_2)^2}(x^i_l - W^2_l)^2 \right) \tag{7}$$

followed by normalization again. We term this generalization of RLVQ and GLVQ *generalized relevance learning vector quantization* or GRLVQ, for short. Note that both updates can be motivated intuitively by the Hebb paradigm taking the normalization into account: those weighting factors are reinforced most which coefficients are closest to the respective data point $x$ if this point is classified correct; if $x$ is classified wrong, those factors are reinforced most which coefficients are far away.

Apart from greater flexibility due to an adaptive metric, GRLVQ provides a simple way of determining which data dimensions are relevant: we can just drop those dimensions with lowest weighting factor until a considerable increase of the classification error is observed. Alternatively, one can use the above mentioned GSOM algorithm. However, because of its remaining hypercubical structure the results may be inaccurate. Another method is to apply a Grassberger-Procaccia-analysis to determine the intrinsic dimension which is unfortunately sensitive to noise [8;26]

## 3 Experiments

We first tested GRLVQ on two artificial data sets from [3] in order to compare it to RLVQ. The data comprise clusters with small or large overlap, respectively, of the clusters in two dimensions as depicted in Fig. 1. We embed the points in $\mathbb{R}^{10}$ as follows: Assume $(x_1, x_2)$ is one data point. Then we add 8 dimensions obtaining a point $(x_1, \ldots, x_{10})$. We choose $x_3 = x_1 + \eta_1, \ldots, x_6 = x_1 + \eta_4$, where $\eta_i$ is distributed according to Gaussian or uniform noise, respectively, with different variances. $x_7$, ..., $x_{10}$ contain pure noise of different variances. GRLVQ obtains about the same accuracy as RLVQ on all data sets and clearly indicates which dimensions are less important via assigning small weighting factors to the less important dimensions which are known in these examples. Typical weighting factors are the vector $(0.49, 0.4, 0.07, 0.02, 0, 0.02, 0, 0, 0, 0)$ for the first data set or the vector $(0.28, 0.36, 0.3, 0.05, 0, 0, 0, 0, 0, 0)$ for the second data set, hence clearly separating the important first two data dimensions from the remaining 8 dimensions of which the first 4 contain some information (as pointed out via a comparably large third weighting term for the second data set) and the remaining four contain no information at all. However, GRLVQ shows a faster convergence and larger stability compared to RLVQ in particular if used for noisy data sets with large overlap of the classes. Concerning RLVQ, pre-training with LVQ and small learning rates were mandatory in order to ensure good results; the same situations turn out to be less critical for GRLVQ, although it is advisable to choose the learning rate for the weighting terms an order of magnitude smaller than the learning rate for the codebook update. These results indicate that GRLVQ is particularly well suited for noisy real life data sets.

In a second test we applied GRLVQ to the well known Iris data set provided in the UCI repository of machine learning [4]. The task is to predict three classes of plants based on 4 numerical attributes in 150 instances, i.e., we deal with data points in $\mathbb{R}^4$ with labels in $\{1, 2, 3\}$. Both, LVQ and RLVQ obtain an accuracy of about 0.95 for a training and test set if trained with 2 codebooks for each class. RLVQ shows a slightly cyclic behavior in the limit, the accuracy changing between 0.94 and 0.96. The computed weighting factors for RLVQ are $(0.02, 0.01, 0.02, 0.89)$

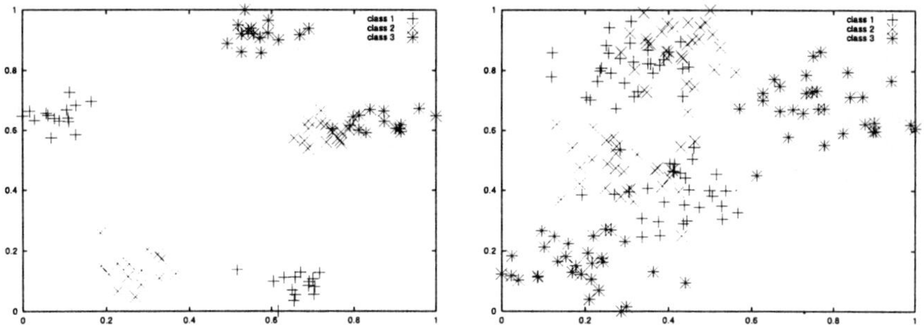


**Fig. 1.** Artificial data sets consisting of three classes with two clusters each and small or large overlap, respectively; only the first two dimensions are depicted.

indicating that based on the last dimension a very good classification would be possible. If more dimensions would be taken into account, a better accuracy of about 1.0 would be possible as reported in the literature, although we could not produce such a solution with LVQ or RLVQ. GRLVQ yields the better accuracy of at least 0.97 on the training as well as the test set and obtains weighting factors of the form $(0, 0, 0.4, 0.6)$, hence indicating that the last dimension is most important as already found by RLVQ, and dimension 3 contributes to a better accuracy which has not been pointed out by RLVQ.

Finally, we applied the algorithm to a large real world data set: a multi-spectral LANDSAT TM satellite image of the Colorado area.[1] Satellites of LANDSAT-TM type produce pictures of the earth in 7 different spectral bands. The ground resolution in meter is $30 \times 30$ for the bands 1-5 and band 7. Band 6 (thermal band) has a resolution of $60 \times 60$ only and, therefore, it is often dropped. The spectral bands represent useful domains of the whole spectrum in order to detect and discriminate vegetation, water, rock formations and cultural features [5,16]. Hence, the spectral information, i.e., the intensity of the bands associated with each pixel of a LANDSAT scene, is represented by a vector in $\mathbb{R}^n$ with $n = 6$. Generally, the bands are highly correlated [1,23]. Additionally, the Colorado image is completely labeled by experts. There are 14 labels describing different vegetation types and geological formations. Thereby, the label probability varies in a wide range [24]. The size of the image is $1907 \times 1784$ pixels.

We trained RLVQ and GRLVQ with 42 codebooks (3 for each class) on 5% of the data set till convergence. The algorithm converged in less than 10 cycles if $\epsilon$ was chosen as 0.1 and $\epsilon_1$ was chosen as 0.01. RLVQ yields an accuracy of about 87% on the training data as well as the entire data set, however, it does not provide a ranking of the codebooks, i.e. all weighting terms are close to their initial value 0.16. GRLVQ leads to the better accuracy of 90% on the training set as well as the entire data set and provides a clear ranking of the several data dimensions. In all experiments, dimension 6 is ranked as least important with weighting factor close to 0. The weighting factors approximate $(0.1, 0.17, 0.27, 0.21, 0.26, 0)$ in several

[1] Thanks to M. Augusteijn (University of Colorado) for providing this image.

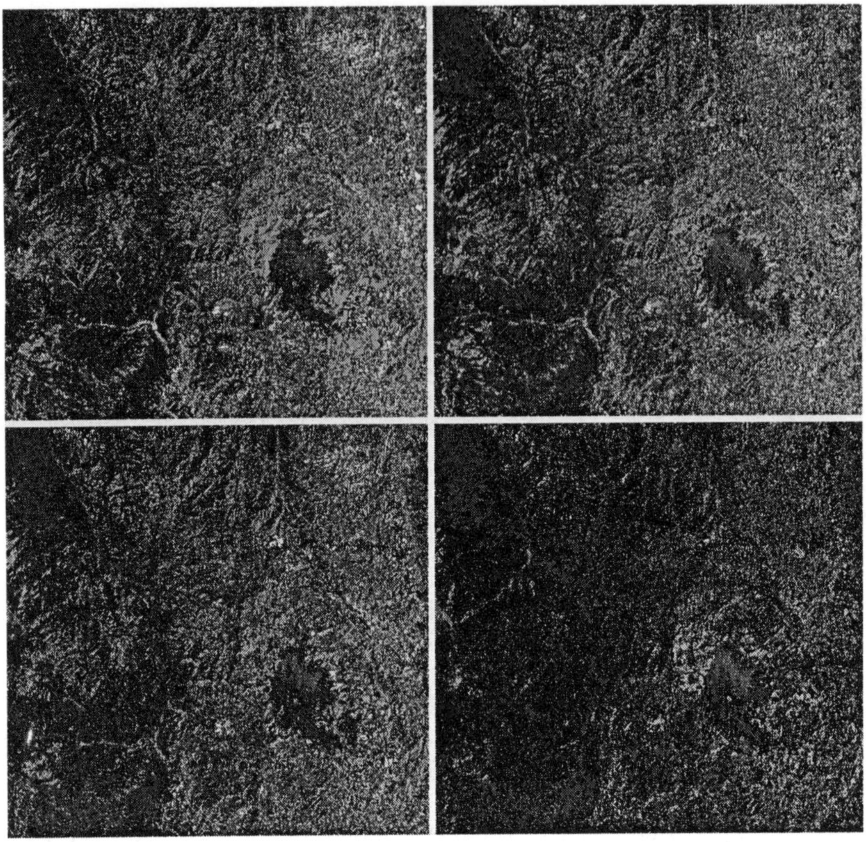

**Fig. 2.** Colorado-satellite-image: the pixels are colored according to the labels; above-left: original labeling; above-right: GRLVQ without pruning; below-left: GRLVQ with pruning of dimensions 1,2,6; below-right: GRLVQ with pruning of 4 dimensions. Misclassified pixels in the GRLVQ-generated images are black colored. (A colored version of the image can be obtained from the authors on request.)

runs. This weighting clearly separates the first two dimensions via a small weighting factor. Pruning dimensions 1 and 2 additionally to dimension 6 yields an accuracy of 84% hence indicating, that the intrinsic data dimension is at most 3. Pruning one additional data dimension still allows an accuracy of more than 50%, hence indicating that the intrinsic dimension may be even lower and the relevant directions are not parallel to the axes or even curved. These results are visualized in Fig. 2 via coloring the misclassified pixels in the respective cases in black.

For comparison we applied the GSOM approach and a Grassberger-Procaccia-analysis. The latter estimates the intrinsic dimension as $d^{\mathcal{GP}} \approx 3.1414$ whereas GSOM generates a lattice of shape $12 \times 7 \times 3$, hence indicating an intrinsic dimension between 1 and 3. These methods show a good agreement with the drastic loss of information if more than 3 dimensions are pruned in the GRLVQ.

## 4 Conclusions

The presented clustering algorithm GRLVQ provides a new robust method for automatically adapting the Euclidian metric used for clustering to the data, determining the relevance of the several input dimensions for the overall classifier, and estimating the intrinsic dimension of data. It reduces the input dimensions onto the essential parameters which is demanded to obtain optimal network structures. The algorithm was successfully tested on real world data, a large and noisy satellite multi-spectral image. A comparison with other approaches validates the results even in real life applications. It should be mentioned that the GRLVQ algorithm can bė adapted to other types of neural vector quantizers as neural gas or SOM, to mention just a few. Further considerations of GRLVQ should include information theory approaches like entropy maximization to improve the capabilities of the network.

## References


1. M. F. Augusteijn, K. A. Shaw, and R. J. Watson. A study of neural network input data for ground cover identification in satellite images. In S. Gielen and B. Kappen, editors, *Proc. ICANN'93, Int. Conf. on Artificial Neural Networks*, pages 1010–1013, London, UK, 1993. Springer.
2. H.-U. Bauer and T. Villmann. Growing a Hypercubical Output Space in a Self–Organizing Feature Map. *IEEE Transactions on Neural Networks*, 8(2):218–226, 1997.
3. T. Bojer, B. Hammer, D. Schunk, and K. Tluk von Toschanowitz. Relevance determination in learning vector quantization. To appear at ESANN'01.
4. C.L. Blake and C. J. Merz, UCI Repository of machine learning databases , Irvine, CA: University of California, Department of Information and Computer Science.
5. J. Campbell. *Introduction to Remote Sensing*. The Guilford Press, U.S.A., 1996.
6. B. Fritzke. Growing grid: a self-organizing network with constant neighborhood range and ada[tation strength. *Neural Processing Letters*, 2(5):9–13, 1995.
7. I. Gath and A. Geva. Unsupervised optimal fuzzy clustering. *IEEE Transactions on Pattern Analysis and Machine Intelligence*, 11:773–791, 1989.
8. P. Grassberger and I. Procaccia. Maesuring the strangeness of strange attractors. *Physica*, 9D:189–208, 1983.
9. D. Gustafson and W. Kessel. Fuzzy clustering with a fuzzy covariance matrix. In *Proceedings of IEEE CDC'79*, pages 761–766, 1979.
10. A. Hyvärinen and E. Oja. A fast fixed-point algorithm for independent component analysis. *Neural Computation*, 9(7):1483–1492, 1997.
11. S. Kaski. Dimensionality reduction by random mapping: fast similarity computation for clustering. In *Proceedings of IJCNN'92*, pages 413–418, 1998.
12. T. Kohonen. Learning vector quantization. In M. Arbib, editor, *The Handbook of Brain Theory and Neural Networks*, pages 537–540. MIT Press, 1995.
13. T. Kohonen. *Self-Organizing Maps*. Springer, 1997.
14. T. Martinetz and K. Schulten. Topology representing networks. *Neural Networks*, 7(3):507–522, 1993.
15. U. Matecki. *Automatische Merkmalsauswahl für Neuronale Netze mit Anwendung in der pixelbezogenen Klassifikation von Bildern*. Shaker, 1999.

16. E. Merenyi. The challenges in spectral image analysis: An introduction and review of ANN approaches. In *Proc. Of European Symposium on Artificial Neural Networks (ESANN'99)*, pages 93–98, Brussels, Belgium, 1999. D facto publications.
17. A. Meyering and H. Ritter. Learning 3D-shape-perception with local linear maps. In *Proceedings of IJCNN'92*, pages 432–436, 1992.
18. E. Oja. Principal component analysis. In M. Arbib, editor, *The Handbook of Brain Theory and Neural Networks*, pages 753–756. MIT Press, 1995.
19. H. Ritter. Self-organizing maps in non-euclidean spaces. In E. Oja and S. Kaski, editors, *Kohonen Maps*, pages 97–108. Springer, 1999.
20. A. S. Sato and K. Yamada. Generalized learning vector quantization. In G. Tesauro, D. Touretzky, and T. Leen, editors, *Advances in Neural Information Processing Systems*, volume 7, pages 423–429. MIT Press, 1995.
21. P. Somervuo and T. Kohonen. Self-organizing maps and learning vector quantization for feature sequences. *Neural Processing Letters*, 10(2):151–159, 1999.
22. M.-K. Tsay, K.-H. Shyu, and P.-C. Chang. Feature transformation with generalized learning vector quantization for hand-written Chinese character recognition. *IEICE Transactions on Information and Systems*, E82-D(3):687–692, 1999.
23. T. Villmann and E. Merenyi. Extensions and modifications of the Kohonen-SOM and applications in remote sensing image analysis. In U. Seiffert and L.C. Jain (eds.): *Self-Organizing Maps. Recent Advances and Applications*, pages 121-145. Springer, 2001
24. T. Villmann. Benefits and limits of the self-organizing map and its variants in the area of satellite remote sensoring processing. In *Proc. Of European Symposium on Artificial Neural Networks (ESANN'99)*, pages 111–116, Brussels, Belgium, 1999. D facto publications.
25. T. Villmann, R. Der, M. Herrmann, and T. Martinetz. Topology Preservation in Self–Organizing Feature Maps: Exact Definition and Measurement. *IEEE Transactions on Neural Networks*, 8(2):256–266, 1997.
26. W. Wienholt. *Entwurf Neuronaler Netze*. Verlag Harri Deutsch, Frankfurt/M., Germany, 1996.

# An Overview on Unsupervised Learning from Data Mining Perspective

Lei Xu *

Department of Computer Science and Engineering, Chinese University of Hong Kong, Shatin, NT, Hong Kong, P.R. China, E-mail: lxu@cse.cuhk.edu.hk

**Abstract.** A number of unsupervised learning studies have been summarized from the perspectives of (a) mining dependence among components (PCA and MCA, LMSER and nonlinear PCA, ICA and independent factor analyses, three layer net and hidden unit number, etc); (b) mining homogeneous or nonhomogeneous groups among samples (MSE clustering and VQ, RPCL learning and elliptic extensions, Gaussian mixture and modified EM algorithms, etc); (c) mining dependences within local groups (localized extensions of part (a) as well as modular supervised learning such as RBF nets, mixture-of-experts and kernel on support vectors); and (d) mining topological structures with organized groups ( self-organization map and others).

## 1 Data Mining and Statistical Learning

In recent years, knowledge database discovery (KDD) becomes an increasingly popular area. Generally, it stands for discovering and visualizing the regularities, structures and rules from data. Its core part [22, 28, 8] consists of dimension reduction, structure mining and visualization, which is, referred as *data mining* (DM) in a broad sense, actually a revisit from a more engineering perspective of those statistical learning studies in the literature of neural networks and pattern recognition [108, 20, 19, 21].

The so called regularities, structures and rules are actually different expressions of various types of dependences among data. We can categorize these types of dependences in five levels based on the complexities of data in different media. Learning from data is actually made through samples of data. Usually, a sample is in the expression of a vector $x = [x^{(1)}, \cdots, x^{(d)}]^T$ with each component $x^{(i)}$ called feature that is either a real number or a symbol. Thus, the basic level of dependences is among features and can be represented in a set of either mathematical functions or logical rules. The next level of dependences is demonstrated via the locations of samples vectors. That is, under such dependences, samples located closely are regarded as same or similar and thus form a group or cluster, such that each sample in a cluster can be represented by either any one sample in this cluster or a representative of this cluster (e.g., by the cluster

* The work described in this paper was fully supported by a grant from the Research Grant Council of the Hong Kong SAR (project No: CUHK4383/99E).

center) [35]. In contrast, samples that locates far away are different and belong to different clusters. In other words, this level of dependence participates samples into a number of equivalence class that is called groups or clusters. Again, such a type of dependences can be represented via either mathematical functions or logical rules. The third level of dependence is indicated via the topological links among clusters, which is usually demonstrated via the topological links of centers of clusters [73, 42]. Moreover, the next level of dependences is temporal dependences among data that come from speech, audio or other time series [30]. Finally, for data from image and video, we have the 5th level of dependences that represents spatial relations of samples. The spatial relations can be in either the form of regular 2D or 3D lattice or of the more general graphic topology such as a probabilistic graphical structure or called Bayesian or Belief networks [68, 36].

A specific learning task may focus on learning or mining one or a combination of the five levels of dependences, and thus we can have a number of learning tasks in different complexities. In this paper, we focus on the first three levels of dependences, which lead to five meaningful combinations that are referred as learning tasks. Shortly we denote them as Task 1, Task 2, Task 1-2, Task 1-3, Task 1-2-3, where Task 1 means mining the first level of dependences and Task 1-2 means mining a combination that consists of the first level and second level of dependences. Specifically, we explore these tasks from statistical perspective. That is, data is regarded to consist of independent and identically distributed (i.i.d.) samples from a underlying unknown distribution $p(x)$ with each sample being a vector $x = [x^{(1)}, \cdots, x^{(d)}]^T$, and each of the tasks actually involves estimating either the distribution $p(x)$ or its certain statistics.

In the sequel, we provide an overview on several typical statistical learning models and algorithms that implement the five learning tasks. There are already certain survey papers in the literature. Instead of providing a complete overview on all the existing studies in the literature, we focus those studies which have not been well discussed yet, and particularly on those studies that can be viewed from a unified perspective. Moreover, we also emphasize the issues of regularization and model selection for the well known problem of a small size of samples.

## 2 Task 1: Mining Dependence among Components

### 2.1 Typical Dependence Structures

We consider three categories of typical dependence structures as follows.

*(1) Forward Mapping* Dependence among components are represented via mapping $x$ forwardly to another vector $y = [y^{(1)}, \cdots, y^{(m)}]^T$. Specifically, this forward mapping $f(x, \theta) = y$ can be further classified in three types according to the role of $y$ in representation, namely, whether the role of $y$ is minimized or maximized:

• *Pattern Matching* The idealistic case that the role of $y$ is minimized happens at $f(x, \theta) = y = 0$ for all the samples of $x$. It means that these samples match the pattern described by $f(x,\theta)=0$ with a specific value of $\theta$, where $f(x,\theta)=0$ is a set of either mathematical functions or logical rules. Usually, $y\neq 0$ and is regarded as random error or residual under this matching. It is natural to assume that the residual $y$ should be unbiased $Ey=0$, and that no dependence relation remains among components, e.g., the components become uncorrelated $Eyy^T = \Lambda = diag[\lambda_1,\cdots,\lambda_m]$ in a sense of independence up to the 2nd order statistics, where $Eu$ denotes the mean of $u$ and $diag[\lambda_1,\cdots,\lambda_m]$ denotes a diagonal matrix that consists of diagonal elements $\lambda_1,\cdots,\lambda_m$. In other words, we can model the residual $y$ by a Gaussian $q(y|\theta_y)=G(y|0,\Lambda)$, where and in the sequel, $G(u|m,\Sigma)$ denotes a Gaussian density with mean $m$ and covariance matrix $\Sigma$. Generally, we consider that the components of residual $y$ are completely independent:

$$q(y|\theta_y)=\prod_{j=1}^{m} q(y^{(j)}|\theta_y^{(j)}). \tag{1}$$

To minimize the role of $y$, we also expect that the variance of each $y^{(j)}$ is minimized. To make this desire meaningful, we usually need certain constraint on $\theta$, shortly denoted by $S(\theta)$. E.g., when $f(x,\theta)$ is linear as given by

$$f(x,\theta)=\theta g(x), g(x)=\begin{cases}[x,1]^T, & \text{(a)},\\ \text{a nonlinear map of } x, & \text{(b)},\end{cases} \tag{2}$$

if we have $Ey=0$ and $Eyy^T=\Lambda$ diagonal for $\theta g(x)=y$, then for any diagonal $D$ and orthogonal $\phi$ with $\phi^T\phi=I$, we still have that

$$Ey'=0,\ Ey'y'^T=D^2\ for\ y'=D\phi\Lambda^{-0.5}y. \tag{3}$$

We can remove this indeterminacy by imposing the constraint $\theta^T\theta=I$ as $S(\theta)$. Then, we can learn $\theta$ via

$$\min_{\theta,\ s.t.\ \theta^T\theta=I}\frac{1}{N}\sum_{t=1}^{N}\|y_t\|^2,\ y_t=\theta g(x_t), \tag{4}$$

or equivalently we can regard the residual $y_t$ being Gaussian and implement the following ML learning

$$\max_{\theta,\Lambda,\ s.t.\ \theta^T\theta=I}\frac{1}{N}\sum_{t=1}^{N}\ln G(y_t|0,\Lambda),\ y_t=\theta g(x_t). \tag{5}$$

Generally, for eq.(1) we further have

$$\max_{\{\theta,\theta_y,\ s.t.\ S(\theta)\}}\frac{1}{N}\sum_{t=1}^{N}\ln q(y_t|\theta_y),\ y_t=f(x_t,\theta). \tag{6}$$

• *Maximum Information Transfer* In contrast, to maximize the role of $y$ means to make $f(x,\theta)$ able to transfer maximum information from $x$. Typically, we expect to maximize the variance of every components of $y$. However, it may also not meaningful, if there is no some extra constraint $S(\theta)$. E.g., when $f(x,\theta)=\theta g(x_t)$, each variance tends to infinity for any $c\theta$ as $|c|\to\infty$. This problem can be removed again by the constraint $\theta^T\theta = I$. Without losing any generality, we can assume $Ey = 0$ (otherwise we can simply let $f(x,\theta) - Ey$ to be a new forward mapping). In this case, we have

$$\max_{\theta,\ s.t.\ \theta^T\theta=I} Tr[\Sigma_y] \ or \ \max_{\theta,\ s.t.\ \theta^T\theta=I} |\Sigma_y|,$$
$$\Sigma_y = \frac{1}{N}\sum_{t=1}^{N} y_t y_t^T, \quad y_t = \theta g(x_t). \tag{7}$$

Generally, we can maximize the information transferred from $x$ via $f(x,\theta)$ by

$$\max_{\{\theta,\theta_y,\ s.t.\ S(\theta)\}} [-\int p(y)\ln p(y)dy], \ p(y) = \int p(y|x)p_0(x)dx, \tag{8}$$
$$y = f(x,\theta) \ or \ p(y|x) = \delta(y - f(x,\theta)), \ p_0(x) = \frac{1}{N}\sum_{t=1}^{N}\delta(x - x_t),$$
$$E.g., \quad S(\theta) \ is \ \theta^T\theta = I, \ for \ y = \theta g(x).$$

Particularly, for the linear mapping

$$y = \theta x + m_y, \quad \theta^T\theta = I, \tag{9}$$

when $x$ is Gaussian, then $y$ is also Gaussian. In this case, we have that eq.(8) degenerates back to eq.(7). Alternatively, changing 'max' into 'min' in eq.(5) will also become equivalent to eq.(7).

In contrast to the case that the role of $y$ is minimized as residuals such that the dependence structure among components are directly represented in $f(x,\theta) = 0$, it looks not so directly how $f(x,\theta) = y \neq 0$ explores the dependence structures when the role of $y$ is maximized. This issue can be understood in two different situations. On one hand, when the mapping by $f(x,\theta) = y$ is information preserving with all the information of $x$ transferred to $y$, there is no need to drive $f(x,\theta)$ to explore the structure of samples of $x$ for this information transfer. A typical example is eq.(9) with $|\theta| = I$. In this case, we simply have $Tr[E(\theta xx^T\theta^T)] = Tr[E(xx^T)]$, $|E(\theta xx^T\theta^T)| = |E(xx^T)|$, and even $-\int p(y)\ln p(y)dy = -\int p(x)\ln p(x)dx$. Thus, all the information is preserved for any $\theta$ with $|\theta| = I$, irrelevant to the structure of samples of $x$. $|\theta| = I$. On the other hand, when $f(x,\theta) = y$ can only implement an information loss mapping, in order to maximize the information transferred $f(x,\theta) = y$ must explore the major structure of $x$ such that the major information part is picked to map. In this case, eq.(7) or eq.(8) will make $y$ become not only independent in components, and but also carry the major information of $x$.

• *Equalization* In the above two mapping types, $f(x,\theta) = y$ acts in two aspects, namely making $y$ become decorrelated or independent in components, and making the role of $y$ either minimized or maximized. Alternatively, we also have a third choice that we only consider the first aspect. In this case, we replace the constraint $S(\theta)$ such as $\theta^T\theta = I$ in eq.(9) by imposing the constraint that every component of $y$ is distributed under a same density, e.g., we impose $Eyy^T = I$ in eq.(9). We call this 3rd type of mapping *equalization*, as it can be regarded as an extension of histogram equalization used in image processing. Clearly, to reach such an equalization, $f(x, \theta) = y$ must be able to explore the major structure of $x$ unless $x$ has no structure among components.

*(2) Backward Mapping* In contrast to a forward mapping, we can also consider a backward mapping $g(y, \phi) = x$ which represents how $y$ is mapped backward to generate $x$. Usually, a Gaussian noise $e$ is taken into consideration, and thus it becomes $g(y, \phi) + e = x$. More generally, this mapping is represented by

$$q(x|\theta) = \int q(x|y, \theta_{x|y}) q(y|\theta_y) dy. \tag{10}$$

E.g., $g(y, \phi) + e = x$ under Gaussian $e$ gives $q(x|y, \theta_{x|y}) = G(x|g(y, \phi), \Sigma_e)$ where the covariance matrix $\Sigma_e$ is either simply $\sigma_e^2 I$ or diagonal. Moreover, $q(x|y)$ can be an other density when $e$ is nonGaussian. Also $q(y|\theta_y)$ is involved in eq.(10).

The key point for learning the parameters of $\theta_{x|y}, \theta_y$ is to make $q(x|\theta)$ fit a given set of samples of $x$ under a learning principle. The typical example is the ML learning

$$\max_{\{\theta_{x|y}, \theta_y, \ s.t. \ S(\theta)\}} \frac{1}{N} \sum_{t=1}^{N} \ln p(x_t|\theta), \tag{11}$$

where $S(\theta)$ is again certain constrain to be impose on $\theta$ to avoid certain indeterminacy that depends what type of $q(y|\theta_y)$ is used. In this paper, we consider the following three types of $q(y|\theta_y)$ that trades off the strengths of $q(x|y, \theta_{x|y}), q(y|\theta_y)$ in modeling the dependence structure of samples of $x$:

• *Homogenous Independent Factors* The simplest type is each component, or called factor, of $y$ being a standard Gaussian, i.e., $G(y|0, I)$, while $q(x|y, \theta_{x|y})$ takes all the responsibility to describe the dependence structure. As to be discussed in Sec.2.4, the disadvantage of the simplest type is leading to certain crucial indeterminacy. Another type is that each component density $q(y^{(j)}|\theta_y)$ in eq.(1) is a same non-Gaussian density, which provides a complementary help to $q(x|y, \theta_{x|y})$ in modeling the dependence structure of data.

• *Nonhomogenous Independent Factors* More generally, $q(y^{(j)}|\theta_y)$ may be different for different $j$ such that $q(y|\theta_y)$ takes more share on modeling the dependence structure of data. E.g., we consider $q(y^{(j)}|\theta_y)$ that is same to every $j$ only in its first two order statistics but different in higher order statistics, as given by an expansion

$$q(y^{(j)}|\theta_y^{(j)}) = G(y^{(j)}|0, 1)[1 + k_3(y^{(j)}) h_3(y^{(j)})/6 + k_4(y^{(j)}) h_4(y^{(j)})/24], \tag{12}$$

where $k_3, k_4$ are the cumulants of $y^{(j)}$, and $h_3, h_4$ are Hermite polynormails. Another example is that each $q(y^{(j)}|\theta_y)$ in eq.(1) is given by a finite mixture as in [88].

• *Gaussian Mixture Factor* Instead of the independent factor model eq.(1), another interesting type of $q(y|\theta_y)$ is either a mixture of gaussians $G(y, \mu_j, I)$ with different location $\mu_j$ or a nonhomogenous Gaussian mixture with each Gaussian with a different covariance matrix.

*(3) Forward Mapping and Backward Mapping* We can combine the features of both the forward mapping and backward mapping. There are two ways for such a combination. One is to make learning on each separately and then put them together. The other is to consider the both mappings coordinately during parameter learning. The details will be further discussed later.

Finally, we add that either $f(x, \theta)$ or $g(y, \phi)$ may also be trained via supervised learning algorithms when a set of paired samples $\{x_t, y_t\}$ is known. Supervised learning tasks are encountered in pattern recognition, prediction, function approximation, discriminant analysis and many others. A number of survey papers on these issues are available in the literature. E.g., an early systematical overview is referred to [108]. This paper will focus on unsupervised learning. However, by considering $\xi_t = \{x_t, y_t\}$ as an augmented sample vector, a supervised learning task can also be regarded as a unsupervised learning task on $\xi_t$ under the structural constraint between its part $x_t$ and its part $y_t$. For this reason, this overview also includes in appropriate cases certain results on supervised learning. Meanwhile, we only need to consider how to get $f(x, \theta)$ based on a training set of $\{x_t, y_t\}$ since the task of getting $g(y, \phi)$ is similar.

## 2.2 Forward Mapping (I): Maximum Information Transfer

• *PCA and Subspace Analyses* Studies on this topic can be traced back as early as [32]. In past decades, Principal Component Analysis (PCA), obtained directly from eq.(7) with $y = Wx$, has been widely studied and used in the field of statistics and many other fields. Its close relation to the well known Hebbian learning in neuroscience was first established in [62] where a simple linear neuron trained by an adaptive modified Hebbian learning rule is shown to perform exactly PCA. Since then, extensive studies have been made to develop adaptive algorithms on linear neural nets of multiple units for PCA. The studies consist of two main streams. One is for adaptive algorithms on asymmetrical architecture that extract the first k-PCA vectors sequentially one by one, with an initial study made in [74] and several sequential studies referred to [45]. The other stream consists of adaptive algorithms on symmetrical architectures that extract the first k-PCA vectors in parallel. Oja subspace rule [63] is one initial effort, which as well as several related efforts, however, do not perform k-PCA as discussed in [103], but a "collective version" of PCA, called *Principal Subspace Analysis (PSA)* that extracts a subspace spanned by the k-PCA vectors.

Two years after [63], two further results have been obtained by the present author in [103]. One is that a global convergence proof on the Oja subspace rule was provided. The other is an adaptive algorithm is given for extracting the m PCA vectors instead of only performing PSA. Moreover, adaptive PCA algorithms for robust performance and for working on missing data have also been proposed [94, 99]. Furthermore, in the past decade, extensive studies have also been made on showing that PCA or PSA can be equivalently implemented by various different learning rules and architectures. A number of such criteria and the corresponding adaptive learning rules are summarized in 1994 [100, 101]. For examples, PCA is performed by either of the following two criteria

$$(a)\ \max_W E\|y^T \Lambda^{-1} y\|^2,\ (b)\ \max_W E\|x - W^t y\|^2,\ y = Wx,\ s.t.\ WW^t = I, \tag{13}$$

where $\Lambda = diag[\lambda_1, \cdots, \lambda_m]$, $\lambda_1 > \cdots > \lambda_m$ is pre-given. The case (a) is called *Maximum Variation of Normalized Output (MVNO)* and the case (b) is called *Least Mean Square Error Reconstruction (LMSER)*. We can get adaptive learning to implement them as shown in [101].

Typical PCA applications relate to compressing and representing as well as visualizing data in a reduced dimension. Also, PCA is used in [106] for retrieving complicated structures via an approximate but fast implementation of attributed graph matching. Moreover, as a dual to [107], the direction of the first principal vector provides a total least square fitting of data by a line. Also, we can approximately fit a curve by transferring it to a line, e.g., $g(x)$ in eq.(2) is set by the same trick in [107].

• *Nonlinear PCA and ICA* PCA has two main features, namely the mapping is linear $y = Wx$ and the components of $y$ are uncorrelated (i.e., independent only in the 2nd order statistics). Extensions have been made on changing each of the features. Under the name of *nonlinear PCA*, the efforts on extending $y = Wx$ into the post-linear $y = S(Wx)$ with $S(y) \doteq [s(y(1)), \cdots, s(y^{(m)})]$, in a sense that each component of linear mapping $Wx$ is followed by a nonlinear scaling by a scalar function (e.g., a sigmoid function). Two typical efforts were made in 1991 from different perspectives. One is to use $y = s(Wx)$ in Oja rule and it turned out that the learning becomes very robust under noise and outliers [64]. The other is made in [103] (published firstly in Proc. IJCNN91), that uses $y = s(Wx)$ in the above $E\|x - W^t y\|^2$ in eq.(13) with an adaptive gradient algorithm provided in its Eq.(9b), which is rewritten as follows:

$$\min_W E\|x - W^t S(Wx)\|^2,\ W^{new} = W^{old} + \eta[S(Wx)e^T + S'(Wx)\varepsilon x^T], \tag{14}$$
$$e = x - W^t S(Wx),\ \varepsilon = y - WW^t S(Wx), S'(y) = diag[s'(y(1)), \cdots, s'(y^{(m)})],$$

where $s'(r) = ds(r)/dr$. Moreover, it has been firstly discovered experimentally that the sigmoid nonlinearity $s(r)$ can automatically break the symmetry in the PSA subspace such that components of $W$ tend to directions that are close to but not equal to the first m orthogonal PCA directions. Also, the other PCA criterion $E\|y^T \Lambda^{-1} y\|^2$ in eq.(13) is also extended into the so called the Nonlinear Maximum Variance (NMV) learning rule [96] by using the post-linear unit $y = s(Wx)$.

Started from Jutten and Herault in 1988 [39], under the name *Independent Component Analysis (ICA)*, efforts have been made on targeting on making the components of $y = Wx$ independent beyond the 2nd order statistics. Due to this clear motivation, it has become a popular topic in the past decade.

Advances on this topic can be roughly classified into four streams. One is based on solving the equation group on the higher order moments of $x$ and of $y$ subject to $y = Wx$ and the independent components of $y$. The second is to minimize a so called contrast function such that the minimum is reached when $y$ becomes independent, with the minimization implemented by adaptive algorithms. The third one is the above mentioned nonlinear PCA. Actually, three years after its publication [103], eq.(15) was directly adopted to implement ICA by authors of [40] with some promising results, which experimentally sets up a link between ICA and nonlinear PCA by eq.(15). The last one is called information theoretic approach, which is further discussed in the sequel.

Typical ICA applications include blind source separation, feature extraction, medical information processing, and many others. Extensive publications on ICA are also available in literatures, e.g., readers are referred to a special issue [27] and a survey paper [33]. Here we only further provide certain advances on the information theoretic approach, roughly summarized into three stages. The first stage is featured by the INFORMAX [5] and the MMI [2]. The two typical algorithms as well as several others update $W$ adaptively after the marginal density of each components of $y$ is heuristically prefixed [5] or estimated via density expansion [2]. These algorithms work well on the cases that the components of $y$ are either all sub-Gaussians or all super-Gaussians, but fail on the cases that the components of $y$ consist of partly sub-Gaussians and partly super-Gaussians. At the second stage, it is realized [91] that the marginal density $q(y^{(j)})$ of each component should also be learned simultaneously during learning on $W$ to automatically fit any combination of super-Gaussian or sub-Gaussian components of $y$. This idea could be implemented by learning the parameters $\theta_y$ in eq.(1). An early effort is the learning parametric mixture based ICA [91,88], where a finite mixture is used as $q(y^{(j)}|\theta_y^{(j)})$ and $\theta_y^{(j)}$ is updated by an EM-like algorithm during updating $W$. Alternatively, efforts in [65,15] estimate the kurtosis of $q(y^{(j)})$. At the third stage, extensions have been made towards various cases, e.g., (a) the dimension of $x$ is larger than that of $y$ instead of that $W$ is invertible [89, 88,86], (b) some specific nonlinear ICA [78,86], (c) the so called temporal ICA that takes temporal relation among samples in consideration [82] and (d) the so called competitive ICA, which will be further discussed in Sec.4.1.

• *Supervised learning: discriminant analysis and three layer net* Forward mapping can also be built via supervised learning on a training set of paired samples $\{x_t, y_t\}$. A typical example of linear mapping $y = Wx$ is given by the Fisher discriminant analysis that maps $x$ into $y$ for a best classification purpose [23, 21].

The key advantage of a forward mapping via supervised learning is able to explore the complicated nonlinear mapping among given samples in pairs $\{x_t, y_t\}$, in help of various forward networks. Readers are referred to [108] for an overview. Here, we only introduce certain new results obtained in recent years from Bayesian Yang-Ying learning [80,82,83] on the following three layer net for mapping $x \to \xi \to y$:

(a) $x \to \xi$ by a post-linear map $s(Ax+a_0)$ and $\xi$ from a multivariate Bernoulli $q(\xi|x) = \prod_{j=1}^m s(z^{(j)})^{\xi^{(j)}}[1-s(z^{(j)})]^{1-\xi^{(j)}}, z = Ax + a_0, s(r) = 1/(1+e^{-r})$;

(b) $\xi \to y$ is made by linear mapping $B\xi + b_0$ subject to a Gaussian noise with covariance $\sigma^2 I$, i.e., $q(y|\xi) = G(y|B\xi + b_0, \sigma^2 I)$.

Thus, the mapping $x \to y$ is given by $q(y|x) = \sum_\xi G(y|B\xi + b_0, \sigma^2)q(\xi|x)$. By the linear Taylor expansion of $G(y|B\xi + b_0, \sigma^2)$ around the mean $E\xi = s(Ax+a_0)$, we approximately have $q(y|x) \approx G(y|Bs(Ax+a_0)+b_0, \sigma^2)$, and thus $E(y|x) = \int yq(y|x)dy = Bs(Ax + a_0) + b_0$ implements the conventional three layer net with sigmoid hidden units.

Instead of using the back-propagation technique to train this three layer net in the sense of least square learning, we make its training via the so called BYY harmony learning [81], which provides the following EM-like algorithm:

$$
\begin{aligned}
&E\ step: \ \hat{\xi} = arg \max_{\xi}\{G(y|B\xi + b_0, \sigma^2 I)\prod_{j=1}^{m} s(z^{(j)})^{\xi^{(j)}}[1-s(z^{(j)})]^{1-\xi^{(j)}}\},\\
&M\ step: \ (a)\ e_y = y - B^{old}\hat{\xi} - b_0^{old},\ B^{new} = B^{old} + \eta e_y \hat{\xi}^T,\\
&\qquad b_0^{new} = b_0^{old} + \eta e_y,\ \sigma^{2\ new} = (1-\eta)\sigma^{2\ old} + \eta \|e_y\|^2, \qquad (15)\\
&\qquad (b)\ e_\xi = \hat{\xi} - s(Ax + a_0),\ A^{new} = A^{old} + \eta e_\xi x^T,\ a_0^{new} = a_0^{old} + \eta e_\xi.
\end{aligned}
$$

Specifically, E-step is equivalent to find $\hat{\xi}$ that maximizes $C(\xi) = -0.5\sigma^{-2}\|y - B\xi - b_0\|^2 + \sum_{j=1}^m[\xi^{(j)} \ln s(z^{(j)}) + (1-\xi^{(j)}) \ln(1 - s(z^{(j)})),$ which is a typical discrete quadratic programming problem. For a fast approximation, we can first solve the linear equation $\nabla_\xi C(\xi) = 0$ by regarding $\xi$ being real and then hard-cutting the solution into a binary one. That is,

$$
\hat{\xi}^{(j)} = \begin{cases} 1, & \text{if } \bar{\xi}^{(j)} > 0.5, \\ 0, & \text{otherwise}, \end{cases} \quad \bar{\xi} = (B^TB)^{-1}[B^T(y-b_0) + \sigma^2\pi],
$$
$$
\pi = [\pi^{(1)}, \cdots, \hat{\pi}^{(m)}]^T,\ \pi^{(j)} = \ln \frac{s(z^{(j)})}{1-s(z^{(j)})}. \qquad (16)
$$

Another issue is to set up the step size $\eta$ in the M-step. It can be two choices:

(a) It is fixed at a small constant $\eta > 0$ when the learning eq.(15) is made in the so called *empirical learning* [81]. Moreover, its specific value can be different for different parameters. E.g., two different sizes $\eta_A, \eta_B$ are used for updating $A, B$, respectively.

(b) When the learning eq.(15) is made in the so called *normalization learning* [81], it is given by $\eta = \eta_t\eta_0$ with $\eta_0 > 0$ and $\eta_t$ given by

$$
\eta_t = \frac{1}{N} - \gamma_0 \frac{\gamma_t}{S_q},\ \gamma_t = G(y|B\hat{\xi}_t + b_0, \sigma^2 I)\prod_{j=1}^{m} s(z^{(j)}(t))^{\hat{\xi}_t^{(j)}}[1-s(z^{(j)}(t))]^{1-\hat{\xi}_t^{(j)}},
$$
$$
S_q = \sum_{t=1}^N G(y|B\hat{\xi}_t + b_0, \sigma^2 I)s(z^{(j)}(t))^{\hat{\xi}_t^{(j)}}[1-s(z^{(j)}(t))]^{1-\hat{\xi}_t^{(j)}}, \qquad (17)
$$

where $1 > \gamma_0 > 0$ is a given constant that compensates the finite sample size in normalization by $\sum_{t=1}^N q(y_t|\hat{\xi}_t)q(\hat{\xi}_t|x_t)$. By this $\eta_t$, after the winner-take-all competition by the E-step, a de-learning is introduced to regularize the learning on the winner for each sample in proportional to the current fitting of the model to the sample. However, it is expensive to compute on all the samples as in eq.(17). We can also approximate the sum $S_q$ adaptively by $S_q(t+1) = (1-\lambda)S_q(t) + \lambda G(y|B\hat{\xi}_t + b_0, \sigma^2 I)\prod_{j=1}^m s(z_t^{(j)})^{\hat{\xi}_t^{(j)}}[1-s(z_t^{(j)})]^{1-\hat{\xi}_t^{(j)}}$ for a suitable $0 < \lambda < 1$. Then, as $t$ varies, we have

$$\eta_t = \frac{1}{t} - \gamma_0 \frac{\gamma_t}{S_q(t+1)}. \tag{18}$$

The detail derivations of eq.(15), eq.(16) and eq.(17) are further referred to [81], from which we also known that eq.(15) is actually implement a so called harmony learning that will push $p(\xi|x)$ into a least complexity form to avoid using extra hidden units. Alternatively, by enumerating a number of $m$ values incrementally, we can also select a best number $m^*$ for hidden units by the following criterion

$$\begin{aligned} &m^* = arg \min J(m),\ J(m) = 0.5m\ln\sigma^2 + J_y(m),\ \hat{\xi} = Ax + a_0, \\ &J_y(m) = -\tfrac{1}{N}\textstyle\sum_{t=1}^N \sum_{j=1}^m \{\hat{\xi}_t^{(j)} \ln s(z_t^{(j)}) + (1-\hat{\xi}_t^{(j)})\ln(1-s(z_t^{(j)}))\}. \end{aligned} \tag{19}$$

## 2.3 Forward Mapping (II): Pattern Matching and Equalization

• *Pattern Matching: MCA-MSA and Surface Fitting* Playing a dual role to PCA, eq.(4) results in the so called minor component analysis (MCA) that minimizes the residuals $y$ such that the samples are fitted in the total least square sense by either the hyperplane with the 1st minor component as its normal direction or the subspace that is orthogonal to the subspace spanned by the minor components. That is, using a $m$-dimensional MSA subspace to represent a data set is equivalent to use a $d-m$ dimensional PSA subspace to represent to represent the same set, where $n$ is the dimension of $x$. Particularly, instead of using $d-1$ principal components to represent a hyperplane, we can equivalently represent it by only the 1st minor component. In [109], MCA or MSA is used together with PCA or PSA to form a dual representation with a more effective dimension reduction. In 1992, an adaptive learning algorithm has been proposed to implement MCA for the total least square fitting of not only lines, planes, and hyperplanes but also circles, curves, planes, surfaces, and hypersurfaces [107]. Moreover, MCA is also used in [87] for the so called co-integration regularity in time series. Furthermore, studies has also been extended to object identification via fitting a general function $g(y, \phi)$ [95].

• *Equalization: ICA, P-ICA and M-ICA* Though the concept of ICA was proposed in parallel to PCA, the existing studies on ICA are made without distinguishing the concepts 'minor' or 'principal'. There may be two reasons. One is that the original purpose of ICA is to recover $x = Ay$ by $y = Wx$ and such a recovery is indeterminacy on the scales of each component of $y$. The other is that $W$ is usually invertible and $y = Wx$ becomes independent but take all the components in consideration, where to distinguish which components are 'minor' or 'principal' is not necessary.

Strictly speaking, the concept of ICA is parallel to de-correlation component analysis (DCA) $y = Wx$ that makes $Eyy^T$ be diagonal, including both PCA and MCA as well as linear equalization by $y = Wx$ with $Eyy^T = I$ as special cases. Thus, when the dimension $m$ of $y$ is lower than the dimension $d$ of $x$, we should also have the principal ICA (P-ICA), minor ICA (M-ICA), and nonlinear equalization. The last one has already discussed in Sec.2.1. The difference of the principal ICA and minor ICA can be understand from the perspective of the so called pre-whitening ICA. After making the covariance of samples of $x$ be orthogonal, we will have $C_d^m$ combinations to select $m$ de-correlated components and then normalize them white. Thus, ICA can be made in two steps. First, we select a combination of $m$ components and then normalize them white. Second, we do the invertible ICA in the $m$ dimension space. That is, we have $C_d^m$ combinations of ICA. Among them, we have either the P-ICA when the $m$ principal components are used to whitening or the M-ICA when the $m$ minor components are used to whitening. This way of defining the P-ICA and the M-ICA depends on the whitening preprocessing. In Sec.2.5, we will give another way to define them without relying on a preprocessing.

## 2.4 Backward Mapping: Three Typical Independent FA

• *Gaussian FA and Independent FA* The typical example of the early efforts on this topic is factor analysis, which can be traced back to the beginning of the 20th century by Spearman [77]. Formulated by Anderson and Rubin in 1956 [3], it considers the simplest linear special case of eq.(8), that is

$$x = Ay + e, \quad e \text{ is independent from } y, \tag{20}$$

where both $e, y$ come from Gaussians with $E(e) = 0, E(y) = 0, E(y_t y_t^T) = I$, and $e$ is uncorrelated among its components with a diagonal covariance matrix $\Sigma_e$. However, this model suffers the problem of not having a unique solution because its indeterminacy on rotation and on the communality estimation [55]. Early studies towards such problems consists of either constraining $A$ to be orthogonal matrix only or imposing heuristics to select a specific rotation such as in Quartimax and Varimax [55].

In the past decade, efforts have been made on considering eq.(20) with the independence assumption on the components of $y$. For clarity, we refer this new type of factor analysis (FA) as *independent FA (IFA)* to avoid being confused with the original one, which should now be more precisely referred as *De-correlating FA (DFA)*. Similar to ICA, when each $q(y^{(j)})$ is nonGaussian or at most only one of them is Gaussian, the rotation indeterminacy can be removed. However, it is much more difficult to implement IFA than ICA. Not only we need to deal with the problem of modeling each component density $q(y^{(j)})$ as in ICA, but also we need to handle the noise $e$. Due to this noise, the ML learning by eq.(11) encounters the computational difficulty on handling the integral over $y$ in eq.(8). Several efforts have been made towards to solving this difficulty.

The most simple way is to approximately regard $e = 0$ such that estimating $A$ in $x = Ay$ becomes equivalent to ICA that gets the inverse mapping $y = Wx$. But it works only in a small noise $e$ case. The other ways include preprocessing for filtering noise, making ML learning via a Monte-Carlo sampling [86], and using heuristic structures [15,27]. Readers are referred to a survey paper [33].

In the sequel, we further add on certain advances obtained from BYY harmony learning in recent years [80,82,83].

• *Bernoulli FA, Independent FA, and BYY harmony learning* We consider eq.(20) with eq.(1). In help of the BYY harmony learning [80], we can get the following EM-like algorithm:

$$\begin{aligned} &E\ step: \ \hat{y} = arg \max_y [G(x|Ay, \Sigma_e) q(y|\theta_y)], \\ &M\ step: \ e = x - A^{old}\hat{y}, A^{new} = A^{old} + \eta e \hat{y}^T, \ \Sigma_e^{new} = (1-\eta)\Sigma_e^{old} + \eta e e^T, \\ &\qquad \theta_y^{new} = \theta_y^{old} + \eta\phi(\hat{y}), \ \phi(y) = \nabla_y \ln q(y|\theta_y). \end{aligned} \tag{21}$$

Specifically, E-step is equivalent to find $\hat{y}$ that maximizes $C(y) = -0.5(x - Ay)^T \Sigma_e^{-1}(x - Ay) + \sum_{j=1}^k \ln q(y^{(j)}|\theta_y^{(j)})$, which can be made by a fast approximation that solves the equation $\nabla_y C(y) = 0$ as shown in [80]. For illustration, we provide two examples:

$$\begin{aligned} &(a)\ for\ \ q(y|\theta_y) = G(y|0, I), \ \ \hat{y} = [I + A^T \Sigma_e^{-1} A]^{-1} A^T \Sigma_e^{-1} x, \\ &(b)\ for\ \ q(y|\theta_y) = \prod_{j=1}^k q_j^{y^{(j)}} (1-q_j)^{1-y^{(j)}}, \ \ \hat{y}^{(j)} = \begin{cases} 1, & \text{if } \hat{y}^{(j)} > 0.5, \\ 0, & \text{otherwise,} \end{cases} \\ &\qquad \hat{y} = (A^T \Sigma_e^{-1} A)^{-1} \{A^T \Sigma^{-1} x + [\pi_1, \cdots, \pi_k]^T\}, \pi_j = \ln \frac{q_j}{1-q_j}. \end{aligned} \tag{22}$$

With the case (a) in eq.(21), there is no need on updating $\theta_y$, and eq.(21) actually is an adaptive algorithm for implementing DFA. For the case (b), the updating on $\theta_y$ is simply given by

$$q_j = 1/(1 + e^{c_j}), \ \ c_j^{new} = c_j^{old} + \eta(\hat{y}^{(j)} - q_j^{old}). \tag{23}$$

In this case, eq.(21) is an adaptive algorithm for implementing a Bernoulli FA. Also a variant of Bernoulli FA is given in [80] for the case that $x$ is also binary from a multivariate Bernoulli. In the literature of neural networks, other efforts have been also made on modeling binary $x$ (e.g., representing a binary image) by interpreting it as generated from binary hidden factor $y$ with mutually independent bits. Typical examples include multiple cause models [75,16] and Helmholtz machine [17,31].

Similar to eq.(15), another issue in eq.(21) is the step size $\eta$ in the M-step. Again, it can be either fixed at constants for implementing *empirical learning* or given by $\eta = \eta_t \eta_0$ with

$$\eta_t = \frac{1}{N} - \gamma_0 \frac{\gamma_t}{S_q}, \ \ \gamma_t = G(x_t|A\hat{y}_t, \Sigma_e) q_j^{\hat{y}_t^{(j)}} (1-q_j)^{1-\hat{y}_t^{(j)}},$$

$$S_q = \sum_{t=1}^{N} G(x_t|A\hat{y}_t \Sigma_e) q_j^{\hat{y}_t^{(j)}} (1-q_j)^{1-\hat{y}_t^{(j)}}, \tag{24}$$

for implementing *normalization learning* [80]. Still, we can use eq.(18) by adaptively updating $S_q(t+1) = (1-\lambda)S_q(t) + G(x_t|A\hat{y}_t, \Sigma_e) q_j^{\hat{y}_t^{(j)}} (1-q_j)^{1-\hat{y}_t^{(j)}}$. Again, similar to eq.(15), the harmony learning of eq.(22) will push $q(y|\theta_y)$ into a least complexity form to avoid using a redundant dimension [80].

• *Principal subspace dimension* From BYY harmony learning [80, 82, 83], we can also get the following criterion for selecting a best dimension $m^*$:

$$m^* = arg\min J(m),\ J(m) = 0.5\ln|\Sigma_e| + J_y(m),$$
$$J_y(m) = \begin{cases} m\ln(2\pi) + m, & \text{(a)} \\ -\frac{1}{N}\sum_{t=1}^{N}\sum_{j=1}^{m}[\hat{y}_t^{(j)}\ln q_j + (1-\hat{y}_t^{(j)})\ln(1-q_j)], & \text{(b)} \\ -\frac{1}{N}\sum_{t=1}^{N}\sum_{j=1}^{m}\ln q(\hat{y}_t^{(j)}|\theta_y^{(j)}), & \text{(c),} \end{cases}$$
$$(a)\ q(y|\theta_y) = G(y|0, I),\ \ (b)\ q(y|\theta_y) = \prod_{j=1}^{m} q_j^{y^{(j)}}(1-q_j)^{1-y^{(j)}},$$
$$(c)\ q(y|\theta_y)\ in\ eq.(1), \tag{25}$$

where the value $J(m)$ is obtained after parameter learning at each $m$ as we enumerate a number of $m$ values incrementally. Specifically, the case (b) determines the number of binary bits that is required for $y$, while the case (a) and case (c) describe the dimension of the subspace dimension spanned by principal components in the sense of the 2nd order independence and of the higher order independence, respectively. E.g., we can use the case (a) together with PCA for determining an appropriate number $m$ of principal components, and we can use the case (c) together with ICA for determining the number of blind sources.

• *Temporal FA and Higher order HMM* Both the DFA and Bernoulli FA have been further extended to taking temporal relation among samples in consideration [80, 82] via adding a state equation $y_t = By_{t-1} + \varepsilon_t$ with $B$ being a diagonal and $\varepsilon_t$ is a white noise. Specifically, DFA is extended into the so called temporal FA (TFA). Interestingly, as shown in [80, 82], the rotation indeterminacy of DFA has been removed due to temporal relation. While Bernoulli FA is extended into the so called independent hidden Markov model. Moreover, the temporal extensions have also been applied to perform generalized APT financial analyses.

## 2.5 Bi-directional Mapping: LMSER-ICA and Helmholtz machine

A backward mapping focuses on how $x$ is generated from $y$ such that not only noise is taken in consideration but also which components of $y$ are principal or minor can be evaluated according to their roles in the reconstruction and its matching to the observed data samples. As a result, it makes the problem of selecting an appropriate $m$ and eq.(25) meaningful. However, the disadvantage of a backward mapping is that it is expensive and also inconvenient, based on the learning results of a backward mapping, to perform the mapping $x \to y$. The disadvantage can be remedied by a bi-directional architecture that combines both a forward mapping and a backward mapping. Actually, the nonlinear LMSER in eq.(15) is a simple example. It combines a forward mapping $y = S(Wx)$ and a backward mapping $x = W^t y$, both are considered during the learning on $W$. As discussed in Sec.2.2, it performs ICA. Moreover, this ICA has a feature that components can be assessed as being principal or minor by evaluating how good its backward mapping $x = W^t y$ matches the observed data samples, similar to PCA. Thus, the ICA performed by this LMSER is more appropriately to be regarded as a P-ICA. Generally, the reconstruction error $J(W) = E||x - W^t S(Wx)||^2$ or even its linear version $J(W) = E||x - W^t(WW^t)^{-1}Wx||^2$ can be used to measure $y = WA$ in implementing ICA, e.g., it is P-ICA when the minimum of $J(W)$ is reached or it is M-ICA when the maximum $J(W)$ is reached.

Several extensions of the LMSER learning have been obtained in help of Bayesian Yang-Ying harmony learning [81]. Here we introduce one example, which is equivalent to minimize

$$J(A, W, \{q_r\}_{r=1}^m) = d \ln \sigma + \frac{1}{N} \sum_{t=1}^N \sum_{r=1}^m [y_t^{(r)} \ln q_r + (1 - q_r) \ln (1 - y_t^{(r)})],$$
$$y_t = Wx_t, \quad \sigma^2 = \frac{1}{N} \sum_{t=1}^N ||x_t - AS(y_t)||^2. \tag{26}$$

This minimization can be implemented by an adaptive EM-like algorithm

$$\begin{aligned} E\ step:\ & y_t = W^{old} x_t, \ \hat{y}_t = arg \min_{y_t} [0.5\sigma^{-2} ||x_t - AS(y_t)||^2 \\ & - \textstyle\sum_{r=1}^m [y_t^{(r)} \ln q_r + (1 - y_t^{(r)}) \ln (1 - q_r)], \\ M\ step:\ & q_r = \frac{1}{1 + e^{c_r}}, \ c_r^{new} = c_r^{old} + \eta(\hat{y}^{(r)} - q_r^{old}), \ e_t = x_t - A^{old} S(\hat{y}_t), \\ & A^{new} = A^{old} + \eta e_t S^T(\hat{y}_t), \ \sigma^{2\ new} = (1-\eta)\sigma^{2\ old} + \eta d^{-1} ||e_t||^2 \\ & W^{new} = W^{old} + \eta S_d(\hat{y}_t) e_t x_t^T, \ S_d(y) = diag[s'(y(1)), \cdots, s'(y^{(m)})], \end{aligned} \tag{27}$$

Again, similar to eq.(15), this adaptive harmony learning will push those redundant $q_j$ towards zero such that model selection is automatically made during learning. Alternatively, we also have the model selection criterion

$$\min_m J(m), J(m) = 0.5 d \ln \sigma^2 + \textstyle\sum_{j=1}^m [q_r \ln q_r + (1 - q_r) \ln (1 - q_r)]. \tag{28}$$

Furthermore, it has been shown in [80, 82, 83] that this LMSER-ICA eq.(15) is a special case of the Bayesian Yang-Ying independence learning, and from which we also get other extensions of the nonlinear LMSER that not only relax $x = W^t y$ to $x = Ay$ without the constrain $A = W^T$ but also take several possible distributions of $y$ in consideration. Particularly, one special case is equivalent to the one layer deterministic Helmholtz machine learning [31, 16, 17]. Furthermore, extensions have also been made towards to temporal situations, resulting in temporal LMSER [80].

## 3 Task 2: Mining Groups among Samples

Mining groups or called clustering are made among samples such that samples within a group are regarded as same or similar while samples in different groups are regarded being different. More specifically, we can classify the studies on the tasks into two categories:

(a) *Homogeneous grouping* Samples are grouped based on a criterion of similarity or distance $d(x_i, x_j)$ that is homogeneous to any sample pair $x_i, x_j$. A typical example is the Euclidean distance, as further discussed in Sec.3.1.

(b) *Nonhomogeneous grouping* Samples are grouped under a measure that is not homogeneous to every sample, but relates to the specific structure of each group, as further discussed in Sec.3.2.

In particular, when $x$ has a high dimension, which is usually the case in many real applications and especially in multimedia processing, samples are grouped based on each group's specific structure in a subspace of much lowered dimension instead of in the full original space of $x$. We call such particular cases *Subspace structure based nonhomogeneous grouping*, which will be discussed in Sec.4.

### 3.1 MSE-VQ Clustering and RPCL Learning

• *MSE Clustering, VQ and KMEAN algorithm* Extensive studies have been made in literature of statistics and pattern recognition for several decades under the name of clustering analysis [4, 21, 35, 19]. The most widely used homogeneous measure is the Euclidean distance $d(x_i, x_j) = ||x_i - x_j||^2$. The grouping tasks in this situation is equivalent to use a number of vectors to represent a data set such that each vector locates at the center of each group or cluster. The existing algorithms for the purpose can be classified into two types. One is usually called incremental/hierarchical/dynamic clustering [21, 19, 35] or competitive learning [29]. The key point is incrementally adding one cluster center once a newly coming sample is regarded to be far beyond a threshold. This type is easy to implement and the number of clusters is decided dynamically. However, the performances highly depend on the initialization and the specific way that those clusters grows up.

The other type of clustering algorithms considers all the possible cluster centers in parallel via minimizing the mean square distances or a global measure on all the samples. However, the complexity of finding the global minimum grows exponentially with the number of clusters, and thus the problem is usually tackled by a heuristic algorithm that usually produces a solution at a local minimum. A typical example is the KMEAN algorithm [21, 19] and variants, which is also called Vector Quantization (VQ) in literature of speech and image processing [51, 60]. Such a task is equivalently tackled under the name of competitive learning in the literature of neural networks [1, 107]. The KMEAN algorithm has also been adopted and modified in the literature of data mining as one of most popular tools for compressing, categorizing, and organizing data, with emphasis on scaleable ability for a large database [22]. Readers are referred to these cited textbooks, survey papers and references, particularly to a recent survey paper [45] on multimedia data processing.

In this paper, we emphasize certain essential issues in real applications and especially in multimedia data processing. Specifically, two issues will introduce below and several other issues in Sec.3.2.

• *Deciding the number of clusters* The KMEAN algorithm and others work well only when a correct number $k$ of clusters are pre-given. We can get a very poor performance with a wrong data structure found if we do not know this number and thus set $k$ inappropriately [104]. Moreover, on a training set of samples of $x$, though using a large $k$ may apparently result in a smaller mean square error, it usually generalizes badly on new samples from data, especially in a changing environment. A possible solution to the problem is to choose a best $k^*$ by a selection criterion. Many heuristic criteria have been proposed in the statistic literature [24, 57, 58, 76, 72]. Recently, based the Bayesian Ying-Yang learning, a simple criterion has been also obtained in companion with the KMEAN algorithm as follow [90]:

$$\min_k J(k), J(k) = \ln k + 0.5d \ln E^2_{MSE}, \tag{29}$$

where $d$ is the dimension of $x$ and $E^2_{MSE}$ is the mean square error.

However, any selection criterion suffers a large computational cost since we need to make clustering at a number of different value of $k$, even though such a process can be organized in a more efficient way, e.g., embedding the evaluation of the selection measure during clustering as did in ISODATA[4]. Alternatively, proposed firstly in 1992 [104], the so called rival penalized competitive learning (RPCL) solves this problem with the correct number $k^*$ determined automatically during learning for an initial $k$ that is large enough, in the sense that extra units are driven far away from data due to appropriately penalized learning on the rival. Later, RPCL has been adopted to various applications, including information retrieval in image databases[48, 41, 46], Plant diagnosis[26], nonlinear financial prediction, and hidden Markov model [13, 12], clustering, vector quautization, object classification in 3-D images, scene segmentation in 2D&3D image as well as multidimensional data [9, 49, 14, 54]. Also, following the initial suggestion in [104] for training RBF net, a number of authors have used or recommended RPCL algorithm for the training of various RBF nets [6, 10, 112, 11, 47, 9]. Subsequently, we will further introduce various RPCL extensions to elliptic clustering and subspace structure based nonhomogeneous grouping as well as its relation to the BYY harmony learning.

• *Fast implementation in a binary tree* In data mining on a large database with data of high dimension, a clustering algorithm that can be fast implemented is preferred. Also, the clusters should be well indexed and thus can be retrieved conveniently. A so called hierarchical PCA [97, 96] for vector quantization provides a solution for such demands. By this technique, a binary clustering tree is formed by recursively splitting a set $D_c$ of samples associated with the current node into two subsets that associate two son nodes such that samples of each part locate on each side of a hyperplane that passes the mean of $D_c$ and is perpendicular to the principal component direction of $D_c$. After each split, the node associated with $D_c$ is marked CLOSED. Moreover, we can compute the value of $J(k)$ by the above eq.(29) and check whether $J(k)$ turns to increase. If not, the two sons are marked OPEN. Otherwise, we stop and discard the two sons. Next we go to another OPEN node in either the depth-first or the breadth-first way [67]. The root node of the tree is associated with the entire data set of samples. The tree grows as such until $J(k)$ turns to increase on all the OPEN nodes. Such a technique of forming a data tree can be used for fast indexing and retrieving in multimedia data processing.

## 3.2 Gaussian Mixture, EM Variants and Eliptic RPCL learning

• *Gaussian Mixture and EM Algorithm* The above discussed algorithms apply to homogeneous data with each being spherical Gaussian and sharing a same or similar portion of samples. Studies in the literatures have also been made on extending the KMEAN algorithm and competitive learning algorithms to clusters of the so called elliptic shapes. Most of these studies can be related to the special cases of the ML learning on Gaussian mixture

$$p(x_t|\theta) = \sum_{j=1}^{k} \alpha_j G(x|m_j, \Sigma_j) \tag{30}$$

in help of the EM algorithm [18, 70, 56] with good convergence properties [92]. E.g., in [90], a simplified EM algorithm on Gaussian mixture is shown to extend the KMEAN algorithm to elliptic clustering. Actually, density estimation by a Gaussian mixture covers the various tasks of clustering with each group represented in a Guassian $G(x|m_j, \Sigma_j)$. It not only provides more accurate clustering results but also sophisticated data structure via $\Sigma_j$. As a popular topic, Gaussian mixture with the EM algorithms has been both extensively studied and widely used in many fields. The readers are referred to [45] for its applications in multimedia data processing and to [18, 70, 56] in a broad scope. Here, we add on several results that improve the generalization ability of learning.

• *Three Variants of The EM algorithm* From BYY learning [90, 83, 81], we get as follows:

(a) *Re-weighted EM* We can replace the likelihood $N^{-1}\sum_t \ln p(x_t|\theta)$ by the general convex likelihood $N^{-1}\sum_t f(p(x_t|\theta))$ with a convex function $f'(r) > 0, f''(r) < 0, r > 0$. We maximize this likelihood on a Gaussian mixture $p(x_t|\theta)$ by a so called re-weighted EM algorithm in [90] since a re-weighting factor is attached to each sample, which was shown empirically to be more robust than the ML learning via the original EM algorithm, especially when $f(r) = r^\beta, \beta < 1$.

(b) *Smoothed EM* The performance of the ML learning will degenerate considerably on a set of finite number of high dimensional samples. To solve the problem, a so called smoothed ML learning is proposed, which replaces the likelihood function $N^{-1}\sum_t \ln p(x_t|\theta)$ with an integral $\int p_h(x) \ln p(x_t|\theta) dx$ and $p_h(x)$ given by a Parzen window estimator [83, 81]. Moreover, we are lead to a modified EM algorithm that simply modifies the original EM algorithm at its M-step with its updating on each $\Sigma_j$ added a smoothing parameter $h$ to its diagonal elements. Furthermore, after each iteration of the E-step and M-step, we can also update the parameter $h$ via a simple one dimensional search [83, 81].

(c) *De-learning EM* Another special case of BYY harmony learning on Gaussian mixture, also taking the effect of finite number of high dimensional samples in consideration, avoids the smoothing parameter $h$ with its role replaced by a de-learning in the M-step.

To illustrate, we provide a unified adaptive EM-like procedure that covers all the above three algorithms:

$$\begin{aligned} &E\ Step: j_c = arg\max_j [\ln G(x|m_j, \Sigma_j) + \ln \alpha_j], \\ &M\ Step: m_{j_c}^{new} = m_{j_c}^{old} + \eta(x - m_{j_c}^{old}), \\ &\Sigma_{j_c}^{new} = (1-\eta)\Sigma_{j_c}^{old} + \eta[hI + (x - m_{j_c}^{old})(x - m_{j_c}^{old})^T]. \end{aligned} \quad (31)$$

Specifically, it implements the smooth EM with a constant step size $\eta > 0$, together with $h > 0$ updated via a simple one dimensional search [83, 81]. Moreover, when $h = 0$, according to different settings of $\eta > 0$, it acts as:

(1) An elliptic adaptive KMEAN algorithm for a constant step size $\eta > 0$;

(2) An adaptive re-weighted EM algorithm as in [90] for $\eta = f'(p(x_t|\theta))\, p(x_t|\theta)\eta_0 > 0$ with $\eta_0$ being a constant;

(3) An adaptive de-learning EM algorithm as in [81] for $\eta = \eta_0\eta_t > 0$, where $\eta_t$ is in the form of eq.(24) but with $\gamma_t = p(x_t|\theta)$ and $S_q = \sum_t p(x_t|\theta)$. Again, $S_q$ can be approximated by adaptively updating $S_q(t+1) = (1-\lambda)S_q(t) + p(x_t|\theta)$.

• *Selection of Gaussians* Similar to the homogeneous clustering, how to decide the number $k$ of Gaussian is an essential issue for a good performance of Gaussian mixture. Again, one solution is to choose a best $k^*$ by a selection criterion. In [90], such a criterion is obtained as follow:

$$\min_k J(k), J(k) = 0.5\textstyle\sum_{j=1}^k \alpha_j \ln|\Sigma_j| - \sum_{j=1}^k \alpha_j \ln \alpha_j. \quad (32)$$

• *Elliptic RPCL learning and BYY harmony Learning* Also, the correct number $k^*$ of Gaussians can be determined automatically during learning in help of extending RPCL learning [104] to the cases of any elliptic shapes and in any portion of samples [95, 85, 81]. As a result, we have the following elliptic RPCL algorithm:

$$\begin{aligned} &Step\ 1: j_c = arg\max_j d_j(x),\ j_r = arg\max_{j \neq j_c} d_j(x), d_j(x) = -\ln[G(x|m_j, \Sigma_j)\alpha_j], \\ &Step\ 2: m_{j_c}^{new} = m_{j_c}^{old} + \eta_c(x - m_{j_c}^{old}),\ m_{j_r}^{new} = m_{j_r}^{old} - \eta_r(x - m_{j_r}^{old}), \\ &\qquad S_{j_c}^{new} = S_{j_c}^{old} + \eta_c \Delta S_{j_c},\ S_{j_r}^{new} = S_{j_r}^{old} - \eta_r \Delta S_{j_r}, \end{aligned} \quad (33)$$

where the learning rate $\eta_c$ is much smaller than the de-learning rate $\eta_r$, e.g., $8 \le \eta_c/\eta_r \le 15$. Also, we indirectly compute $\Sigma_j = S_j S_j^T$ via updating $S_j$ in order to guarantee that $\Sigma_{j_c} = S_{j_c} S_{j_c}^T$ remains semi-positive definitive. In eq.(33), $\Delta S_j$ is the gradient direction given as follows:

$$\Delta S_j = \nabla_{S_j} \ln G(x|m_j, \Sigma_j) = \{\Sigma_j^{-1}[hI + (x - m_j)(x - m_j)^T]\Sigma_j^{-1} - \Sigma_j^{-1}\}S_j.$$

In implementing, we can always keep $S_j$ in storage or get it at each updating by decomposing $\Sigma_j$ that is always keep $S_j$ in storage. In the latter case, the updating on $S_{j_c}$ can also be replaced by $\Sigma_{j_c}^{new} = (1-\eta_c)\Sigma_{j_c}^{old} + \eta_c[hI + (x - m_{j_c}^{old})(x - m_{j_c}^{old})^T]$, and eq.(33) degenerates back to eq.(31) when $\eta_r = 0$. Moreover, we can also let $\eta_r$ to be different for $m_j$ and for $\Sigma_j$, even when $\eta_r = 0$ for $\Sigma_j$ but $\eta_r > 0$ for $m_j$.

The role of $h$ is same as above discussed. When $h > 0$, we get the smoothed RPCL learning. Also, we can let $\eta_c = f'(p(x_t|\theta))p(x_t|\theta)\eta_{0,c} > 0$ and $\eta_r = f'(p(x_t|\theta))p(x_t|\theta)\eta_{0,r} > 0$ to get the robust feature of the Re-weighted EM algorithm. Though, RPCL is originally proposed heuristically [104], it has been shown that it is qualitatively equivalent to a special case of the general RPCL learning algorithm obtained from the BYY harmony learning [82, 83] and thus get a guide for determining the learning rate $\eta_c$ and de-learning rate $\eta_r$.

# 4 Task 3: Mining Dependences within Local Groups

We can get the dependence structure among components locally on each cluster in data. One way to do so is anyone of the algorithms in Sec.3.2 for nonhomogeneous grouping to get every covariance matrix $\Sigma_j$ and then get local dependence structures based on $\Sigma_j$. However, since $\Sigma_j$ contains only the 2nd order statistics, based $\Sigma_j$ we can not implement local ICA or find a local nonlinear dependence structure. Moreover, even theoretically we can get a linear dependence structure based on $\Sigma_j$, not only it wastes many computing costs on getting $\Sigma_j$, but also it may result in a bad estimate when the dimension $d$ of $x$ is high, because each $\Sigma_j$ contains $d(d+1)/2$ parameters to be specified, which needs a large number of samples to avoid the resulted $\Sigma_j$ to be singular.

A better alternative is to make a nonhomogeneous clustering based on mining local dependence structures that are locally within much lower dimensional subspaces. Specifically, we can get the local extensions of those algorithms in Sec.2.

## 4.1 Local PCA, Competitive ICA and Modular Models

• *Local PCA and Local PSA* In [109], PCA and PSA are used for local subspace representation of data in pattern recognition. In [100, 101, 96], local PCA is also used for fitting a number of lines and hyperplanes.

Provide that $m_j$ is the center of the $j$-th cluster, the diagonal elements of the diagonal matrix $\Lambda_j$ are the $d_j$ largest eigen-values, and the $d_j$ row vectors of $W_j$ are the corresponding eigen-vectors, we define the following subspace based distance

$$d_j(x) = [W_j(x - m_j)]^T \Lambda_j^{-1} W_j(x - m_j), \tag{34}$$

and then use eq.(33) for implementing local PCA with

$$\Lambda_j = S_j S_j^T, \; \Delta S_j = S_j^{-1} diag[(hI + W_j(x - m_j)(x - m_j)^T W_j^T) - \Lambda_j], \tag{35}$$

where $S_j$ is diagonal and $diag[A]$ means a diagonal matrix that takes the diagonal part of $A$. Moreover, at each location we update as follows

$$Step\ 3: \quad W_{j_c}^{new} = W_{j_c}^{old} + \eta_c \Delta W_{j_c}^{pca}, \; W_{j_r}^{new} = W_{j_r}^{old} + \eta_r \Delta W_{j_r}^{mca}, \tag{36}$$

where $\Delta W_{j_c}^{pca}$ can use one existing stable adaptive PCA learning rule on $x'_{j_c} = x - m_{j_c}$ with a linear net $W_{j_c}^{pca} x'_{j_c}$, and $\Delta W_{j_r}^{mca}$ can use one existing stable adaptive MCA learning rule on $x'_{j_r} = x - m_{j_r}$ with a linear net $W_{j_r}^{mca} x'_{j_r}$. Particularly, we can get local PSA as a special case by simply setting $\Lambda_j = \lambda$.

• *Competitive ICA* Instead of exploring local de-correlation structures by Local PCA, we can also explore local independent structures via a so called competitive ICA algorithm:

$$Step\ \ 1: j_* = arg \max_j [0.5 \ln |W_j W_j^T| + \ln q(W_j x + \mu_j | \theta_y)], \tag{37}$$

$$\begin{aligned} Step\ \ 2: W_{j.}^{new} &= W_{j.}^{old} + \eta [I + \phi(y^*)(W_{j.}^{old} x_t)^T] W_{j.}^{old}, \\ \phi(y) &= \nabla_y \ln q(y|\theta_y), \; y^* = W_{j.}^{old} x + \mu_{j.}^{old}, \\ \mu_{j.}^{new} &= \mu_{j.}^{old} + \eta \phi(\hat{y}_t), \; \theta_y^{new} = \theta_y^{old} + \eta \nabla_{\theta_y} \ln q(W_j x + \mu_j | \theta_y). \end{aligned}$$

For a fixed $j$, Step 2 is the same as the learning parametric mixture based ICA [88, 86], where a finite mixture is used as $q(y^{(j)}|\theta_y^{(j)})$, with $\theta_y$ updated by an EM-like algorithm during updating $W_j$. Moreover, we can also generalize it to competitive temporal ICA for handling temporal situation [80].

• *Modular Supervised Learning: RBF net, mixture-of-experts, and support vectors* Local dependence structures can also be built via supervised learning on modular models such as the radial basis function (RBF) [59, 61, 102], nonparametric kernel regression [20, 102], the mixture-of-expert (ME) models [34, 37, 38], and support vector machine [79]. Similar to the previously discussed unsupervised learning examples, these supervised modular models also build dependence structures among components based on local properties of data. E.g., the conventional learning on RBF nets is made usually in two sequential steps. The first step decides the centers of basis functions usually via certain clustering algorithm, and the second step determines the parameters of the output layers by the least square learning. Such a two-step algorithm actually provides a suboptimal solution. Extensive literatures are available on these supervised learning models. For a more detailed introduction, readers are referred to [108] for an early survey and to [81] for a recent discussion on the relation between these models.

Here we summarize several results on training these models by either adaptive EM-like algorithms or RPCL related algorithms:

(a) The mixture-of-expert (ME) model [34, 37, 38] implements forward mapping by a number of local experts that are engaged in via a probabilistic controlling of a so called gating net, with each individual expert being a three layer net. Moreover, an alternative ME model is further proposed [98, 69, 84] such that learning can be made completely by the EM algorithm in the case that each expert is described by a Gaussian with a linear regression, while the training on the gating net of the original ME is trained by a gradient-based algorithm but not by the EM algorithm.

(b) The normalized RBF nets and the extended normalized RBF nets are shown in [84] to be regarded as special cases of the alternative mixture-of-expert (ME) model, and thus can be trained by ML learning by the EM algorithm, instead of the conventional two-step method. Moreover, the hard-cut EM algorithm and adaptive EM-like algorithms have been proposed for fast learning on not only these RBF nets but also both the original and alternative ME models in help of a so called coordinated competition [84].

(c) In [82,81], all the above studies are be related to the Bayesian Ying-Yang harmony learning as special cases. As a result, their learning algorithms can be replaced by their corresponding RPCL-type learning algorithms that perform parameter learning with automated model selection on experts or basis functions. Also, criteria are obtained in a way similar to eq.(19).

(d) In [102], nonparametric kernel regression [20] is shown to be a special case of the normalized RBF nets such that several previous results on kernel regression can be brought to provide certain understandings on the normalized RBF nets. Recently in [81], such a link is revisited from the perspective of using the above discussed learning algorithms on generalizing kernel regression technique, resulting in not only an easily implemented approach for determining the smoothing parameter in kernel regression, but also an alternative approach to select supporting vectors in the popular supporting vector machines for a better generalization.

### 4.2 Local MCA-MSA and Curve Detection

As a dual to local PCA-PSA discussed in Sec.4.1, local MCA-MSA can be used for local subspace representation of data [109], for fitting curve, hyperplane and hypersurface [107] and detecting a number of curves, hyperplanes and hypersurfaces at different locations [104,100,101,96].

Moreover, as a dual to the RPCL algorithm in Sec.4.1, local MCA-MSA can also be implemented by RPCL learning. Specifically, we can replace eq.(34) by

$$d_j(x) = \|(I - W_j^T W_j)(x - m_j)\|^2, \tag{38}$$

and then use eq.(33) with eq.(35) for learning. The difference is that eq.(36) is replaced by

$$Step\ 3: W_{j_c}^{new} = W_{j_c}^{old} + \eta_c \Delta W_{j_c}^{mca},\ W_{j_r}^{new} = W_{j_r}^{old} + \eta_r \Delta W_{j_r}^{pca}, \tag{39}$$

where the positions of using a PCA rule and a MCA rule are swapped.

Taking curve detection as an example, this technique provides an alternative to the Hough transform-like technique on detecting curves on image in noisy environment [111,105]. Such tasks may be implemented in two ways. One is to use the trick in [107] to transfer a curve into a form such that the above local MCA can be used directly. E.g., for detecting a quadratic curve such as circles or ellipses, we consider the equation $a_j x^2 + b_j xy + c_j y^2 + d_j x + e_j y +$

$f_j = 0$ and rewrite it into $w_j^T(x - m_j) = 0$ with $x = [x^2, xy, y^2, x, y]^T$ and $w_j = [a_j, b_j, c_j, d_j, e_j]^T$, and then. Therefore, we can perform a local MCA by the above algorithm at a special case that each $W_j$ consists of only one vector of $w_j$. After learning, we turn each resulted $w_j$ into parameters of curve with $f_j = -w_j^T m_j$.

Another way is to define $d(x, \theta_j)$ as the shortest distance from $x$ to the $j$-th curve represented by $\theta_j$, and then use RPCL learning as follows

$$\begin{aligned} &Step\ 1: j_c = arg \max_j d(x, \theta_j),\ j_r = arg \max_{j \neq j_c} d(x, \theta_j), \\ &Step\ 2: \theta_{j_c}^{new} = \theta_{j_c}^{old} + \eta_c \nabla_{\theta_{j_c}} d_j(x, \theta_{j_c}),\ \theta_{j_r}^{new} = \theta_{j_r}^{old} - \eta_r \nabla_{\theta_{j_r}} d_j(x, \theta_{j_r}). \end{aligned} \tag{40}$$

More generally, in the so called Multi-sets modeling [95], $d(x, \theta_j)$ can be the shortest distance from $x$ to the $j$-th object described in a general set, and we use eq.(40) for learning.

### 4.3 Local Backward Mapping and Competitive LMSER

• *Backward mapping: Local DFA and Local Independent FA* The advantage of using a Gaussian mixture for mining groups and the advantage of using the factor model eq.(20) for mining the dependence structure can be combined. We consider the following two possibilities:

$$p(x_t|\theta) = \sum_{j=1}^{k} \alpha_j \begin{cases} \int G(x|A_j y + m_j, \Sigma_j) G(y|0, I) dy, & \text{(a)}, \\ \int G(x|Ay, \Sigma) G(y|\mu_j, \Lambda_j) dy, & \text{(b)}. \end{cases} \tag{41}$$

In the case (a), at each location $m_j$, each Gaussian is decomposed into a local DFA model

$$x = m_j + A_j y_j + e_j, \tag{42}$$

where $e_j$ is a Gaussian noise of zero mean and covariance matrix $\Sigma_j$. Thus, similar to a DFA previously discussed in Sec.2.4, there is still indeterminacy on rotation and scale at every location. This situation is removed by in the case (b) in eq.(41) where a Gaussian mixture factor $\sum_{j=1}^{k} \alpha_j G(y|\mu_j, \Pi_j)$ is mapped to $x$ via a common $x = Ay + e$ or equivalently each local DFA locates at $A\mu_j$. Due to the constraint of this common mapping, the indeterminacy on rotation at every location is removed except the singular case that the distribution is same at each location.

Moreover, the local structure based Gaussian mixture in eq.(41) can be further extended to nonGaussian finite mixture $p(x|\theta) = \sum_{j=1}^{k} \alpha_j p(x|\theta_j)$ by letting $G(y_j|0, I)$ or $G(y|\mu_j, \Lambda_j)$ replaced by the independent factor eq.(1), which leads to two corresponding local independent FA models.

Furthermore, we can combine eq.(31) and eq.(21) to get a double loop EM-like adaptive algorithm for implementing learning on local DFA and independent

FA. Taking the case (a) as an example, we have

$$
\begin{aligned}
&E\ \ Step: j_c = arg\max_j[\ln p(x|\theta_j) + \ln \alpha_j],\\
&\quad p(x|\theta_j) = \begin{cases} \int G(x|A_j y_j + m_j, \Sigma_j)G(y_j|0, I)dy_j, & \text{Gaussian},\\ \int G(x|A_j y_j + m_j, \Sigma_j)q(y|\theta_{y,j})dy, & q(y|\theta_{y,j}) \text{ given by eq.(1)}, \end{cases}\\
&M\ \ Step: \text{ implement the inner E step and M step in eq.(21) once, with}\\
&\quad A_{j_c}\ as\ A,\ \ x - m_{j_c}\ as\ x,\ \ \Sigma_{j_c}\ as\ \Sigma_e,\ \ \theta_{y,j}\ as\ \theta_y,\\
&\quad Then,\ update\ m_{j_c}^{new} = m_{j_c}^{old} + \eta(x - m_{j_c}^{old}).
\end{aligned} \tag{43}
$$

• *Bi-directional Mapping: Competitive LMSER* We can extend the LMSER learning eq.(26) to the local models at different $m_j$. Similar to eq.(43), we can get the following double loop EM-like adaptive algorithm for learning, with its E step in the outer loop implementing competition each time a sample comes:

$$
\begin{aligned}
&E\ \ Step: j_c = arg\max_j J(A_j, W_j, \{q_{j,r}\}),\ \ y_t = W_j(x_t - m_j),\\
&\quad J(A_j, W_j, \{q_{j,r}\}) = 0.5d\ln\sigma_j^2 + 0.5\sigma_j^{-2}||x_t - A_j S(y_t)||^2\\
&\quad -\textstyle\sum_{r=1}^{d_j}[y_t^{(r)}\ln q_{j,r} + (1 - q_{j,r})\ln(1 - y_t^{(r)})],\\
&M\ \ Step: \text{ implement the inner E step and M step in eq.(27) once, with}\\
&\quad A_{j_c}\ as\ A,\ \ W_{j_c}\ as\ W,\ \ x_t - m_{j_c}\ as\ x_t,\ \ \sigma_{j_c}^2\ as\ \sigma^2,\ \ q_{j,r}\ as\ q_r,\\
&\quad Then,\ update\ m_{j_c}^{new} = m_{j_c}^{old} + \eta(x - m_{j_c}^{old}).
\end{aligned} \tag{44}
$$

## 5 Tasks 4 & 5: Topologically Organized Groups and Local Dependence Structures

Another important data structure consists of topological relations among groups or clusters. Early efforts can also be traced back several decades. Roughly, these studies originated along two lines. One is for reducing high dimensional data into lower dimension such that topological relations among samples can be reserved for statistic data analysis and engineering purpose. Several heuristic techniques were proposed. Among them, a typical representative is called Samon mapping [73]. The other line is motivated by mathematical modeling biological striate cortex. A typical work is Malsburg's self-organization of orientation sensitive cells [52]. A breakthrough advance, that is able both to model self-organized formation of topological feature maps in biological system and to apply to data analysis with efficient computing, is the well known Kohonen map [43,42].

In the past decade, several efforts have been further made in the literature of neural networks for implementing Samon-type mapping in help of nonlinear architecture of neural networks [66,53,71]. Moreover, very extensive studies on topological map have been made on Kohonen map, which actually forms a major stream of unsupervised learning in the literature of neural networks, as shown by the main theme of this series of workshops. Readers are referred to papers in the workshops' proceedings and other vast volumes on this theme in the neural network literature.

Here, we only incompletely mention three lines of developments that relate to the local dependence mining methods discussed in Sec.4. One is extending the lattice map architecture to more sophisticated architectures. An early attempt is made in 1990 for training a number of Kohonen maps that are automatically organized in a pipe-line architecture during learning [110]. Further developments along this direction include Kohonen map tree [44] and the Growing Grid or gas [25]. The another line is to combine the feature of Kohonen map with Gaussian mixture such as given in [7]. Another line that deserves to mention is applying certain advanced mathematical results on deformation analyses to interpret and evolve self-organization map [50].

It may deserve to mention that few efforts has been made in the existing literature on combining the tasks of mining local dependence structures into the formation of topological structure yet, which should be a promising direction of developments of studies on self-organizing map.

## References


1. Ahalt, S.C., et al, "Competitive learning algorithms for vector quantization", *Neural Networks*, **3**, 277-291, 1990.
2. Amari, S.-I., Cichocki, A., & Yang, H.H., "A new learning algorithm for blind separation of sources", in D. S. Touretzky, et al, eds, *Advances in Neural Information Processing 8,* MIT Press, 757-763, 1996.
3. Anderson, T.W., & Rubin, H., "Statistical inference in factor analysis", *Proc. Berkeley Symp. Math. Statist. Prob. 3rd 5*, UC Berkeley, 111-150, 1956.
4. Ball, G.H., & Hall, D.J., "ISODATA: A novel method of data analysis and pattern classification', *Tech. Rep. No. AD 699616*, Stanford Research International, 1965.
5. Bell, A.J. & Sejnowski, T.J., "An information-maximization approach to blind separation and blind de-convolution", *Neural Computation 7*, 1129-1159, 1995.
6. Billings, S. A., & Zheng, G. L., "Radial basis function network configuration using genetic algorithms", *Neural Networks 8*, 877-890, 1995.
7. Bishop, C.M., Svensen, M. and Williams, C.K.I., " GTM: the Generative Topographic Mapping", *Neural Computation 10*, 215-234, 1998.
8. Bradley, P.S., et al, "Data mining: overview and optimization opportunities", *Microsoft Research Technical Report*, TR-98-04, 1998.
9. Bors, A. G. & Pitas, I., "Object classification in 3-D images using alpha-trimmed mean radial basis function network," *IEEE Trans. on Image Process 8*, 1744-1756, 1999.
10. Bors, A. G. & Pitas, I., "Median radial basis function neural network", *IEEE Trans. on Neural Networks 7*, 1351-1364, 1996.
11. Chang, P. R. & Yang, W. H., "Environment-adaptation mobile radio propagation prediction using radial basis function neural networks", *IEEE Trans. on Vehicular Technology 46*, 155-160, 1997.
12. Cheung, Y. M. & Xu, L., "A RPCL-based approach for Markov model identification with unknown state number," *IEEE Signal Processing Letters 7*, 284-287, 2000.
13. Cheung, Y. M., et al, "A RPCL-CLP architecture for financial time series forecasting," in *Proc. of 1995 IEEE ICNN 2*, 829-832, 1995.
14. Chiarantoni, E., et al, "Scene segmentation in video sequences by a RPCL neural network," *Proc. of 1998 IEEE WCCI 3*, 1877-1882, 1998.

15. Cichocki, A., Douglas, S.C., & Amari, S., " Robust techniques for independent component analysis (ICA) with noisy data", *Neurocomputing 22*, 113-129, 1998.
16. Dayan, P., & Zemel, R.S., "Competition and multiple cause models", *Neural Computation 7*, 565-579, 1995.
17. Dayan, P. & Hinton, G., E., "Varieties of Helmholtz machine", *Neural Networks 9*, 1385-1403, 1996.
18. Dempster, A.P., et al, "Maximum- likelihood from incomplete data via the EM algorithm", *J. of Royal Statistical Society B39*, 1-38, 1977.
19. Devijver, P. A. & Kittler, J., *Pattern Recognition: A Statistical Approach*, Prentice-Hall, 1982.
20. Devroye, L., et al, *A Probability Theory of Pattern Recognition*, Springer, 1996.
21. Duda, R.O., & Hart, P.E., *Pattern classification and Scene analysis*, Wiley, 1973.
22. Fayyad, U.M., et al " Knowledge discovery and data mining: towards a unifying framework", *Proc. 2nd Intl. Conf. KDD-96*, 1996.
23. Fisher, R.A., "The Use of multiple measurements in taxonomic problems", *Ann. Eugen 7*, 178-188, 1936.
24. Friedman, H.P., & Rubin, J., " On Some invariant criteria for grouping data", *J. Amer. Statis., Assoc. 62*, 1159-1178, 1967.
25. Fritzke, B, "Growing Grid- A self-organizing network with constant neighborhood range and adaptation strength", *Neural Processing Letters 2*, (5), 1995.
26. Furukawa, H., et al, "A systematic method for rational definition of plant diagnostic symptoms by self-organizing neural networks", *Neurocomputing 13*, 171-183, 1996.
27. Fyfe, C., ed., Special issue on *Independence and artificial neural networks*, Neurocomputing 22, No.1-3, 1998.
28. Glymour, C. et al, " Statistical Themes and lessons for data mining", *Data Mining and Knowledge Discovery 1* , 25-42, 1996.
29. Grossberg, S., "Competitive learning: from interactive activation to adaptive resonance", *Cognitive Science 11*, 23-63, 1987.
30. Hamilton, J.D., *Time Series Analysis*, Princeton University Press, New Jersy, 1994.
31. Hinton, G. E., et al, "The wake-sleep algorithm for unsupervised learning neural networks", *Science 268*, 1158-1160, 1995.
32. Hotelling. H, "Simplified calculation of principal components", *Psychometrika 1*, 27-35, 1936.
33. Hyvarinen, A. "Survey on Independent Component Analysis", *Neural Computing Surveys 2* , 94-128, 1999.
34. Jacobs, R.A., et al, "Adaptive mixtures of local experts", *Neural Computation 3*, 79-87, 1991.
35. Jain, A.K., & Dubes, R.C., *Algorithm for Clustering Data*, Prentice-Hall, 1988.
36. Jensen, F.V., *An introduction to Bayesian networks*, University of Collage London Press, 1996.
37. Jordan, M. I., & Jacobs, R.A., " Hierarchical mixtures of experts and the EM algorithm", *Neural Computation 6*, 181-214, 1994.
38. Jordan, M. I., & Xu, L., "Convergence results for the EM approach to mixtures of experts", *Neural Networks 8*, 1409-1431, 1995.
39. Jutten, C. & Herault, J., " Independent Component Analysis versus Principal Component Analysis", *Proc. European Signal Processing Conf EUSIPCO88*, 643-646, 1988.
40. Karhunen, J. & Joutsensalo, J., "Representation and separation of signals using nonlinear PCA type Learning," *Neural Networks 7*, 113-127, 1994.

41. King, I., et al, "Using rival penalized competitive clustering for feature indexing in Hong Kong's textile and fashion image database," *Proc. of 1998 IEEE WCCI 1*, 237-240, 1998.
42. Kohonen, T, *Self-Organizing Maps* , Springer-Verlag, Berlin, 1995.
43. Kohonen, T., "Self-organized formation of topologically correct feature maps", *Biological Cybernetics 43*, 59-69, 1982.
44. Koikalainen, P, " Progress with the Tree structure Self-Organizing Map", Proc. ECAI94: 11th European Conference on Artificial Intelligence, Ed. A.Cohn, John Wiley&Son, 211-215, 1994.
45. Kung, S. Y., & Hwang, J.N., "Neural Networks for Intelligent Multimedia Processing," *Proceedings of the IEEE, 86*, No. 6, 1244-1272, 1998.
46. Lau, T.K. & King, I., "Performance analysis of clustering algorithms for information retrieval in image databases," *Proc. of 1998 IEEE WCCI 2*, 932-937, 1998.
47. Lee, J., et al, "A practical radial basis function equalizer," *IEEE Trans. on Neural Networks 10*, 450-455, 1999.
48. Li, X. Q., & King, I., "Regression analysis for rival penalized competitive learning binary tree," *Proc. of IJCNN2000 6*, 290-295, 2000.
49. Li, R., et al, "Fast image vector quantization using a modified competitive learning neural network approach" *Intl J. of Imaging Systems and Technology 8*, 413-418, 1997.
50. Liou, C.Y. and Tai, W.P., "Conformality in the self-organization network", *Artificial Intelligence, 116*, 265-286, 2000.
51. Makhoul, J., Rpucos, S., & Gish, H., "Vector quantization in speech coding," *Proc. IEEE*, **73**, 1551-1558, 1985.
52. von der Malsburg, Ch. "Self-organization of orientation sensitive cells in the striate cortex", *Kybernetik 14*, 85-100, 1973.
53. Mao, J, and Jain, A.K., " Artificial neural networks for feature extraction and multivariate data projection", *IEEE Trans. Neural Networks 6*, No. 2, 296-317, 1995.
54. Marazzi, A., et al, "Automatic selection of the number of clusters in multidimensional data problems", *Proc. of Intl. Conf. on Image Processing 3*, 631-634, 1996.
55. McDonald, R, *Factor Analysis and Related Techniques*, Lawrence Erlbaum, 1985.
56. McLachlan, G.J., & Basford, K.E., *Mixture Models: Inference and Application to Clustering*, Dekker, 1988.
57. Millgan, G. W., " A monte carlo study of thirty internal criterion measures for cluster analysis", *PSYCHOMETRIKA 46*, 187-199, 1985.
58. Millgan, G. W., & Copper, M.C., " An examination of procedures for determining the number of clusters in a data set", *PSYCHOMETRIKA 50*, 159-179, 1985.
59. Moody, J. & Darken, J., "Fast learning in networks of locally-tuned processing units", *Neural Computation 1*, 281-294, 1989.
60. Nasrabadi, N. & King, R. A., "Image coding using vector quantization: a review," *IEEE Trans. Commun. 36*, 957-971, 1988.
61. Poggio, T., & Girosi, F., "Networks for approximation and learning", *Proc. of IEEE 78*, 1481-1497, 1990.
62. Oja, E., "A simplified neuron model as a principal component analyzer", *J. Mathematical Biology, 16*, 267-273, 1982.
63. Oja, E., "Neural networks, principal components, and subspaces", *Int. J. Neural Systems 1*, 61-68, 1989.
64. Oja, E., et al, "Learning in nonlinear constrained Hebbian networks" , *Proc. ICANN'91*, Helsinki, 385-390, 1991.

65. Oja E. & Hyvarinen,A., "Blind Signal Separation by Neural Networks", *Proc. ICONIP96*, Springer-Verlag, 7-14, 1996.
66. Ornes, G and Sklansky, J, " A neural network that visualizes what it classifies", *Pattern Recognition Letters 18*, No.11-13, 1307-1316, 1997.
67. Pearl J. *Heuristics: Intelligent Search Strategies for Computer Problem Solving*, Addison-Wesley, 1984.
68. Pearl, J, *Probabilistic reasoning in intelligent systems: networks of plausible inference*, San Fransisca, CA: Morgan Kaufman, 1988.
69. Ramamurti, V., & Ghosh, J., "Regularization and error bars for the mixture of experts network", *Proc. of IEEE ICNN97*, 221-225, 1997.
70. Redner, R.A., & Walker, H.F., "Mixture densities, maximum likelihood, and the EM algorithm", *SIAM Review 26*, 195-239, 1984.
71. de Ridder, D., and Duin, R.P.W., " Sammnon's mapping using neural networks: a comparision", *Pattern Recognition Letters 18*, No.11-13, 1301-1306, 1997.
72. Rissanen, J. et al, "Unsupervised classification with stochastic complexity in multivariate statistical modeling", *Proc. The 1st US/Japan Conf. on the Frontiers of Statistical Modeling: An In formational Approach*, 1994.
73. Samon, John W., Jr.. "A Nonlinear Mapping for Data Structure Analysis," *IEEE Trans. on Computers, Vol.C-18*, No.5, 1969.
74. Sanger, T.D., " Optimal unsupervised learning in a single-layer linear feed forward neural network", *Neural Networks 2*, 459-473, 1989.
75. Saund, E., "A multiple cause mixture model for unsupervised learning", *Neural Computation 7*, 51-71, 1995.
76. Scott, A.J., & Symons, M.J., "Clustering methods based on likelihood ratio criteria", *Biometrics 27*, 387-397, 1971.
77. Spearman, C., "General intelligence objectively determined and measured", *Am. J. Psychol. 15*, 201-293, 1904.
78. Taleb, A.. & Jutten, C., "Non-linearity source separation: the post-nonlinear mixtures", *Proc.ESANN97*, Bruges, April 16-18, 279-284 , 1997.
79. Vapnik, V.N., *The Nature Of Statistical Learning Theory*, Springer-Verlag, 1995.
80. Xu, L., " BYY Harmony Learning, Independent State Space and Generalized APT Financial Analyses ", in press, *IEEE Trans on Neural Networks*, 2001.
81. Xu, L., "Best Harmony, Unified RPCL and Automated Model Selection for Unsupervised and Supervised Learning on Gaussian Mixtures, Three-Layer Nets and ME-RBF-SVM Models", in press, a special issue on *International Journal of Neural Systems*, 2001. A part of its preliminary version on *Proc. IJCNN99, Vol.1: 540-545* and on *Proc. WCNN96: 193-200.*
82. Xu, L., "Temporal BYY Learning for State Space Approach, Hidden Markov Model and Blind Source Separation", *IEEE Trans on Signal Processing 48*, 2132-2144, 2000. A part of its preliminary version on *Proc. IJCNN99, Vol.2: 949-954* and on *Proc. ICONIP98-Kitakyushu, Vol.2:877-884.*
83. Xu, L., "BYY System and Theory for Statistical Learning: Best Harmony, Data Smoothing, and Model Selection", *Neural, Parallel and Scientific Computations 8*, 55-82, 2000. Its preliminary version on *Proc. 1999 Chinese Conf. on NNSP*, 12-29, 1999.
84. Xu, L., "RBF Nets, Mixture Experts, and Bayesian Ying-Yang Learning", *Neurocomputing 19*, No.1-3, 223-257, 1998.
85. Xu, L., "Rival penalized competitive learning, finite mixture, and multisets clustering," *Proc. of 1998 IEEE IJCNN 3*, 251-2530, 1988.
86. Xu, L., "Bayesian Kullback Ying-Yang Dependence Reduction Theory", *Neurocomputing 22*, No.1-3, 81-112, 1998.

87. Xu, L., and Leung, W.M., "Cointegration by MCA and modular MCA", *Proc. IEEE/IAFE 1998 Intl. Conf. on Computational Intelligence for Financial Engineering (CIFEr)*, March 29-31, NY City, 157-160, 1998.
88. Xu, L., Cheung, C.C., & Amari, S.-I., "Learned Parametric Mixture Based ICA Algorithm", *Neurocomputing 22*, No.1-3, 69-80, 1998. A part of its preliminary version on *Proc. ESANN97*, Bruges, April 16-18, 291-296, 1997.
89. Xu, L., "Bayesian Ying-Yang Learning Based ICA Models" , *Proc. 1997 IEEE Signal Processing Society Workshop*, Sept. 24-26, Florida, 476-485, 1997.
90. Xu, L., "Bayesian Ying-Yang Machine, Clustering and Number of Clusters", *Pattern Recognition Letters* **18**, *1167-1178, 1997.*
91. *Xu, L., Yang, H.H., & Amari,S.-I., "Signal Source Separation by Mixtures Accumulative Distribution Functions or Mixture of Bell-Shape Density Distribution Functions ", Research Proposal, presented at* FRONTIER FORUM (speakers: D. Sherrington, S. Tanaka, L.Xu & J. F. Cardoso), *organized by S.Amari, S.Tanaka & A.Cichocki, RIKEN, Japan, April 10, 1996.*
92. *Xu, L., & Jordan, M. I., "On Convergence Properties of the EM Algorithm for Gaussian Mixtures", Neural Computation 8, 129-151, 1996.*
93. *Xu, L., "A unified learning scheme: Bayesian-Kullback YING-YANG machine",* Advances in NIPS 8, *eds., D. S. Touretzky, et al, MIT Press, 1996, 444-450. A part of its preliminary version on* Proc. ICONIP'95, *1995, 977-988.*
94. *Xu, L., & Yuille, L., "Robust Principal Component Analysis by Self-Organizing Rules Based on Statistical Physics Approach",* IEEE Tr. Neural Networks *6, 131-143, 1995.*
95. *Xu, L., "A Unified Learning Framework: Multisets Modeling Learning", Proc. of 1995 World Congress on Neural Networks 1*, 35-42, 1995. A part of its preliminary version on *Proc. 1994 IEEE Intl. Conf. on Neural Networks 1*, 315-320, 1994.
96. Xu, L., "Advances on three streams of PCA studies" *Proc. 1995 IEEE Intl Conf. on NNSP'95*, 480-483, 1995.
97. Xu, L., "Vector Quantization by Local and Hierarchical LMSER", *Proc. 1995 Intl Conf. on Artificial Neural Networks II*, Paris, 575-579, 1995.
98. Xu, L., Jordan, M.I., & Hinton, G.E., "An Alternative Model for Mixtures of Experts", *Advances in Neural Information Processing Systems* 7, 633-640, 1995. Its preliminary version on *Proc. of WCNN'94 2*, 405-410, 1994.
99. Xu, Y. & Xu, L., " An Approach to Missing Data Principal Component Analysis", *Proc. ICONIP'95 II*, 1013-1016, 1995.
100. Xu, L., " Beyond PCA Learning: From Linear to Nonlinear and From Global Representation to Local Representation ", *Proc. ICONIP'94*, 943-949, 1994
101. Xu, L., " Theories for Unsupervised Learning: PCA and Its Nonlinear Extensions", *Proc. IEEE ICNN'94 II*, 1252-1257, 1994.
102. Xu, L., Krzyzak, A., & Yuille, A.L., "On Radial Basis Function Nets and Kernel Regression: Statistical Consistency, Convergence Rates and Receptive Field Size", *Neural Networks* 7, 609-628, 1994.
103. Xu, L., "Least mean square error reconstruction for self-organizing neural-nets", *Neural Networks 6*, 627-648, 1993. Its early version on *Proc. IJCNN91'Singapore*, 2363-2373, 1991.
104. Xu, L., Krzyzak, A., & Oja, E., " Rival Penalized Competitive Learning for Clustering Analysis, RBF net and Curve Detection", *IEEE Trans. on Neural Networks 4*, 636-649, 1993. Its early version on *Proc. of 11th Intl. Conf. on Pattern Recognition 1*, 672-675, 1992.

105. Xu, L, & Oja, E., "Randomized Hough Transform (RHT): Basic Mechanisms, Algorithms and Complexities", *Computer Vision, Graphics, and Image Processing : Image Understanding 57*, No.2, 131-154, 1993.
106. Xu, L., & Klasa, S. (1993), " A PCA Like Rule for Pattern Classification Based on Attributed Graph",*Proc. IJCNN93*, 1281-1284, 1993.
107. Xu, L, Oja, E. & Suen, C. Y., "Modified Hebbian learning for curve and surface fitting", *Neural Networks 5*, 441-457, 1992.
108. Xu, L., Klasa, S. & Yuille, L., "Recent Advances on Techniques of Static Feed-forward Networks With Supervised Learning", *Int. J. Neural Systems 3*, 253-290, 1992.
109. Xu, L, Krzyzak, A., and Oja, E., "A Neural Net for Dual Subspace Pattern Recognition Methods", *Intl. J. Neural Systems, 2*, No.3, 169-184, 1991.
110. Xu, L., "Adding Learned Expectation into The Learning Procedure of Self-Organizing Maps", *Intl. J. Neural Systems, 1*, No.3, 269-283, 1990.
111. Xu, L, Kultanen. P., & Oja, E., "A New Curve Detection Method: Randomized Hough Transform (RHT)", *Pattern Recognition Letters 11*, 331-338, 1990.
112. Zheng, G. L., & Billings, S. A., "Radial basis function network configuration using mutual information and the orthogonal least squares algorithm", *Neural Networks 9*, 1619-1637, 1996.

# Recursive Self-Organizing Maps

Thomas Voegtlin and Peter F. Dominey

Institut des Sciences Cognitives,
CNRS UMR 5015,
69675 Bron cedex, France

**Abstract.** A temporal extension of the Self-Organizing Map (SOM) is presented. The learning algorithm is based on self-reference, and is called Recursive SOM. This network learns local representations of the temporal context associated with a time series, and extends classical properties of SOM to time.

## 1 Introduction

The problem of adequately representing temporal information is of crucial importance. In a neural network, time can be represented with delayed inputs (spatialization of time), with leaky integrators, or with recurrent connections. Recurrent representations are powerful and robust, while representations based on delays are in general too sensitive to temporal deformations of the signal [2]. Representations using leaky integrators are also smooth and robust, but they are limited due to exponential loss of context.

Thus far, a number of methods have been proposed for representing temporal information with the Self-Organizing Map (SOM), with various successes. This includes both distributed and local types of representations. These methods use time-delays [6, 11], recurrent connections [3, 4, 8], leaky integrators [1] or sometimes combine several of those principles [5, 9]. However, it is not clear in what sense these models correctly extend the SOM properties to time. In general, classical notions like vector quantization and quantization error are not adapted to their evaluation.

In the present paper we present the Recursive Self Organizing Map, a SOM that learns adapted representations of temporal sequences. This model uses self-reference, in the sense that the original SOM algorithm is applied to its own representations. This model was first presented in [12], in the context of discrete binary inputs, and it was demonstrated that it generates near-optimal discrete quantization of temporal data. Here we investigate its properties in the case of continuous inputs, measuring quantization error on sequences.

## 2 Algorithm

In the original SOM algorithm [7], each unit $i$ of the map compares its weight vector $w_i$ to the input vector $x(n)$ at step $n$. The learning rule consists in updating the weights of neurons that belong to the neighborhood of the best-matching

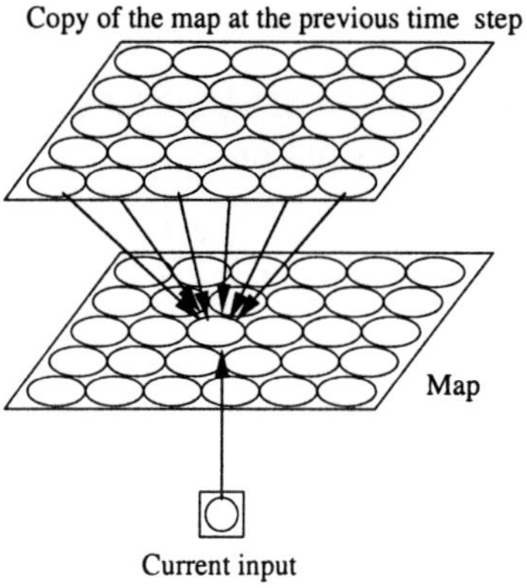


**Fig. 1.** Architecture of the recursive SOM. The original SOM algorithm is used recursively, on both the input and the representation derived at the previous time step. In the present paper the input is one-dimensional

unit, $k$, in the direction of the input vector:

$$\Delta w_i = \gamma h_{ik}(x(n) - w_i) \tag{1}$$

where $\gamma$ is a learning rate and $h_{ik}$ is a decreasing function of the distance between units $i$ and $k$ on the map.

In the Recursive SOM, each unit $i$ has a receptive field defined by two weight vectors, $w_i^x$ and $w_i^y$, that are compared to the input vector, $x(n)$, and to the vector of activities in the map at the previous time, $y(n-1)$, respectively. The activity, $y_i(n)$, of unit $i$ at step $n$ is:

$$y_i(n) = \exp\left(-\alpha \,||x(n) - w_i^x||^2 - \beta \,||y(n-1) - w_i^y||^2\right) \tag{2}$$

where $\alpha$ and $\beta$ are constant coefficients, and the norm is Euclidean. The best matching unit is the unit that has the highest activity. If $k$ is the index of the unit maximizing $y_i(n)$, the learning rules used for the weights are:

$$\Delta w_i^x = \gamma h_{ik}(x(n) - w_i^x) \tag{3}$$
$$\Delta w_i^y = \delta h_{ik}(y(n-1) - w_i^y) \tag{4}$$

where $\gamma$ and $\delta$ are learning rates, and the neighborhood function $h_{ik}$ is a Gaussian function of the Euclidean distance, $d(i,k)$, between units $i$ and $k$ on the map:

$$h_{ik} = \exp(-d(i,k)^2/\sigma^2) \tag{5}$$

where $\sigma$ controls the width of the Gaussian.

Note that this algorithm is nothing more than the original SOM algorithm, applied to both vectors $w^x$ and $w^y$ simultaneously. During training, each unit learns to associate an input vector, $x(n)$, with a pattern of activity, $y(n-1)$, representing its temporal context. The non-linearity in (2) is necessary for stability of learning.

## 3 Evaluation

In testing our algorithm, we used the Mackey-Glass time-series [10], a dynamic system defined by the differential equation:

$$\frac{dx}{dt} = \beta x(t) + \frac{\alpha x(t-\delta)}{1 + x(t-\delta)^{10}} \tag{6}$$

For $\delta > 16.8$, the series $x(t)$ is chaotic and highly sensitive to initial conditions. In this paper, we used the parameter values $\alpha = 0.2$, $\beta = -0.1$ and $\delta = 17$. A sequence of values of $x(t)$ is plotted on figure 2, starting from random initial conditions. In our experiments, the time-series we used as input to the Self-Organizing Map was the above Mackey-Glass series sampled every 3 time units: $[x(0), x(3), x(6), x(9), x(12), ...]$.

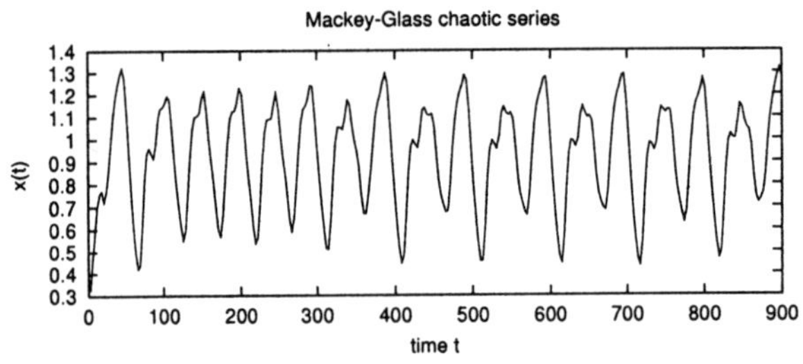


**Fig. 2.** The Mackey-Glass chaotic time series was used as input to the network.

The original SOM is a neural vector quantization method. A classical measure of performance is its quantization error, which is the mean distance between the input vector and the receptive field of the best-matching unit. In order to evaluate our temporal model, we need to generalize this measure to sequences of input vectors. The receptive field, or preferred input sequence, of each unit, is defined as the mean value of the input sequence that triggers its selection. The quantization error is equal to the standard deviation associated to this mean. Note that we need to restrict the computations of receptive fields and quantization error to a time window. However, this time window is defined for monitoring only, and its size does not affect the function of our model.

The Mackey-Glass time series (see figure 2) has regularities that are immediate; that is, past or future values of the series are partially predictable from the current value of the series, with no information about other events. In order to evaluate how temporal information is stored in our model, it is therefore necessary to evaluate these immediate regularities, by comparison with a model that has no sensitivity to time.

For this reason, we compared the Recursive SOM to the original SOM. We also included the Recurrent SOM [9] in our comparison, a temporal SOM that is

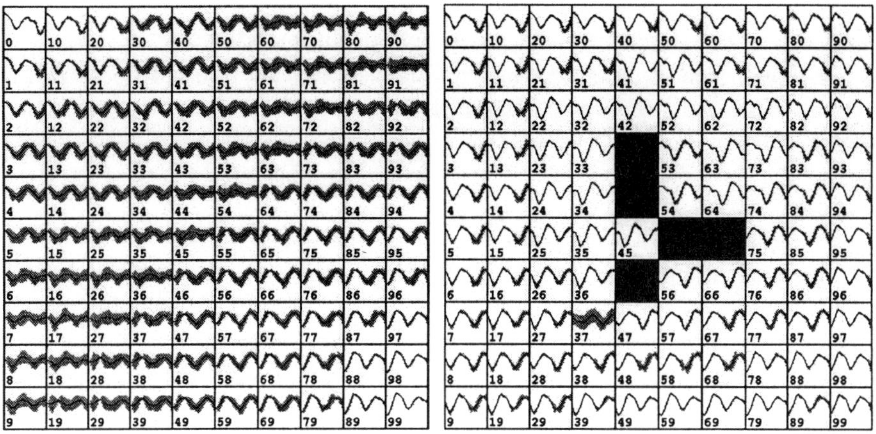

**Fig. 3.** Comparison of the temporal receptive fields of the units of SOM (left) and Recursive SOM (right). Maps were trained on the Mackey-Glass time series. The preferred input sequence of each unit is in black, with the most recent input on the left. The gray intervals reflect the standard deviation, or quantization error. Receptive fields are displayed for 30 past events. Five units are not exploited due to topology constraints.

based on recurrent connections and leaky integrators. In the Recurrent SOM, the best matching unit minimizes the norm of a vector $y_i$ called the leaked difference vector, defined by: $y_i(n) = (1-\alpha)y_i(n-1)+\alpha(x(n)-w_i)$, $0 < \alpha < 1$. This vector is used in place of $x(n)$ in the learning rule: $\Delta w_i = \gamma h_{ik}(y_i(n) - w_i)$. All maps were of size 10 by 10 and were trained during 150000 iterations. Parameters of the Recursive SOM were $\alpha = 2$ and $\beta = 0.06$. The Recurrent SOM used a leak constant $\alpha = 0.1$.

The receptive fields on the surface of the SOM and of the Recursive SOM are displayed on figure 3. Black lines represent the preferred input sequence of each unit. The gray regions below and above the curves correspond to confidence intervals of size plus and minus one standard deviation. The standard deviation of the sequence preceding the selection of a unit is equal to the quantization error, assuming that the receptive field is used as a letter of a code-book for the sequence of events.

For the SOM, the quantization error is minimal for units 0 and 99, which correspond to extreme values of the input. This is not surprising, since it is easy to predict that the series will decrease once it has reached its maximal value. In contrast, the receptive fields corresponding to intermediate values of the input are much less specific, which is reflected by a high standard deviation. This is because the SOM cannot perform temporal aliasing.

The receptive fields of the Recursive SOM are more structured; their diversity reflects the temporal profiles observed in the time series. Spatio-temporal topological organization is observed. For most units of the Recursive SOM, the standard deviation of past events is low; this demonstrates that temporal information is retained, and that the responses of cells is highly specific.

A comparison of quantization errors is displayed on figure 4, for SOM, Recurrent SOM and Recursive SOM. Quantization error has been estimated for 30 past events.

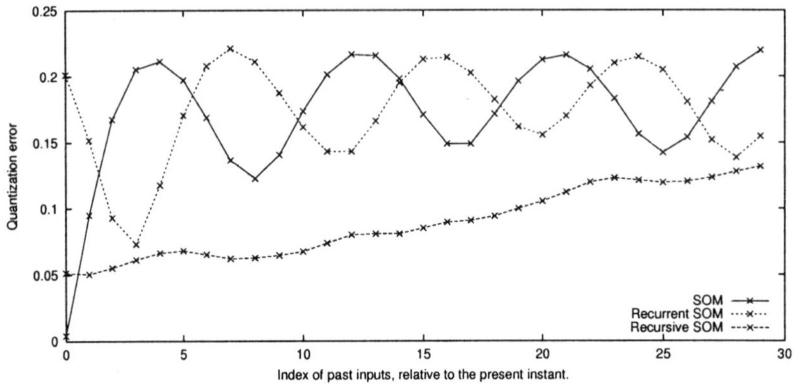


**Fig. 4.** Temporal quantization error. This error is defined as the mean difference between the input sequence and the receptive field of the selected unit. Quantization error reflects the uncertainty about past events. 30 past inputs are considered here.

Since the SOM does not learn to represent temporal information, it provides a reference to which performance of other models can be compared. The SOM develops a very accurate representation of the present event, with a quantization error close to zero. The error curve then oscillates, reflecting temporal regularities of the time series.

The Recurrent SOM demonstrates the same overall level of error as the SOM; however, oscillations are phase-shifted. Experimentally we observed that the leak constant of the Recurrent SOM determines the value of the phase shift. This shift means that temporal information is represented in the Recurrent SOM. However, the representation of certain parts of the signal (e.g. at $n - 3$) is performed at the expenses of other parts of the signal; indeed, the global performance is not significantly better than for the SOM.

In contrast, our model demonstrates significantly reduced quantization error over time. The sole exception is the current input, for which the classical SOM makes a much smaller quantization error. This is because the resources of the Recursive SOM are used for representing past events, while the SOM can use all its neurons for representing the current input with high precision. The curve corresponding to the Recursive SOM does not oscillate; this reflects adaptation of our model to the temporal regularities of the signal.

## 4 Conclusion

In [12] we demonstrated that the representations developed in Recursive SOMs are nearly optimal, in the context of discrete inputs. That is, the quantization of discrete sequences maximizes the mean number of known past events. Although optimality is more difficult to define in the case of continuous inputs, the experiments presented here suggest that the local representation developed in our algorithm achieves an optimal tradeoff between the mean quantization errors of past and present inputs.

The Recursive SOM constitutes a robust temporal extension of SOM, that successfully generalizes the properties of the original SOM to time series, *ie* topology preservation and vector quantization. Moreover, the Recursive SOM is very close to the original SOM algorithm; the difference is that representations in Recursive SOMs involve self-reference.

## References

1. G. J. Chappell and J. G. Taylor. The temporal Kohonen map. *Neural networks*, 6:441–445, 1993.
2. J. L. Elman. Finding structure in time. *Cognitive Science*, 14:179–211, 1990.
3. N.R. Euliano and J.C. Principe. Spatio-temporal self-organizing feature maps. *Proceedings of the ICNN*, pages 1900–1905, 1996.
4. A. Hoekstra and M. Drossaers. An extended Kohonen feature map for sentence recognition. *Proceedings of the ICANN*, pages 404–407, 1993.
5. D. L. James and R. Miikkulainen. SARDNET: a self-organizing feature map for sequences. *Advances in Neural Information Processing Systems*, 7:577–584, 1995.
6. J. Kangas. *On the Analysis of Pattern Sequences by Self-Organizing Maps*. Ph.D. Thesis, Helsinki University of Technology, 1994.
7. T. Kohonen. *Self-Organizing Maps*. Springer-Verlag, Berlin, 2nd edition, 1997.
8. K. Kopecz. Unsupervised learning of sequences on maps with lateral connectivity. *Proceedings of the ICANN*, pages 431–436, 1995.
9. T. Koskela, M. Varsta, J. Heikkonen, and K. Kaski. Time series prediction using recurrent SOM with local linear models. *Int. J. of Knowledge-Based Intelligent Engineering Systems*, 2(1):60–68, 1998.
10. M.C. Mackey and Glass L. Oscillations and chaos in physiological control systems. *Science*, pages 197–287, 1977.
11. J. Vesanto. Using the som and local models in time-series prediction. In *Proceedings of WSOM'97*, pages 209–214, 1997.
12. T. Voegtlin. Context quantization and contextual self-organizing maps. In *Proceedings of the IJCNN'2000*, volume 6, pages 20–25, 2000.

# An Investigation into Catastrophic Interference on a SOM Network

A. Indrayanto and N. M. Allinson
Department of Electrical Engineering and Electronics, UMIST
Manchester, United Kingdom

Abstract

*Catastrophic interference* (CI) is a common problem facing many artificial neural network models in learning new data sequentially. Some studies show that the catastrophic interference problem is largely a consequence of the overlap of internal distributed representations, and neural networks with local representations may suffer less. One particular local representation network under study is the Kohonen self-organizing map (SOM). Little is known whether the SOM network may also exhibit this problem. The experiments use *focused* and *cumulative* learning with two different sets of data. Results show that SOM may well suffer from the catastrophic interference problem for some particular selections of parameters and data sets.

## 1 Introduction

One common problem facing many artificial neural network models in learning new data sequentially is how to accommodate the new data without disrupting the previous learned data. In other word, the network needs to have some sensitivity to the new data but still preserve some stability of the previous learned data. This often referred to the "sensitivity-stability" dilemma [1] or the "stability-plasticity" problem [2]. One particular manifestation of this problem, known as *catastrophic interference (CI)*, had been exposed in [3] and [4]. Catastrophic interference is the tendency of a neural network to completely forget previous learned information when it learns the new data.

The problem has been first studied intensively in [3] and [4], while a comprehensive analysis of the catastrophic interference has been done in [5]. Several proposed solutions can be found in [6], [7], [8], and [9]. However, most of the studies and the proposed solutions referred to the distributed neural network architectures with a single set of shared (or partially shared) multiplicative weights, such as backpropagation.

From these studies, it is argued that the CI problem is largely a consequence of the overlap of internal distributed representations, and that reducing this overlap would reduce the problem [8]. One way to do this is through localizing the internal representations such as localizing the hidden units. French also argues that neural networks with local representations may suffer less from the CI problem compare to the neural networks with distributed representations. Moreover, a very local representation will not exhibit catastrophic interference because there is little interaction among representations [6]. This kind of network may face some gradual-forgetting problem instead. However, it is not clear under what circumstances a neural network with local representations may suffer from the CI problem. Does the CI problem never occur in this type of network? Is there any specific condition that may affect the response of the network to this problem? This investigation is an attempt to answer these questions.

Since there are many neural network models that belong to the local representation networks, this investigation will only focus primarily on a particular local representation network, namely the Kohonen self-organizing map (SOM) [10].

# 2 Self Organizing Map

The Kohonen Self-Organizing Map (SOM) network belongs to a local representation network category since it employs a competitive learning procedure that allows only a single neuron to become most active at one time. The SOM network in this experiment uses a set of neurons to form a one-dimensional topology map. Let $\mathbf{C}$ represents the map with every neuron $c$, $c \in \mathbf{C}$, is arranged in a one-dimensional space, and let $\mathbf{X} \in \mathbf{R}^2$ represents a two-dimensional input spaces. Every neuron $c$ is connected to all dimensional components of the input sample $\mathbf{x} = [x_1,\ x_2]^T$, where $\mathbf{x} \in \mathbf{X}$, then the connection weights are $\mathbf{w}_c(n) = [w_{c1}(n),\ w_{c2}(n)]^T, \forall c \in \mathbf{C}$, where $n \geq 0$ is the discrete learning time.

The competitive learning procedure searches for the best matching neuron through comparing an input sample $\mathbf{x}$ with all the weight vectors $\mathbf{w}_c$ of the network. The smallest of the *Euclidean distances* $\|\mathbf{x} - \mathbf{w}_c\|$ is defined as the best matching node or the winner. This can also be expressed as

$$v(n) = \arg\min_{c \in \mathbf{C}} \|\mathbf{x}(n) - \mathbf{w}_c(n)\|,\ \forall c \in \mathbf{C},$$

where $v(n)$ is the winning neuron given an input sample $\mathbf{x}$. The weights are then updated according to

$$\mathbf{w}_c(n+1) = \mathbf{w}_c(n) + \alpha(n) h_{cv(n)}(n)[\mathbf{x}(n) - \mathbf{w}_c(n)],\ \forall c \in \mathbf{C},$$

where $\alpha(n)$ is the learning function and $h_{cv(n)}(n)$ is the neighbourhood function. The learning function used in the experiment is a linearly decreased function expressed as

$$\alpha(n) = \alpha(0)\left(1.0 - \frac{n}{cycles}\right),$$ $n \geq 0$ and *cycles* is the total learning cycles.

While the radius of the winner's neighbourhood decreases linearly following the step function, i.e.

$$h_{cv(n)}(n) = \begin{cases} 1 & c \in \aleph_v(n) \\ 0 & c \notin \aleph_v(n) \end{cases}$$, where $\aleph_v(n)$ is the winner's neighbourhood at time $n$.

# 3 Experimentation

The focus of the experiments is to examine the interference effects of the previous learning data (*early-acquired data*) in the presence of the new data (*late-acquired data*) given to the network sequentially. To simulate similarity and discrimination among the data, we use different data positions in the Euclidean space. The data that are located close to each other will *look similar* compared to the data that are located further apart. Therefore, the investigation also examines the effects of different data positions in Euclidean space that may initiate a *catastrophic interference* (*catastrophic forgetting*) or a *gradual forgetting* problem in the SOM network.

The data are arranged into two separate data sets and presented to the network in consecutive order. With this arrangement, the first data set is defined as the *early-acquired* data while the second data set is defined as the *late-acquired* data. There are two kinds of arrangement to learn the sets, i.e. the *focused* learning and the *interleaved* (*cumulative*) learning. In the *focused* learning, the *late-acquired* data is presented to the network without interleaving it with the *early-acquired* data, while in the *interleaved* (*cumulative*) learning the *late-acquired* data is presented together with the *early-acquired* data.

## 3.1 Data

The data sets consist of uniformly distributed random numbers in two-dimensional Euclidean space. Let $\mathbf{X} \in \mathbf{R}^2$ represents the input space, and $\mathbf{u} = [u_1, \ u_2]^T$, $\mathbf{u} \in \mathbf{X}$, is the *early-acquired* data sample, while $\mathbf{v} = [v_1, \ v_2]^T$, $\mathbf{v} \in \mathbf{X}$ is the *late-acquired* data sample. Each data set is a cluster of samples located within an ellipse in two-dimensional space, i.e. $\exists \mathbf{u} \in D_i$, $D_i \subset \mathbf{X}$ where $\{D_i, i = 1,2,\ldots N\}$ is $N$ data sets used in the experiments.

Two separate experiments using different combination of data sets were completed. The first experiment uses two sets of data where one of them is a subset of the other; hence, it is called the *subset data test*. The first data set covers a larger data space and contains the second data set, i.e. $D_1 \subset \mathbf{X}$ and $D_2 \subset \mathbf{X}$ are the two data sets such that $D_2 \subset D_1$, $D_2 \neq D_1$. These data sets are designed to simulate the insertion of new data sample within the boundary (cluster) of the previous learned (*early-acquired*) data sample. From the Euclidean-distance point of view, the data samples in the two data sets are close to each other and hence they may *look similar*. The data sets are shown in Figure 1a.

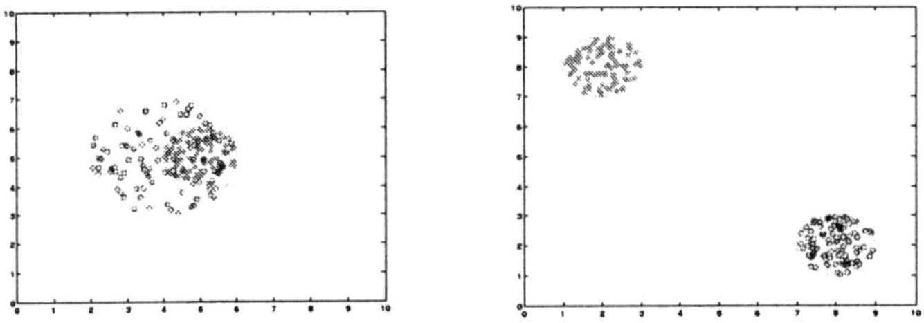

(a) (b)

Figure 1 Data sets for (a) the subset data test; (b) the mutually exclusive data test.

The first data cluster is shown as a large ellipse with 100 hollow points that represents each data sample, i.e. $\forall \mathbf{u} \in D_1$. The second data cluster is shown as a small ellipse inside the large one with 100 grey points as the data samples, i.e. $\forall \mathbf{v} \in D_2$.

The second experiment uses two sets of data that are separated from each other, i.e. $D_1 \subset \mathbf{X}$ and $D_2 \subset \mathbf{X}$ are the two data sets such that $D_1 \cap D_2 = \emptyset$, $D_1 \cup D_2 \neq \mathbf{X}$. Hence $D_1$ and $D_2$ are *mutually exclusive*. This experiment is called the *mutually exclusive data test*.

The network is expected to distinguish the *late-acquired* data set as a totally new data set outside the boundary (cluster) of the *early-acquired* data set. These data sets are shown in Figure 1b. The first data set is shown at the bottom-right corner represented as 100 hollow points inside the ellipse, i.e. $\forall \mathbf{u} \in D_1$, while the second data cluster is shown at the top-left corner represented as 100 grey points inside the ellipse, i.e. $\forall \mathbf{v} \in D_2$, and $\mathbf{u} \neq \mathbf{v}$.

## 3.2 Procedure

There are four arrangements for the simulations, i.e. the *subset data test* with *focused* learning, the *subset data test* with *cumulative* learning, the *mutually exclusive data test* with *focused* learning, and the *mutually exclusive data test* with *cumulative* learning.

The experiments use networks with different network sizes, initial learning rates, number of learning cycles and initial neighbourhood radii. Each trial uses a specific parameter setting, but this setting is kept unchanged for the complete trial. Only one parameter is varied in every trial.

Every weight vector for each network is initialised with uniform random values and then trained with the *early-acquired* data set for a specific number of learning cycles before learning the first batch data. This is to provide some bias toward the *early-acquired* data set in the beginning of experiment. After initialisation and pre-training, the network first learns the *early-acquired* data set at some particular parameter settings, and the final *average quantization error (AQE)* is recorded. The trained network then learns the *late-acquired* data set, and similarly its final average quantization error is recorded. Both learning sequences are restricted to a specified number of learning cycles. After the second learning, the network is tested using the *early-acquired* data, while its average quantization error is recorded. During the testing phase, there is no weight vector adaptation; i.e. it is merely a recognition phase and not a learning phase.

# 4 Results and Discussions

The AQE results show that whether a network with a particular parameter setting, experiences catastrophic forgetting or a gradual one. If the network converges after the first training (i.e. *AQE* of the *early-acquired* data is relatively small), and the network also converges after the second training (i.e. *AQE* of the *late-acquired* data is also comparatively small), yet it gives a relatively high *AQE* when tested again with the *early-acquired* data after the second training, then we can say that the network may experience a CI problem. Conversely, when the second test of the *early-acquired* data after the second training shows a relatively small *AQE* then the network does not experience a CI problem.

Figure 2a and 2b show the results for 10, 50, and 100 1-D maps with *focused* learning. The first input data is the *early-acquired* data set and the second input data is the *late-acquired* data set. After initialization the network is pre-trained with the first data set for 10,000 learning cycles, and with the initial learning rate is set to 0.3 while the initial radius of neighbourhood is set to the map dimension. During this pre-training the learning rate decreases linearly until it reaches zero and the radius decreases proportionally toward radius size one (i.e. only the winner is updated). Then the network learns the first data set once again for another 10,000 cycles, and it continues learning with the second data set for another 10,000 cycles. The figures show the *AQE* after the pre-training phase. If we look closely, the initial *AQE* on cycle 0 is around 0.4, which is the final *AQE* of the pre-training phase. From Figure 2a, we can see that although the network has been pre-trained with the first data set, it starts to *forget* the first data set when the learning rate and the radius size are reset to their initial values. This conforms to the proof found in [11]. However, when the radius size is fixed to size one (i.e. only the winner is updated) during learning, some networks with more neurons continue to preserve the first data set, as shown in Figure 2b.

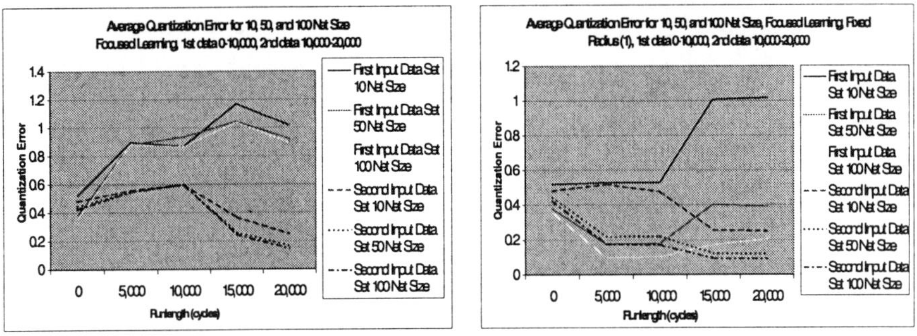

(a) (b)

Figure 2 *AQE* for networks with initial radius size (a) equal to network size; (b) fixed to radius one

To see what is actually happening in the network, in particular to its topological map, Figure 3a and 3b show typical results. In these figures, the first data set are the hollow points distributed within the large ellipse while the second data set are the grey points distributed within the smaller one. These are the same data points as in Figure 1a, except that the picture is rotated 90 degrees. The string with connected dots is the one-dimensional topological map. Figure 3 shows the final topological map at the end of the learning process. It is shown in Figure 3a that the final map only covers the points within the smaller ellipse, that is the network only learns the second data set and forgets almost all of the first data set. On the other hand, the network with fixed radius size still preserves some of the previous learning but with a non-optimal map. A better result with a near optimal map is shown in Figure 4a. This map is a result from the *cumulative* learning experiment. This is understandable since the *cumulative* learning uses both the *early-acquired* and the *late-acquired* data as its second data set. The result also shows that the final map totally depends on the second data set.

Some results that also correspond to the CI problem are also found in the second experiment of *focused* learning using *mutually exclusive data test*, as shown in Figure 4b. This shows significant AQE differences for the *early-acquired* data set before and after the introduction of the *late-acquired* data set. This strongly suggests that the SOM network has also experienced the CI problem, especially if the *late-acquired* data samples are significantly different to the previous *early-acquired* data.

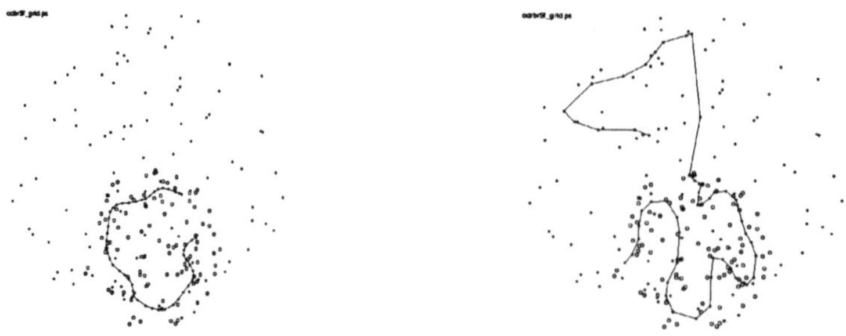

(a) (b)

Figure 3 Topological map for network with radius (a) equal to network size; (b) fixed to radius one

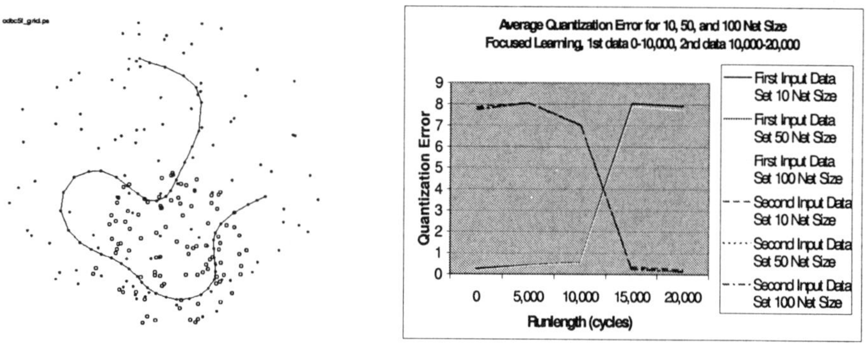


(a) (b)

Figure 4 (a) Topological map after *cumulative* learning, (b) *AQE* in *mutually exclusive* data test.

# 5 Conclusion

From the experimental results, it appears that a neural network with internal local representations such as the Kohonen self-organizing map (SOM) also experiences the catastrophic interference (CI) problem, especially if the *late-acquired* data samples are significantly different to the previous *early-acquired* data samples. However, when the *late-acquired* data is a subset of the *early-acquired* data then the network exhibits *gradual-forgetting* instead.

The catastrophic interference problem in SOM is not because of the overlap of internal distributed representations as in the distributed neural network model such as backpropagation, but it depends on the selection of the initial radius of neighbourhood. This parameter along with the learning parameter plays as a *maturation* factor for the network structure.

Choosing a small value for the initial radius size may reduce the catastrophic interference effect. However, this may result to a non-optimal topological map. From the point of view of memorisation, this may be applicable despite of its non-optimal map. However, for the purpose of generalisation with a necessity to noise tolerance this condition is not preferable.

The selection of the network learning parameters in certain circumstance is critical. It is especially important to select proper learning parameters when the network size is barely enough to accommodate the new learning. However, if the network size is significantly enough, then the selection of the learning parameters becomes less critical.

# 6 References


1. Hebb DO. Organization of Behavior. Wiley & Sons, New York, 1949
2. Carpenter G, Grossberg S. ART 2: Self-organization of stable category recognition codes for analog input patterns. Applied Optics 1987; 26(23):4919-4930
3. McCloskey M, Cohen NJ. Catastrophic interference in connectionist networks: the sequential learning problem. Psychology of Learning and Motivation 1989; 24:109-165
4. Ratcliff R. Connectionist models of recognition memory: constraints imposed by learning and forgetting functions. Psychological Review 1990; 97(2):285-308
5. Sharkey NE, Sharkey AJC. An Analysis of catastrophic interference. Connection Science 1995; 7(3&4):301-329
6. French RM. Using semi-distributed representations to overcome catastrophic forgetting in connectionist networks. In: Proceedings of the 13th Annual Cognitive Science Society Conference. LEA, NJ, 1991, pp 173-178
7. French RM. Pseudo-recurrent connectionist networks: an approach to the "sensitivity-stability" dilemma. Connection Science 1997; 9(4):353-379
8. French RM. Catastrophic forgetting in connectionist networks: causes, consequences and solutions. Trends in Cognitive Sciences 1999; 3:128-135
9. Frean M, Robins A. Catastrophic forgetting in simple networks: an analysis of the pseudorehearsal solution. Network: Comput. Neural Syst 1999; 10:227-236
10. Kohonen T. Self-organizing maps. Springer-Verlag, Berlin, 1995
11. Yin HJ. Self-organizing maps: statistical analysis, treatment, and applications. PhD thesis. University of York, York, 1996

# A Topography-Preserving Latent Variable Model with Learning Metrics

Samuel Kaski and Janne Sinkkonen

Helsinki University of Technology
Neural Networks Research Centre
P.O. Box 5400, FIN-02015 HUT, Finland
{samuel.kaski,janne.sinkkonen}@hut.fi

**Abstract.** We introduce a new mapping model from a latent grid to the input space. The mapping preserves the topography but measures local distances in terms of auxiliary data that implicitly conveys information about the relevance or importance of local directions in the primary data space. Soft clusters corresponding to the map grid locations are defined into the primary data space, and a distortion measure is minimized for paired samples of primary and auxiliary data. The Kullback-Leibler divergence-based distortion is measured between the conditional distributions of the auxiliary data given the primary data, and the model is optimized with stochastic approximation yielding an algorithm that resembles the Self-Organizing Map, but in which distances are computed by taking into account the (local) relevance of directions.

## 1 Introduction

Topograhy-preserving latent variable models like the Self-Organizing Map (SOM) [2,3] are valuable tools especially for descriptive data analysis tasks, creating overviews of the data. Such models form an organized mapping from the latent space, usually a two-dimensional discrete map grid, to the input space. The map grid can be used as a graphical display whereon close-by locations represent similar data. Additional properties of the data, such as the density (cluster) structure and the distribution of the values of data variables, can be visualized on the display.

The mapping characterizes the probability density $p(\mathbf{x})$ of the multivariate (vectorial) data $\mathbf{x}$, but it depends on the metric of the data space. The metric in turn depends on the feature extraction: it is usually selected by first choosing the data variables and their relative scales, and then using a simple global measure such as the Euclidean distance. Feature extraction is often far from trivial. The variables may be of diverse nature, have different units of measurement, and their relative importance may be unknown. Moreover, the relative importance may be different in different locations of the data space.

We have earlier studied methods for learning suitable local distance measures. The task is impossible unless more information is brought to the setup. Our assumption has been that there is available some auxiliary data $c$ whose conditional distribution implicitly conveys information about the relative importance of the available variables: the more the distribution $p(c|\mathbf{x})$ of the auxiliary data changes when $\mathbf{x}$ changes, the more important the change of $\mathbf{x}$ is. The auxiliary data occurs as paired samples $\{(\mathbf{x}^k, c^k)\}_k$ with the primary data. For simplicity we have so far assumed that the auxiliary data is discrete, multinomially distributed. In our case studies the auxiliary data have indicated a future bankruptcy in bankruptcy analysis, functional classes of genes in gene expression analysis, and keywords selected by experts in content-based text clustering.

We wish to measure distances between distributions, and so a natural choice is the Kullback-Leibler divergence $D$. A new (local) metric $d$ is defined by the divergence between distributions at nearby points:

$$d^2(\mathbf{x}, \mathbf{x} + d\mathbf{x}) \equiv D(p(c|\mathbf{x}) || p(c|\mathbf{x} + d\mathbf{x})) . \tag{1}$$

In an earlier work [1], $p(c|\mathbf{x})$ was approximated with a conditional density estimator and the distances (1) with the Fisher information matrix, and then computed a SOM was computed in the resulting metric. Since this otherwise promising approach involves several steps of approximations and relatively demanding computation, we have started to investigate other alternatives as well.

One approach is to directly incorporate the new distance measure (1) into the cost function of the mapping, as was earlier done for clustering purposes [4]. In the present paper we generalize this work to a latent variable model, in which the cluster centers are arranged onto a SOM-like grid, and their parameters are tied to each other through hidden variables and a neighborhood function: The cluster centers are sums of the hidden variables, weighted by the neighborhood function. This forces the grid to be organized, or "topography-preserving," at the minimum of the cost function.

## 2 The Method

We assume that there is a set of paired data $\{(\mathbf{x}_k, c_k)\}_k$ available, where $\mathbf{x} \in \mathbb{X} \subset \mathbb{R}^n$ and $c$ is multinomially distributed. We postulate a grid on which a set of cluster centers $\mathbf{m}_j$ is attached, and we wish to form a soft clustering of the data space that uses differences between the distributions $p(c|\mathbf{x})$ as the distortion measure. Moreover, the cluster centers that are close-by on the grid are constrained to be relatively similar.

Our cost function should therefore have two properties: it shoud minimize a distortion measure, measured between the conditional distributions $p(c|\mathbf{x})$, and it should promote the smoothness of the map. We look at the distortion first, by starting from a generalized version of vector quantization, and impose a smoothing condition later.

*A cost function measuring distortion between distributions.* In "soft" vector quantization the goal is to find prototypes or codebook vectors $\mathbf{m}_j$ which minimize the average distortion $E$ between the data and the codebook vectors. The average distortion is

$$E = \sum_j \int y_j(\mathbf{x}; \mathbf{m}_j) D(\mathbf{x}, \mathbf{m}_j)\, p(\mathbf{x})\, d\mathbf{x}\,, \tag{2}$$

where $D(\mathbf{x}, \mathbf{m}_j)$ is the distortion made when representing $\mathbf{x}$ by $\mathbf{m}_j$, and $y_j(\mathbf{x}; \mathbf{m}_j)$ is the "soft" cluster membership function that fulfills $0 \leq y_j(\mathbf{x}) \leq 1$ and $\sum_j y_j(\mathbf{x}) = 1$ for all $\mathbf{x}$. (Note that we have simplified the notation here for clarity: $y_j$ may actually depend on all the $\mathbf{m}_j$.)

Because we are interested in differences between the distributions $p(c|\mathbf{x})$, we attach a distributional prototype $\boldsymbol{\psi}_j$ to each cluster $j$, and replace the usual squared Euclidean distortion measure with the Kullback–Leibler divergence $D_{KL}(p(c|\mathbf{x}), \boldsymbol{\psi}_j) \equiv \sum_i p(c_i|\mathbf{x}) \log(p(c_i|\mathbf{x})/\psi_{ji})$. Note that unlike the ordinary prototype of the standard vector quantization, $\boldsymbol{\psi}_j$ is not directly tied to a certain point of the input space (we will later point out an analytical form for it at the optimum of the cost function).

With this distortion measure the cost function of the vector quantization becomes

$$E_{KL} = \sum_j \int y_j(\mathbf{x}; \mathbf{m}_j) D_{KL}(p(c|\mathbf{x}), \boldsymbol{\psi}_j)\, p(\mathbf{x})\, d\mathbf{x}\,. \tag{3}$$

Instead of computing the distortions between the vectorial samples and vectorial prototypes as in (2), we now compute distortions between the distributions $p(c|\mathbf{x})$ and the distributional prototypes $\boldsymbol{\psi}_j$. The cost function (3) is to be minimized with respect to both the $\mathbf{m}_j$ and the $\boldsymbol{\psi}_j$.

*Parameterization and imposing smoothness.* As in the earlier work [4] with independently parameterized clusters, the cluster membership functions $y_j(\mathbf{x})$ will be of the form $y_j(\mathbf{x}) = G_j(\mathbf{x}) / \sum_k G_k(\mathbf{x})$, where the $G_j(\mathbf{x}) = e^{-\|\mathbf{x}-\mathbf{m}_j\|^2/2\sigma^2}$ are Gaussians. The dispersion parameter $\sigma$, implicit in the cost function 3, is selected *a priori* or by using a validation set. Note that because $\sum_j y_j(\mathbf{x}) = 1$, the membership functions may be interpreted as the conditional pdf of a random variable, here denoted by $V$ and the values denoted by $v_j$.

In the present work our intention is to generate a solution where the clusters $j$ reside on a grid and the nearby clusters are relatively similar by their parameters. To achieve such a smoothness, we force nearby clusters to be similar by making the cluster location parameters $\mathbf{m}_j$ dependent on each other through a set of *hidden variables* $\mathbf{n}_k$: $\mathbf{m}_j = \sum_k h_{jk}\mathbf{n}_k$. The coefficients $h_{jk}$ are positive, fulfilling $\sum_k h_{jk} = 1$, and they define the topology of the grid; they resemble the neighborhood function of the SOM. We have used (normalized) Gaussian neighbourhoods, with the standard deviation denoted by $\beta$, and a rectangular grid.

*Minimizing the cost function.* The gradient of the cost function with respect to the hidden variables is

$$\nabla_{\mathbf{n}_j} E_{KL} = -\frac{1}{\sigma^2} \sum_{i,k,l} h_{lj} \log \frac{\psi_{li}}{\psi_{ki}} \int (\mathbf{x} - \mathbf{m}_l) p(v_l, v_k, c_i, \mathbf{x}) d\mathbf{x} , \tag{4}$$

where the random variable interpretation of the membership functions was used to write $y_l(\mathbf{x}) y_k(\mathbf{x}) p(c_i, \mathbf{x})$ as $p(v_l, v_k, c_i, \mathbf{x})$.

A very simple algorithm follows by applying stochastic approximation to the optimization problem. The (factorizable) joint distribution $p(v_l, v_k, c_i, \mathbf{x})$ can be used as the sampling function: an input $\{(\mathbf{x}(t), c_i)\}$ is drawn, and then two clusters $k$ and $l$ from the $\mathbf{x}$-conditioned multinomial distribution $\{y_k(\mathbf{x}(t))\}$. The adaptation rule can be formulated on the basis of the cluster centers $\mathbf{m}_j$ alone, without referring to the hidden variables:

$$\mathbf{m}_j(t+1) = \mathbf{m}_j(t) + \alpha(t) \bar{h}_{lj} \log \frac{\psi_{li}}{\psi_{ki}} (\mathbf{x} - \mathbf{m}_l) , \tag{5}$$

where $\bar{h}_{lj} \equiv \sum_s h_{ls} h_{js}$.

The other free parameters besides the membership function locations are the prototype distributions $\psi_{ji}$. After the reparameterization $\log \psi_{ji} \equiv \gamma_{ji} - \log \sum_r \exp \gamma_{jr}$, to keep $\psi_{ji}$ positive and summed up to unity, the distributional prototype can be optimized with the stochastic approximation step

$$\gamma_{lr}(t+1) = \gamma_{lr}(t) - \alpha(t)(\psi_{lr} - \delta_{ri}) , \tag{6}$$

where $\delta_{ri}$ is the Kronecker delta. (For more details cf. [4].)

It is possible to show that the optimal values for the prototypes are $\psi_{ji} = p(c_i|v_j)$, i.e., they converge to the cluster-wise averages of the conditional distributions $p(c_i|\mathbf{x})$. Plugging this into the distortion function (3) gives a connection to mutual information: minimizing the Kullback-Leibler based distortion function is in fact equivalent to maximizing the mutual information between the random variables $C$ and $V$, i.e. essentially between the guiding auxiliary variable and the cluster membership functions.

## 3 Demonstration

We demonstrate the method by mapping a seemingly simple toy data set, in which we know what the solution should approximately be. The data is uniformly distributed inside a 3D cube (the thin lines in Fig. 1**a**). The class distribution changes only in the direction of the horizontal $xy$-plane, approximately uniformly everywhere. Assuming that the changes of the class distribution are the important aspects in the data, the vertical $z$-direction is irrelevant and resources will be wasted if it is modelled.

*Data.* The data consists of four equally likely classes $c_i$, of which two change linearly along the one side ($x$-direction) of the horizontal square, and the other two along the other side ($y$-direction). If the length of the side of the cube is 1.6 and the origo is in the middle of the cube, the class distributions are $p(c_0|\mathbf{x}) = 0.1x + 0.25$, $p(c_1|\mathbf{x}) = 0.5 - p(c_0|\mathbf{x})$, $p(c_2|\mathbf{x}) = 0.1y + 0.25$, and $p(c_3|\mathbf{x}) = 0.5 - p(c_2|\mathbf{x})$.

*Optimization.* During preliminary simulations it turned out that the algorithm described in Section 2 is fairly sensitive to the initial conditions, and hence we optimized the cost function in two phases. The first phase ensures the organization of the map and the second is a fine-tuning phase.

The map consisting of 7 by 7 units was initialized randomly by setting the model vectors to the first 49 data points. In the *first learning phase* of 2 million iterations the width of the neighborhood $\beta$ was gradually decreased from 4.0 to its final value of 1.0. The width decreased first with a piecewise-linear approximation to an exponential curve, and then linearly to one in the end. The learning coefficient $\alpha(t)$ was simultaneously decreased from 0.5 to zero with the same schedule. The cluster membership functions were rather wide, with $\sigma = 0.3$.

During this organization phase we applied another rule in addition to (5) to reduce exessive degrees of freedom in the mapping. Note that the sole purpose of this first learning phase is to force the mapping to a favourable, organized initial state. All clusters in the neighborhood of the cluster $s$ closest to the data point, $s = \arg\min_j \|\mathbf{x} - \mathbf{m}_j\|$, were adapted towards $\mathbf{m}_s$:

$$\mathbf{m}_j(t+1) = \mathbf{m}_j(t) + (\alpha(t)/C)^2 \bar{h}_{sj}(\mathbf{m}_s(t) - \mathbf{m}_j(t)) \; . \tag{7}$$

In our simulations $C = 4$. Note that as $\alpha$ goes to zero, $\alpha^2$ goes to zero much faster, and hence (7) has an effect only in the beginning of the learning process.

In the second, fine-tuning phase of another 2 million iterations, the width of the neighborhood remained at 1.0, and $\alpha(t)$ decreased with the same piecewise-linear schedule from 0.05 to zero. The cluster membership functions were sharpened during the first 100,000 iterations by changing $\sigma$ linearly from 0.3 to 0.1. The adaptation step (7) was not used.

*Results.* As can be seen in Figure 1**a**, the converged map represents only the two directions $x$ and $y$ that are relevant in the sense of the classes. The map represents these two dimensios in an ordered fashion (Fig. 1**b**). By contrast, a SOM would try to fold itself to fill the whole 3D cube (cf. [3]).

## 4 Conclusions

We have introduced a new mapping method that preserves the topography of the input space but implicitly measures the similarity of close-by data

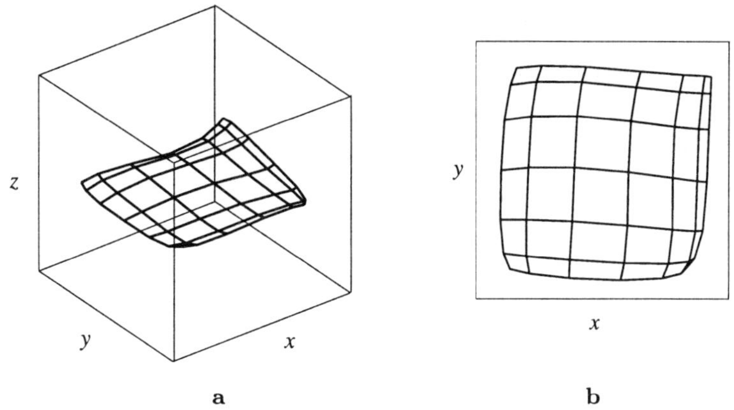


**Fig. 1.** **a** In a three-dimensional cube filled with uniformly distributed data, the map represents only the relevant two-dimensional $xy$-subspace, in which the conditional distribution of the auxiliary data changes. **b** In the relevant subspace the representation is ordered ("topography-preserving"). Model vectors that are neighbors on the map grid have been connected by lines, and the model vectors are represented by the nodes of the resulting "network."

points by differences in the conditional distributions of associated auxiliary data. The auxiliary data can therefore be used to define what is important in the primary data. Assuming such auxiliary information is available, the new method solves the problem of how the (originally arbitrary) distances should be computed, common in many exploratory multivariate analysis tasks.

The proposed optimization method is still fairly sensitive to the initial state and details of the optimization procedure. We will next concentrate on improving the optimization method.

## References

1. S. Kaski, J. Sinkkonen, and J. Peltonen. Bankruptcy analysis with self-organizing maps in learning metrics. *IEEE Transactions on Neural Networks*, 2001. Accepted for publication.
2. T. Kohonen. Self-organized formation of topologically correct feature maps. *Biological Cybernetics*, 43:59–69, 1982.
3. T. Kohonen. *Self-Organizing Maps.* Springer, Berlin, 1995. (Third, extended edition 2001).
4. J. Sinkkonen and S. Kaski. Clustering based on conditional distributions in an auxiliary space. *Neural Computation*, 2001. Accepted for publication.

# Vector Quantization with γ-Observable Neighbors

Michaël Aupetit Pierre Couturier Pierre Massotte

LGI2P – EMA – Site EERIE
Parc Scientifique Georges Besse - 30035 Nîmes - France

Abstract

We define a new neighborhood called "γ-Observable" and investigate its use in Vector Quantization tasks. Considering a datum $\mathbf{v}$ and a set of n units $\mathbf{w_i}$ in a Euclidean space, let $\mathbf{v_i}$ be a point of the segment $[\mathbf{vw_i}]$ whose position depends on $\gamma$ a real number between 0 and 1, the γ-Observable neighbors (γ-ON) of $\mathbf{v}$ are the units $\mathbf{w_i}$ for which $\mathbf{v_i}$ is in the Voronoï region of $\mathbf{w_i}$, i.e. $\mathbf{w_i}$ is the closest unit to $\mathbf{v_i}$. For $\gamma=1$, $\mathbf{v_i}$ merges with $\mathbf{w_i}$, all the units are γ-ON of $\mathbf{v}$, while for $\gamma=0$, $\mathbf{v_i}$ merges with $\mathbf{v}$, only the closest unit to $\mathbf{v}$ is γ-ON of $\mathbf{v}$. The size of the neighborhood decreases from n to 1 while $\gamma$ goes from 1 to 0. For $\gamma$ lower or equal to 0.5, the γ-ON of $\mathbf{v}$ are also its natural neighbors, i.e. their Voronoï regions share a common edge with that of $\mathbf{v}$. We experimentally show that this neighborhood used in VQ gives faster convergence and similar final distortion than the "Neural-Gas" onto synthetic data distributions.

## 1 Introduction

Vector Quantization (VQ) consists in representing a data distribution using a set of units (or "codebook" vectors) such that a distortion measure (energy function) is minimized. In "Hard" Competitive Learning (HCL), only the closest "winning" unit (using euclidean distance) to the current input datum is adapted and moves toward it at each iteration. The problem is that the process rapidly reaches a local minimum of the energy function. In order to reach lower local minima (for lack of the optimum), a neighborhood structure is defined so that all the units in the neighborhood of the winning unit are also adapted but generally to a lower extent than the winner itself. In Self-Organizing Maps (SOMs) [1][2], the neighborhood structure is defined onto an abstract map uncoupled with the data space. The higher the distance on the map, between the winner and a unit, the lower the extent of its adapting step in the data space. The topology of the map is generally predefined and of low dimension, which makes SOMs efficient tools to project and to visualize the data distribution. SOMs have been used for VQ in several works [3][4][5][6] where they provide at the same time a topological representation and a quantization of the data distribution.

However, if one only wants to quantify a distribution without a need for visualization, SOMs are not ideal tools because during the quantization, the units are constrained to respect their predefined topology which might be different from the *a priori* unknown topology of the data distribution.

Another algorithm dedicated to VQ and called "Neural-Gas" (NG) has been presented by Martinetz [7]. In an NG, each time a datum is presented, the units are ranked in the increasing order of their euclidean distance to the input datum. This rank is used to weight the adaptation rule of the units such that the lower their rank, the higher their move toward the input datum in the data space. An NG can be viewed as a one-dimensional SOM changing its structure at each time step and having at one extremity the winner and all along the ranked units. In [7], it is shown that NG gives better results in term of mean distortion obtained than SOM in a VQ task.

The dynamic topology of the NG defines a neighborhood structure similar to a "smooth" k-Nearest Neighbors (k-NN) neighborhood. It is different from that of a SOM (Figure 1a-b) w.r.t three properties which might characterize a neighborhood:

- The structure of the neighborhood: based on euclidean distance in the data space in NGs, based on a topological structure (Delaunay triangulation [8] dual of Voronoï polytopes [9]) in the space of the map in SOMs.
- The number of units at a given activation level: except on the borders, 1 in NGs, 2 in 1-dimensional SOMs, 4 or 6 in 2-dimensional SOMs.
- The center of the neighborhood: the datum in NGs, the closest unit to the datum in SOMs.

These characteristics and the shape of the activation function (usually exponential or binary) might affect the speed of convergence and the final organization of the units. One can think to other neighborhood structures for VQ, based on the Delaunay triangulation of the units *and* possibly the input datum (Figure 1c-d). The Kohonen rule (used in SOMs) might be applied onto the graph structure obtained, each unit being weighted according to its distance on the graph to the winning unit or to the input datum. However, these approaches need heavy computation and memory to define the Delaunay graph at each time step especially in high dimension [8][10]. We propose an alternative neighborhood called the $\gamma$-Observable neighborhood which is closely related to the Delaunay graph (see Section 2) but which needs simple distances and minimum-search or sorting computations. We compare this approach to the Neural-Gas in Section 3.

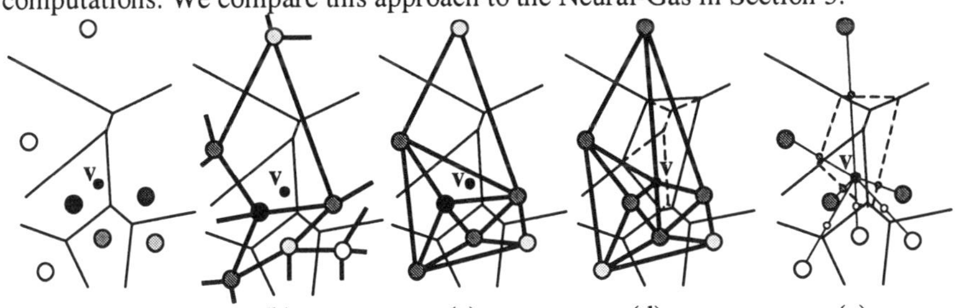


Figure 1: Neighborhood of an input datum **v** for different approaches. Thin plain lines are the edges of the Voronoï regions of the units (large circles) whose dark/bright colors represent high/low activations. Bold plain lines are the explicit neighborhood links between the units. (a) NG. (b) 2-dimensional SOM. (c) Delaunay triangulation of the units. (d) Delaunay triangulation of the units *and* the datum **v**. (e) 0.5-Observable Neighborhood. In (d-e) the neighbors surround **v** which is the center of the neighborhood, while in (b-c) the center is the winning unit and in (a) neighbors of the NG may cluster on one side of **v**.

## 2 The γ-Observable Neighborhood

Considering $W=(\mathbf{w}_1;\ldots;\mathbf{w}_n)$ a set of n vectors in a euclidean space $R^d$ that we call units in the following, the Voronoï region of a unit i over the set W is defined as:

$$V_W(\mathbf{w}_i) = \{\mathbf{v} \in R^d \,|\, \forall j \in [1,n], \|\mathbf{v}-\mathbf{w}_i\| \le \|\mathbf{v}-\mathbf{w}_j\| \}$$

We define the γ-Observable Neighborhood of a vector **v** in $R^d$ euclidean as:

$$N\gamma_W(\mathbf{v}) = \{\mathbf{w}_i \in W \,|\, \mathbf{v}_i \in V_W(\mathbf{w}_i)\} \text{with} \quad \mathbf{v}_i = \gamma \cdot \mathbf{w}_i + (1-\gamma)\cdot \mathbf{v}$$

where γ is a real number between 0 and 1.

This means that a unit $\mathbf{w}_i$ is part of the γ-Observable Neighborhood (γ-ON) of a datum **v** iff $\mathbf{v}_i$, a point of the segment $[\mathbf{v}\mathbf{w}_i]$, is part of the Voronoï region of $\mathbf{w}_i$ (Figure 2). The parameter γ sets the position of $\mathbf{v}_i$ onto the segment $[\mathbf{v}\mathbf{w}_i]$: if $\gamma=1$, $\mathbf{v}_i=\mathbf{w}_i$ and all the units of W are activated (i.e. they are part of the γ-ON of **v**); while if $\gamma=0$, $\mathbf{v}_i=\mathbf{v}$ and only the closest unit $\mathbf{w}_{win}$ to **v** is activated.

At last, if γ is lower or equal to 0.5, we have shown in [11] that the units which are activated are all natural neighbors of **v** (compare Figure 1d and 1e), i.e. **v** would be linked to them if the Delaunay triangulation of W *and* **v** were built.

It appears that the γ-ON may be used for VQ with a dynamic setting of γ during the learning phase which allows to pass from all the units activated to only the closest one to **v**, as in NG and SOM algorithms.

The γ-ON is characterized w.r.t the properties mentioned in the previous section:

- Midway between k-NN and natural neighbors, all the units activated surround the datum (except if the datum is outside the convex hull of the units or possibly if γ is lower than 0.5).
- All the units of the neighborhood are activated at the same level (binary neighborhood).
- The neighborhood is centered onto the datum.

For a given value of γ and a given datum **v**, the number of activated units depends on their relative position in the data space. The name "Observable" comes from the analogy with particular "visibility" problems encountered in computational geometry [12] and described with more details in our previous work [11].

Figure 2 illustrates the γ-ON for different values of γ.

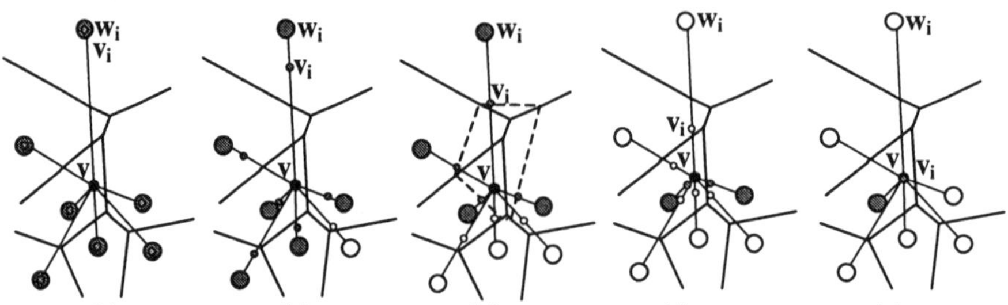


Figure 2: the γ-Observable Neighborhood of a datum **v** for different values of γ: (a) 1; (b) 0.75; (c) 0.5; (d) 0.25 and (e) 0. In (c-d-e), all the units activated are natural neighbors of **v**.

# 3 Experiments

## 3.1 Algorithm

The algorithm to compute the $\gamma$-Observable neighborhood is straight forward following its definition: considering an input datum **v**, the $\mathbf{v_i}$ are computed and going all over W, if $\mathbf{w_i}$ is the closest unit to $\mathbf{v_i}$ then $\mathbf{w_i}$ is $\gamma$-ON of **v**.
The brute force algorithm described above has a $O(d.n^2)$ time complexity. However, the complexity may drop if the algorithm takes into account the following geometrical properties:
A unit $\mathbf{w_i}$ cannot be $\gamma$-ON of a datum **v**, iff it exists another unit $\mathbf{w_j}$ closer to $\mathbf{v_i}$ than $\mathbf{w_i}$, hence if $\mathbf{w_j}$ is inside the hyper-sphere $S_{Vi}$ of radius $R_{Vi}=\|\mathbf{v_i}-\mathbf{w_i}\|=(1-\gamma)\|\mathbf{v}-\mathbf{w_i}\|$ centered on $\mathbf{v_i}$. No unit outside the hyper-sphere $S_{Vout}$ of radius $R_{Vout}=\|\mathbf{v}-\mathbf{w_i}\|$ centered on **v**, can be inside $S_{Vi}$ (Figure 3a-b). Moreover, if $\gamma>0.5$, no unit inside the interior hyper-sphere $S_{Vin}$ centered on **v** and of radius $R_{Vin}=(2\gamma-1)\|\mathbf{v}-\mathbf{w_i}\|$ can be inside $S_{Vi}$ (Figure 3a). If the units are ranked in increasing order of their distance to **v**, then it is easy to find the units which are inside or outside $S_{Vin}$ or $S_{Vout}$ and the number of units to test may be dramatically lower than n, lowering the time complexity (see Section 3.3). The optimized algorithm is then:

```
Compute the distance of each unit to v;(O(d.n))
Rank the units in increasing order of their distance to v;( O(n.log(n)) )
For each unit wi in that increasing order: (O(n))
   Compute di=||vi-wi||=(1-γ)||v-wi||; (O(1))
   Find the rank rkwstart of the first ordered unit wstart outside Svin; (binary search O(log(n)))
   Assume wi is γ-ON of v; (O(1))
   For each unit wj whose rank is in [rkwstart, rkwi[ : (O(n))
      compute dj=||vi-wj|| ; (O(d))
      if (dj<di) break this loop, wi is not γON of v;
   Endfor
Endfor
```

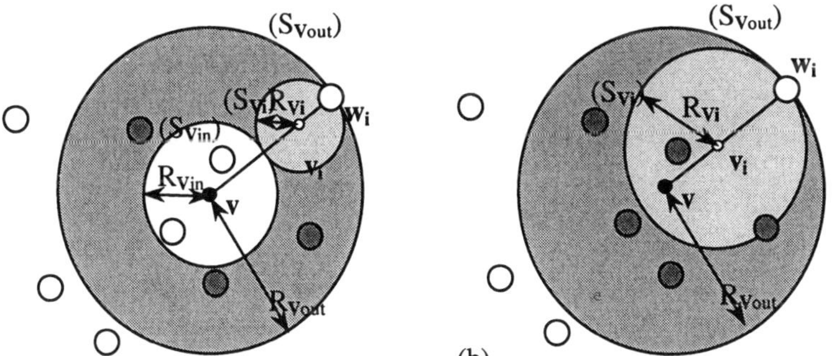


(a) (b)

Figure 3: Considering $\mathbf{w_i}$ part of a set of units (large circles) and a datum **v** (black point), (a) for $\gamma>0.5$, only the units (dark gray) in the gray area (inside $S_{Vout}$ and outside $S_{Vin}$) need to be tested: if $S_{Vi}$ (light gray circle) does not contains another unit than $\mathbf{w_i}$ , then $\mathbf{w_i}$ is $\gamma$-ON of **v**, else $\mathbf{w_i}$ is not $\gamma$-ON of **v**. (b) for $\gamma<=0.5$, all the units inside $S_{Vout}$ need to be tested. In (a), $\mathbf{w_i}$ is $\gamma$-ON of **v**, but not in (b).

## 3.2 Neural-Gas *vs* γ-Observable Neighbors

We compare the effect of the neighborhood structure of Neural-Gas and γ-ON algorithms onto the speed of convergence over four different distributions represented on Figure 4.

The learning rule is of the form: $\forall i \in \{1,\ldots,n\}, \Delta \mathbf{w}_i = \alpha \cdot h_i(\xi, \mathbf{v}, W) \cdot (\mathbf{v} - \mathbf{w}_i)$ where $h_i(.)$ is the neighborhood function which differs for NG and γ-ON.

In order to make both approaches comparable (their structure being different), we define a binary NG to compare with the binary γ-ON and an exponential γ-ON to compare with the exponential NG, and we optimize the width of the neighborhood for all the approaches, all other parameters being fixed.

The exponential NG (NGexp) uses $h_i(.)=e^{(-k_i(\mathbf{v},W)/\lambda)}$ where λ set the width of the neighborhood such that the higher λ, the larger the neighborhood, and for λ=0 only the winner is activated. $k_i$ is a rank equal to j-1 for the $j^{th}$ closest unit to **v** [7]. The binary NG (NGbin) uses a threshold T such that if $e^{(-k_i(\mathbf{v},W)/\lambda)}>T$, $h_i(.)=1$ else $h_i(.)=0$. The binary γ-ON (GONbin) is such that $h_i(.)=1$ if $\mathbf{w_i}$ is part of the γ-ON of **v** ($\mathbf{w_i}$ is activated), else $h_i(.)=0$. The exponential γ-ON (GONexp) uses the sorting of the optimized algorithm such that $h_i(.)=0.01^{ki(\mathbf{v},W)/N}$ where $N=\mathrm{card}(N\gamma_W(\mathbf{v}))$ is the number of γ-ON of **v** and $k_i$ is the rank of the γ-Observable unit $\mathbf{w_i}$ according to its distance to **v** among all the γ-ON of **v**. We get $h_{win}(.)=1$ for the closest unit $\mathbf{w_{win}}$ to **v** ($k_{win}=0$) and $h_i(.)=0.01^{(N-1)/N}$ for the farthest γ-ON unit to **v** ($k_i=N-1$).

We run these algorithms over each distribution using 50 different initial positions of the units in the d-dimensional hyper-cube $[0, 10]^d$, the initial position being the same for each algorithm. The parameters $\xi=\{\lambda,\gamma\}$ decrease exponentially during the learning phase from $\xi_i$ to $\xi_f$ following the rule: $\xi=\xi_i*(\xi_f/\xi_i)^{t/t_{max}}$ where t is the current and $t_{max}$ the total number of training steps. We set α=0.1, T=0.5, $t_{max}$=15000, n=16, and took $\gamma_i$ in {1,0.9,0.8,0.7,0.6,0.5,0.4,0.3,0.2,0.1}, $\lambda_i$ in {500,100,50,10,5,1}, and $\lambda_f$ and $\gamma_f$ in {0.1,0.05,0.01}. For each algorithm onto each distribution with a given couple ($\xi_i$,$\xi_f$), we compute the mean over the 50 trials of the MSE evaluated every 500 steps over 1000 samples **v** randomly chosen in the distribution. Figure 5 shows the best results (i.e. giving the lowest final distortion and the fastest convergence toward it) we obtained for each algorithm. The captions give their corresponding optimal set of parameters ($\xi_i$,$\xi_f$) tested.

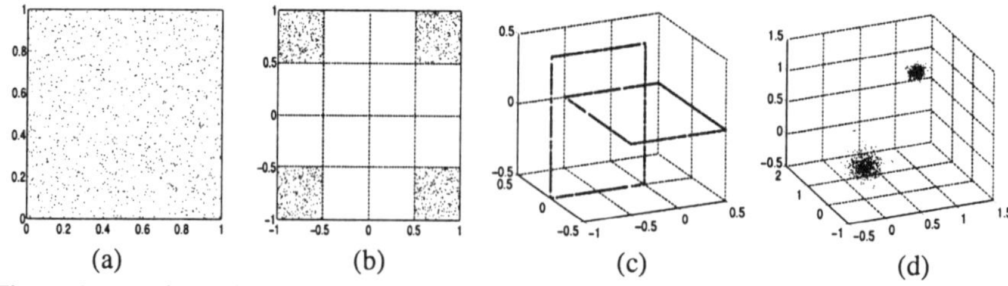

(a) (b) (c) (d)

Figure 4: quantized distributions: (a) 1 uniform square (d=2), (b) 4 uniform squares (d=2), (c) two interlaced square rings (d=3), (d) 3D view of two gaussian distributions in a 10-dimensional space with 0.1 and 0.05 standard deviations and (0,…,0) and (1,…,1) means respectively (d=10).

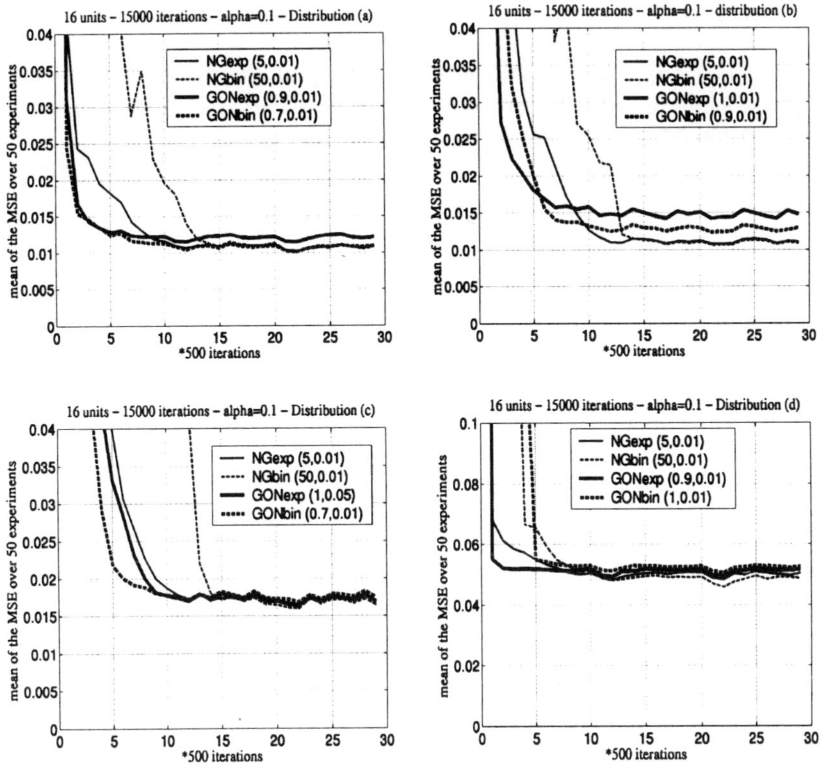


Figure 5: comparison between the best algorithms tested: NGexp, NGbin, GONexp and GONbin. The first and second numbers which appear after the name of the algorithm are $(\lambda_i,\lambda_f)$ for NGs, and $(\gamma_i,\gamma_f)$ for $\gamma$-ONs respectively.

Onto the distribution a) and c), GONbin performs clearly better than the other algorithms. Onto the distribution b), although both GONs converge faster than NGs, they do so toward a higher distortion error. Onto the last distribution d), GONexp converges faster than the other algorithms. A faster convergence is important not only because it increases the speed of the learning phase, but especially in cases where the data arc not stored but need to be processed on the fly, i.e. in cases where the vector quantizer must rapidly catch the structure of the data distribution with less data.

Both types of neighborhoods are not so trivial to compare because the size of the NG's neighborhood only depends on the $\lambda$ parameter, while in $\gamma$-ON, it depends on $\gamma$ and on the relative position of the units (our previous work [11] studies these dependencies with more details).

We have some clues about the reason why $\gamma$-ON gives quite good results: the $\gamma$-ON is not "convex" w.r.t W in the sense it may exist some units inside the convex hull of the activated units (i.e. part of the $\gamma$-ON of a datum) which are not themselves activated. This property seems to homogenize the distribution of the units during the learning phase, i.e. to *self-organize* it with another meaning than in SOMs. Yet, this hypothesis need experimental proving.

## 3.3 Time complexity

Now, we compare the time complexity of NG, and brute force and optimized γ-ON approaches onto the distribution a) (d=2) running[1] for 10000 iterations with n={5,10,50,100,500,1000,5000} units.

For all approaches, the time only depends on the dimension d (appearing in distances computations) and on the number of units n (appearing in loops, sorting and minimum-search computations). For the optimized γ-ON, the time also depends on the value of γ. For γ=0, the complexity is O(d.n) (only the closest unit to **v** is activated: distances in O(d.n) and minimum-search in O(n)), while for γ=1, the complexity is O(1) (all the units are activated: no distance and no search). In both cases the complexity is lower than that of the NG but these cases are marginal appearing at the very beginning or end of the learning phase.

Figure 6 shows that the brute force algorithm is in $O(n^2)$ while for the optimized algorithm, the complexity is in the worst case tested (γ close to 1 and n=5000 units) one order of magnitude higher than for NG and between twice and three times higher in the best cases (less than 100 units). The worst cases correspond to high values of γ, but γ being decreased exponentially toward low values, the complexity is likely to be closer to its lower bound than to its higher one.

Compared with the results of the previous section, the lead of γ-ONs over NGs in term of number of iterations to converge, approximately compensates for its higher time complexity. However, γ-ONs still interesting when the computation time is not a determining factor. Moreover, the operations involved in the brute force γ-ON algorithm are simply distances and minimum-search computations, which lets foresee the possibility of hardware implementations [13] which could drop the time complexity to O(d.n).

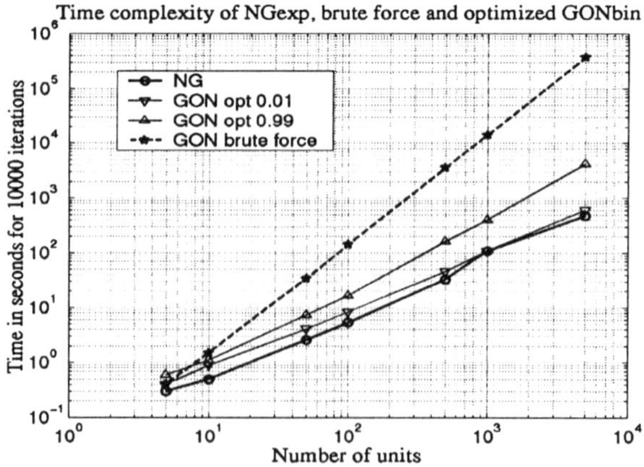


Figure 6: time complexity of NG compared to brute force and optimized γ-ON run onto the distribution a) during 10000 steps. The time curves of the optimized γ-ON are only drawn for γ=0.01 and γ=0.99, the curves for other γ between 0.01 and 0.99 are between these two curves with higher positions for higher γ.

---

[1] in C onto a sparc station ultra-5_10, with a sparcv9 processor 360MHz under SunOS 5.8.

# 4 Conclusion

We present the γ-Observable neighborhood which is closely related to the natural neighborhood but remains simpler to compute and needs not to store a Delaunay graph structure which may be huge in high dimensions. Used in VQ, the γ-ON performs better in term of speed of convergence on several distributions than NGs, and gives similar results in term of the distortion obtained.

In our recent work [11], we use the γ-ON to approximate the natural neighborhood and investigate its behavior in dimensions up to 20. In addition to its properties in Vector Quantization, the γ-ON is a way to explore topological properties of data distributions in high dimensions.

**References**

[1] Kohonen T. Self-organization and associative memory. Springer Verlag, 1988.

[2] Ritter H, Martinetz T, Schulten K. Neural computation and self-organizing maps. Addison&Wesley, 1992.

[3] Kohonen T, Mäkisara K, Saramäki T. Phonotopics maps – insightful representation of phonological features for speech recognition. In Proc. of 7th Int. Conf. On Pattern Recognition, Montréal, 1984, pp. 182-185.

[4] Makhoul J, Roucos S, Gish H. Vector quantization in speech coding. Proc. IEEE, 1985, vol. 73, pp.1551-1588.

[5] Nasrabadi N M, Feng Y. Vector quantization of images based upon the Kohonen self-organizing feature maps. In IEEE Int. Conf. On Neural Networks, San Diego, CA, 1988, pp. 1101-1108.

[6] Ritter H,Martinetz T, Schulten K. Topology-conserving maps for learning visuomotor-coordination. Neural Networks, 1988, vol.2, pp. 159-168.

[7] Martinetz T, Berkovitch S, Schulten K. A "neural-gas" network for vector quantization and its application to time-series prediction. In IEEE Trans. on Neural Networks 1993, vol. 4, 4:558-569.

[8] Fortune S. Voronoï diagrams and Delaunay triangulations. Computing in Euclidean Geometry, D.Z.Du, F.Hwang eds. World Scientific 1992, pp.193-233.

[9] Okabe A, Boots B, Sugihara K. Spatial tessellations: concepts and applications of Voronoï diagrams. John Wiley, Chichester 1992.

[10] Barber C B, Dobkin D P, Huhdanpaa H. The Quickhull algorithm for convex hulls. ACM Trans. on Mathematical Software, December 1996, vol. 22, 4:469-483. Url:http://www.geom.umn.edu/locate/qhull.

[11] Aupetit M, Lepetz D, Nemoz-Gaillard M, Couturier P, Massotte P. Réseaux de neurones et traitement de données: la notion de voisinage γ-observable. In Valgo 2001, ACTH, April 2001. Url:http://www.supelec-rennes.fr/acth/valgo/Valgo_Numero-01-01.html.

[12] Eidenbenz S, Stamm C, Widmayer P. Positioning guards at fixed height above a terrain, an optimum inapproximability result. Proc. of Euro. Symposium on Algorithms, 1998.

[13] De Trémiolles G, Tannhof P, Plougonven B, Demarigny C, Madani K. Visual probe mark inspection, using hardware implementation of artificial neural networks, in VLSI production. Proc. of IWANN1997, Lanzarote, Canary Island, Spain 1997.

# Dynamic Vector Quantization Of Speech

Shayan Garani, José C. Principe

Computational Neuro Engineering Laboratory, University Of Florida
Gainesville, FL 32611-6130, U.S.A
Email: {shayan, principe}@cnel.ufl.edu

Abstract

In this paper we describe a more robust vector quantization model for speech based on spatiotemporal clustering. The principle of self-organization in time is employed to create traveling waves, which are gradually attenuated over time and space to diffuse temporal information and create localized temporal neighborhoods for classification of speech. We also define performance functions for quantification of this method, based on the time varying property of Voronoi cells. This method is better than conventional static vector quantizers built on the same data set and similar training conditions.

## 1 Introduction

Traditional Neural Networks for vector quantization of speech have been mainly concerned with static pattern mappings. Such vector quantizers can act as content addressable memories for static pattern processing. Most of the real life applications like object motion recognition/tracking; automatic speech recognition, time series prediction etc. benefit from paradigms that are capable of processing spatio temporal signals. Spatio temporal signals such as speech have an implicit time dimension. Hence neural network models capable of incorporating both the spatial properties and temporal information for learning, memorizing and recognizing patterns should perform better than their static counterparts.

Temporal Self Organizing Networks with internal Short Time Memory (STM) mechanisms are important due to their biological plausibility and several models have been proposed for handling sequential patterns for recognition and recall [2][3][7]. The key concept in most of the models is that the temporal order within input patterns is captured and the system is trained using unsupervised Hebbian learning rules to form self-organized memories.

In this paper, the concept of Reaction Diffusion equations [5] has been applied to create unique spatio-temporal vector quantizers for speech recognition. The key concept involved here is the creation of traveling waves [6] as a result of the excitation and later relaxation of a processing element in the network. This traveling wave progressively decays over time with the presentation of temporal patterns while the spatial information is diffused through the ranking of the neurons based on Euclidean distance, without the imposition of a pre-defined neighborhood in accordance with the Neural Gas Algorithm [4]. Thus time information is directly brought into the competitive mechanism. The coupling of the spatial information with temporal activity is controlled by a spatio temporal coupling parameter. We define performance functions based on the spatio temporal distance and volume functions suited for the dynamics of time varying Voronoi tessellations, to quantify the accuracy of the classifier. The discussion shall be outlined in the section describing the system model.

# 2 Reaction Diffusion Mechanism

Biological process involved in the generation of iterated structures from undifferentiated entities is a type of temporal sequential process. A quantitative explanation of this process is *Reaction-Diffusion*, initially proposed by *Alan Turing* [5]. Reaction Diffusion is a spatiotemporal process consisting of a set of discrete partial differential equations in time and space governing the space-time concentration of chemical morphogens responsible for the stimulation of growth. The Reaction-Diffusion (R-D) equations permit calculation of growth, diffusion and decay of each morphogen in turn, which effectively propagates the effect of previous activity forward in time while distributing the information spatially by diffusion. This biological mechanism can be exploited to engineer a memory model for spatio temporal processing. A generic spatio temporal model comprising of a lattice of cells suffused in a medium of morphogens is characterized by the following partial differential equation

$$\frac{\partial A}{\partial t} = F(A) + D\nabla^2 A \qquad (1)$$

where, $A$ is the vector of morphogen concentrations, $F$ represents the reaction kinetics and $D$ is the diagonal matrix of positive diffusion coefficients.

The concept of R-D equations, coupled with the use of leaky integrators and activity wave fronts can be used to create space-time memories that can store information in the activity of a distributed neural network. Unlike the STM architectures, which transform time information to spatial information, this model creates self-organization in time and space by creating temporally correlated neighborhoods in the output space.

# 3 Spatio Temporal VQ Model

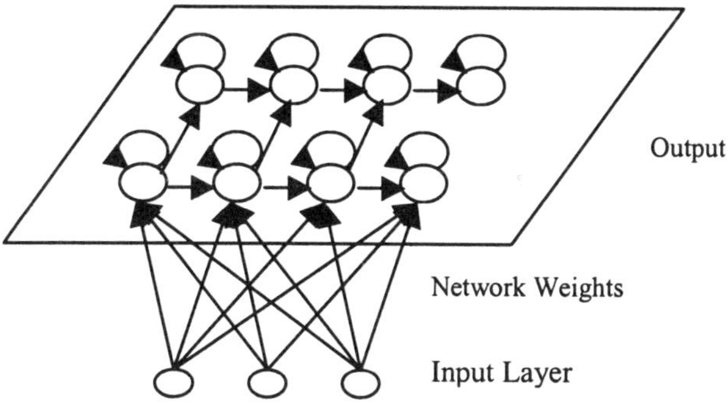


Figure 1. Network Model with spatio temporal connections

In this section we shall outline the VQ algorithm based on spatiotemporal processing. A schematic of the model representation is shown in Figure 1. This serves as a preprocessor engine for a simple speech recognition system.

## Outline of the VQ Algorithm

Given a set of *M* feature vectors $S \in R^D = \{X_1, X_2 .......... X_M\}$ of a signal *S*, we have the following algorithm. Let the codebook size be *N* where *N* is the total number of neurons in the network.

*Step 1:* Determine the winning node *k* according to the equations

$$d = \|X_n - W\| - \mu \times \overline{a} \quad (2a)$$

$$k = \arg\min(d) \quad (2b)$$

where, $\mu$ is the spatiotemporal coupling parameter, *W*- weight vector and $\overline{a}_n$ - activity diffusion at time instant $'n'$.

*Step 2:* Update the activity function as

$$\overline{a_{n+1}} = \overline{a_n} + \frac{[\alpha + (1-\alpha)a_{n,k}]}{\max(C)} \overline{c_{kn}} \quad (3)$$

where, $\alpha$ denotes the rate of decay of the temporal wave front, $a_{n,k}$ represents the activity corresponding to the winning node $k$ at time instant $'n'$, $\overline{c}_{kn}$ denotes temporal connectivity strength from the winner to all other nodes, $C$ is the temporal connectivity matrix (Matrix of dimension $N \times N$ )

The connection strength between the previous winning node $l$ and the current winning node $k$ is updated according to the equations (4a) and (4b)

$$c_{l,k} = c_{l,k} + \Delta \quad (4a)$$

The connectivity strengths of other nodes $j$ with respect to the previous winning node $l$ are normalized.

$$c_{l,j} = \frac{c_{l,j}}{\sum_{j=1}^{N} c_{l,j}} \qquad \forall j = 1, ....... N \quad (4b)$$

where, $\Delta$ is a small positive constant, typically $\Delta = \frac{1}{N}$ where $N$ is the number of neurons.

*Step 3:* Update the weights in accordance with the Neural Gas algorithm [4]

The above three steps are involved in the formation of the self-organized memory. The clustering now needs a performance measure for quantification. The choice of winner based on the distance to weights is not a proper estimation to quantify the dynamics of this model. This is because the Voronoi regions associated with the spatiotemporal model are time varying due to the input signal dynamics even when the weights are fixed. So we present a theoretical discussion on a new methodology to quantify this 'Temporal Plasticity' effect [2] as follows:

- From a set of uniformly distributed random numbers V in $R^D$, determine the nearest neighbors associated with each vector based on the parameters $a$, $W$, $C$ and the winning node $k$ using the spatiotemporal distance measure in equation (2a).
- Estimate the pillbox density $\phi_n(X, k_n)$ associated with the Voronoi cell of the winner $k_n$ at time $'n'$. In the discrete case this is nothing but the fraction of the number of vectors, which fall in the Voronoi cell of the winning node to the total number of vectors spanning the input space.
- Compute the incremental accumulated volume increase associated with successive winner nodes over time according to the equation

$$V = \sum_{n=2}^{n=M} \int_{R^D} \phi_n(\overline{X}, k_n) - \phi_{n-1}(\overline{X}, k_n) d\overline{X} \tag{5}$$

where $k_n$, is the winning node at time $'n'$

- Evaluate the overall cost metric, we define as the volume function to declare the index of the class.

$$S \in \omega_j \quad if \quad V_j(S) = \max\{V_i(S)\} \qquad \forall i = 1, \ldots\ldots C \tag{6}$$

where $C$, is the total number of classes and $V_i(S)$ is the volume function associated with class $w_i$.

The problem now is how do we estimate the volume functions during operation. Letting uniformly distributed random vectors spanning $R^D$, for the estimation of the probability density function associated with each cell, the time complexity is approximately $O(N^D)$ where $D$ is the dimensionality associated with the input space. Since computing this is a huge burden and is impossible for all sets of patterns, we have to search for another variable related to the volume functions and available in the model. We can therefore propose to estimate the increase in the Voronoi cells with the value of $a_k$ in equation (3). The quality of this parameter is experimentally verified by the correlation coefficient between the volume functions and the enhancement in the network for a few test cases.

## 4 Experiments on Speech and Simulation Results

Speaker Independent Speech Recognition [8] is a challenging task. Speech parameters are most often obtained using some type of spectral analysis that concentrates on the short time characteristics of the speech signal. The non-stationary nature of the signal enforces to make spectral measurements sequentially over time, producing a sequence of spectral feature vectors, which need to be effectively vector quantized for recognition purposes. In our example, the speech recognizer comprises of a front end feature extractor constructed from Partial Correlation coefficients approximating the reflection coefficients in the vocal tract model for speech, followed by spatio temporal vector quantizer and a performance function to classify the digits. The crux of the model is in the construction of these vector quantizers and the performance function.

Template training by clustering [8] is required to achieve high word recognition rates for practical purposes. However in the spatio temporal based approach, we choose the index of the class that provides the maximum temporal enhancement for a given speech pattern as a criterion for the recognition index.

## Simulation Example #1 – Speaker Independent Recognition

The data set is chosen from TI-46 database containing Isolated Words. The database comes with a division as training set and testing set. 1600 words (10 different words × 16 different speakers) from 10 different sessions were chosen in the training set and all the words roughly 2560 (10 different words of the same class as the training set × 16 different speakers) in 16 different sessions were chosen in the test set. For each class, isolated code books are developed based on spatio-temporal dynamic Vector Quantization.

In order to train an individual codebook, we choose 800 adaptation steps and selected 20 neurons as a design parameter. The initial learning rate was chosen as 0.1 and exponentially decayed to 0.001 over 800 epochs. Similarly the annealing parameter was chosen as 10 initially and exponentially decayed to 0.001. The spatio temporal coupling parameter was chosen as a raised sinusoid with peak amplitude 0.2. However in general we can consider functions increasing in time, reaching a peak and then decaying to zero as coupling functions. This allows the network to learn the patterns without any rigid distribution. The temporal enhancement in the network reaches a peak and decays slowly for fine-tuning at the end of learning.

Feature vectors belonging to a particular class, from the training set were interspersed with random vectors from other classes. Figure 2. shows the activity as a function of time for the utterance of the digit 'eight'. It has to be noted that the temporal enhancement plot is evolved from unsegmented speech and is derived purely from unsupervised learning. It can also be noted from the plot that one of the speakers just met the threshold for the integrated average activity. The overall recognition accuracy for this approach is 93.08% on testing set and 96.81% on the training set. However with similar training conditions and MSE as a criterion for distortion, the static VQ had an overall recognition rate of 91% on the training set and 86.64 % on the testing set.

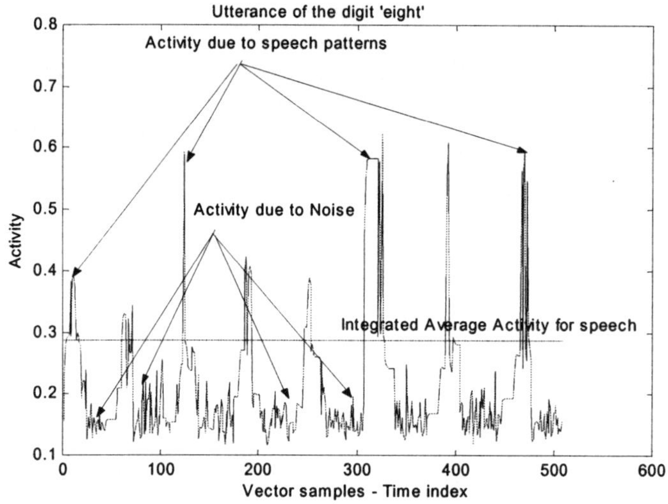


Figure 2. Temporal Activity for utterances of eight

The following plot shows the correlation between the volume functions computed by equation (5) to the associated enhanced activity in the network. The x-axis in Figure 2. corresponds to time index and the y-axis corresponds to the normalized magnitudes. The correlation coefficient between the volume function and the enhancement was 0.9481. In this computation we have neglected consecutive winners, as this is a detection problem.

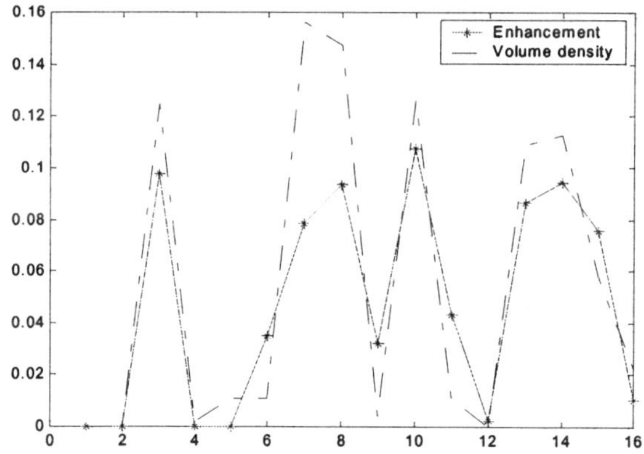


Figure 3. Plots showing volume function and activity enhancement

## Simulation Example #2 - Effect of Noise

In a second experiment we consider a speaker dependent recognition problem in the presence of babble noise [9]. We choose babble noise (Figure 3) as it is correlated over time and resembles the spectrum of speech and this makes it very difficult for automatic speech recognizers. To study the effect of noisy speech recognition, we focused on one of the speakers in the database. Ten different words spoken in eight different sessions formed the training set and a same number was randomly chosen for the testing set. The variance of noise was varied to get different SNRs. The following results show the effect of SNR versus recognition accuracy for the speaker.

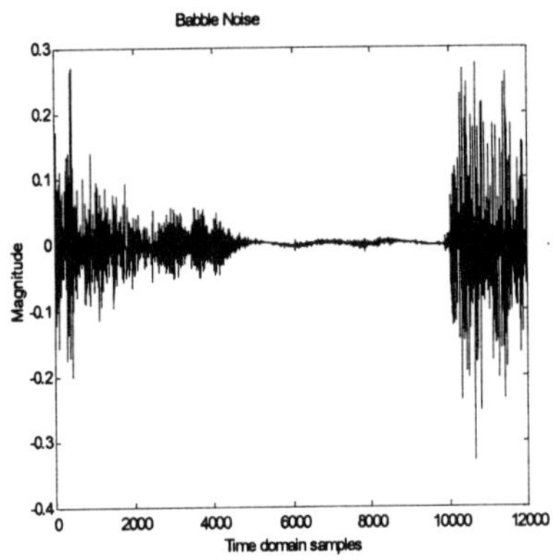


Figure 3. Time domain response

| SNR (In dB) | Recognition Rate based on Static VQ | Recognition Rate based on Dynamic VQ |
|---|---|---|
| ∞ | 88.50 | 95.50 |
| 12 | 80.75 | 85.50 |
| 6 | 70.50 | 77.25 |
| 0 | 56.25 | 68.75 |

Table [I] Recognition Rate

It is interesting to note that the recognition accuracy is fairly good for low SNRs. This shows that temporal enhancement is able to handle noisy speech.

# 5 Conclusions

In this paper we discuss a spatiotemporal vector quantizer and quantify the model for speaker independent recognition of isolated words. The effect of babble noise on speaker dependent recognition is also studied. The approach is unique in a way that it works with temporal information in self-organizing networks. The recognition accuracies are better than other static vector quantization models and are more robust to noise. The improvement in the recognition rate is due to the temporal information that is available with the spatiotemporal model. However better recognition accuracies can be obtained by a careful choice of the energy thresholds to eliminate any outliers during the preprocessing stage of end point detection and feature extraction.

**Acknowledgements** - This work was partially was partially supported by a NSF grant ECS-9900394.

**References**

1. Euliano N. R, Principe J. C. A spatio-temporal memory based on SOMs with activity diffusion, Workshop on Self-Organizing Maps, Helsinki, Finland, 1999, 253-266.
2. Euliano N. R. Temporal Self-Organization for Neural Networks, PhD Dissertation, Graduate School of University of Florida, Gainesville, USA, 1998.
3. Euliano N. R, Principe J. C. Spatiotemporal self-organizing feature maps, in Proc IJCNN 1996.
4. Martinetz T. M, Berkovich S. G, Schulten K. J. Neural Gas Network for Vector Quantization and Application to Time Series Prediction, IEEE Transaction on Neural Networks, July 1993, 4:558-569.
5. Turing A. The Chemical basis of Morphogenesis, in Phil Transactions of the Royal Society of London, Ser.B, 1952; 237:37-72.
6. Murray J. D. Mathematical Biology, Springer-Verlag, New York, 1989.
7. Baretto G. A, Araujo A. F. Time in Self-Organizing Maps: An overview of models.
8. Rabiner L, Juang B. H. Fundamentals of Speech Recognition, Prentice Hall Signal Processing Series, 1993.
9. ICRA Noise Signals ver.0.3 CDROM, International Collegium of Rehabilitative Audiology, Feb 1997.

# Optimization of Electronic Parts Mounting Machines using SOM-TSP Method with 5 Dimensional Data

Kikuo Fujimura[1], Shin-ichi Fujiwaki[1], Obu-Cann Kwaw[1], and Heizo Tokutaka[1]

Tottori University, Koyama Minami 4-101, Tottori 6808552, JAPAN

**Summary.** The SOM-TSP method can easily handle increased dimensions as compared with other TSP methods, such as Hopfield method, Chaotic method ... etc. This offers the possibility where the method can be applied to various optimization problems. In this paper, we propose the application of SOM-TSP method in optimizing the efficiency of surface mounting of electronic parts on the printed circuit board.

## 1 Introduction

The SOM-TSP(**S**elf-**O**rganizing **M**ap for **T**raveling **S**alseman **P**roblem) method is one of the applications of one-dimensional SOM (Self-Organizing Map)[1]. We can obtain the sub-optimum (sub-shortest) tour-length (not the shortest tour-length) in a short time on TSP (Travelling Salesman Problem) by SOM-TSP method as compared to other methods. Although, the SOM-TSP method is useful for industrial application, the theoretical minimum solution is not needed to solve the problem.

Here, we introduce the novel algorithm for the optimization of mounting electronic parts on a printed circuit board using SOM-TSP method[2],[3] with 5 dimensional data. It was found that the required time for mounting electronic parts will be decreased by our proposed method as compared to the built-in method on the mounting-system, through our numerical experiment.

## 2 Problem

Figure 1 is a schematic diagram of the Automatic Electronic Parts Surface-Mounting Machine which is used to mount electronic parts on the printed circuit board. The machine is composed of one XY stage, rotary type handler (with multi head), reels, and one feeder axis.

The printed circuit board is fixed on the XY stage. Electronic chip parts rolled in one reel are used. The reels are set to the feeder axis. The XY stage and feeder axis are movable in the machine.

A rough mechanism that electronic parts are mounted on the printed circuit board is explained here. Electronic parts are picked up at point (a) by

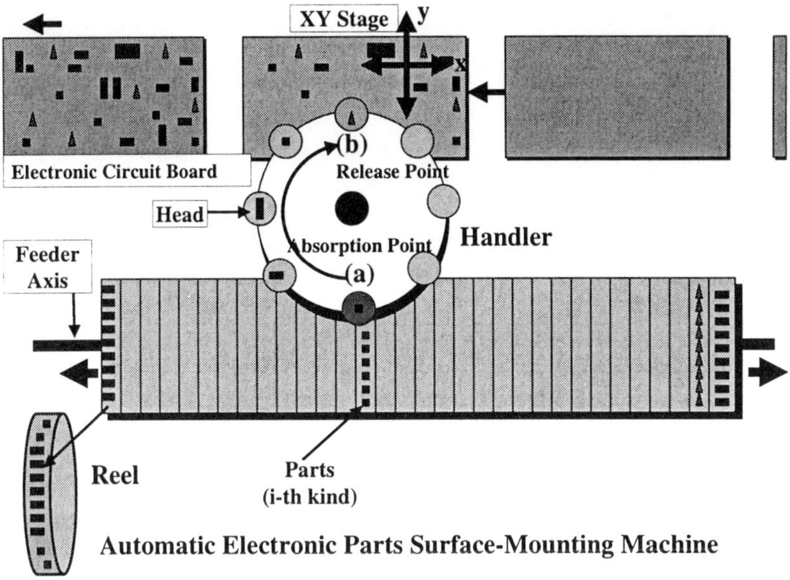


**Fig. 1.** A schematic diagram of the Automatic Electronic Parts Surface-Mounting Machine.

the head in which the handler is installed. The head moves to release point (b) by the rotating handler. As for the XY stage, the position on the board in which elctronic parts will be installed is controlled to agree with the release point (b) (positioning). After positioning, the parts that were picked up by the head are dropped (released).

The minimum Mounting-time can not be obtained by only calculating the shortest tour length for the position of the electronic parts on the circuit board. Moreover, the movement of the feeder axis is slower than the XY stage, because the feeder axis which includes reels is very heavy. These factors complicate this problem further. If we can find the optimum combination of the reel order and the mounting order of electronic parts, the mounting-time will be minimized.

## 3 Algorithm

We used two processes in order to solve this problem. The notation below will help explain our algorithm.

- Kind of electronic parts $S_i$ $(i = 1, 2, \cdots, n)$:
  ("i-th" means that the kind of parts, take the value $i = 1, 2, \cdots, n$);

- Total number of electronic parts $q$ :
  "k-th" means that the parts found in the k-th record in the original *data file*, take the value $k = 1, 2, \cdots, q$ ;
- Each location of electronic parts: $\mathrm{p}_k$
  $\mathrm{p}_k = (x_{i,j}, y_{i,j})$ ;
  where index j means j-th parts in i-th kind of parts $S_i$, where $j = 1, 2, \cdots, m_i$ and $m_i$ is the maximum number of i-th kind of parts.

The algorithm for mounting electronic parts is as shown in the flow chart in figure 2

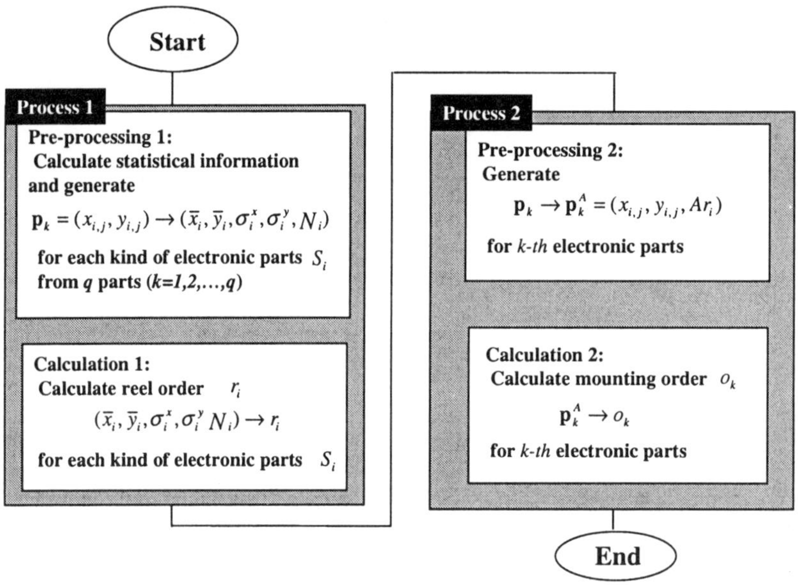


**Fig. 2.** The algorithm for the problem of mounting electronic parts on the printed circuit board.

Process 1: Calculate Reel Order $\{r_i\}$

The order of installing the reel on the feeder axis is calculated. We calculate center of gravity $(\bar{x}_i, \bar{y}_i)$ for each kind of electronic parts $S_i$ as

$$\bar{x}_i = \frac{x_{i,1} + x_{i,2} + \cdots + x_{i,m_i}}{m_i}. \text{ and } \bar{y}_i = \frac{y_{i,1} + y_{i,2} + \cdots + y_{i,m_i}}{m_i}. \quad (1)$$

where,

$x_{i,j}$ : x coordinates and $y_{i,j}$ : y coordinates of j-th parts$S_i$.

The standard deviations of $\sigma_i^x$ and $\sigma_i^y$ for each electronic parts $S_i$ is calculated as

$$\sigma_i^x = \sqrt{\frac{(x_{i,1}-\bar{x}_i)^2+(x_{i,2}-\bar{x}_i)^2+\cdots+(x_{i,m_i}-\bar{x}_i)^2}{m_i}} \quad (2)$$

$$\sigma_i^y = \sqrt{\frac{(y_{i,1}-\bar{y}_i)^2+(y_{i,2}-\bar{y}_i)^2+\cdots+(y_{i,m_i}-\bar{y}_i)^2}{m_i}}. \quad (3)$$

The number of elements $N_i$ was introduced as the fifth dimension (eq.(4)). The optimum value for $\beta$ in this equation was identified to be 750.

$$N_i = \beta log_e\{\ n(S_i)\} \quad (4)$$

The average, the standard deviation and the number elements of each electronic parts $S_i$ are treated as five-dimensional data $(\bar{x}_i, \bar{y}_i, \sigma_i^x, \sigma_i^y, N_i)$.

All the data of electronic parts were merged, and one data set was generated. This data set is used for the calculation of the SOM-TSP method with supplementary cities (described in later section). The order of the visit obtained as a result is used for the order of reels $r_i$.

Process 2: Calculate Mounting Order $\{o_k\}$

The mounting order of the parts on the circuit board is calculated. Data $\mathrm{p}_k^A = (x_{i,j}, y_{i,j}, Ar_i)$ is generated for each k-th parts from the following three elments; x and y cordinates from $\mathrm{p}_k$ for k-th parts and $Ar_i$ which is $r_i$ multiplied by a parameter $A$ . The parameter A is used to control the allowed range of reel skippings. If the value of parameter A is large, the allowed range of reel skippings becomes narrow. In the opposite case, the range becomes wide. Parameter A can be adequately selected according to the performance of the mounting-machine which is used. The data set $\{\mathrm{p}_k^A\}$ is generated and it is used for calculation of SOM-TSP with supplementary cities (described in later section). The order which was obtained from the calculation becomes the set of the mounting order $\{o_k\}$, for the given problem.

### Supplementary cities

In principle, only a closing looped route can be obtained by original SOM-TSP method. The order cannot be obtained even if the route is found, because the beginning point and the ending point are not defined ( as shown in figure 3b). The SOM-TSP method with supplementary cities[5] is introduced to improve this defect ( as shown in figure 3a).

## 4 Simulations and Results

Performance of our algorithm was evaluated using actual printed circuit boards. The performance was finally evaluated on the required time for mounting (this time is known as "**mounting-time**"; as shown in several figures in this paper). The mounting time was measured using the simulator developed for the targeted Mounting Machine.

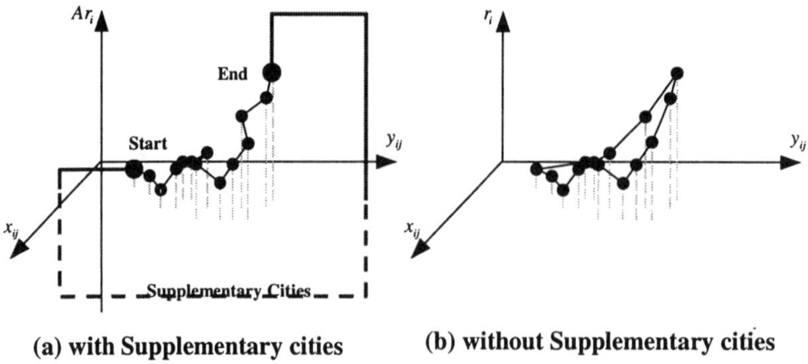


**Fig. 3.** (a) with Supplementary cities and (b) without Supplementary cities.

### 4.1 Case of 109 kinds and 716 parts; max. 0.2[sec/parts] machine

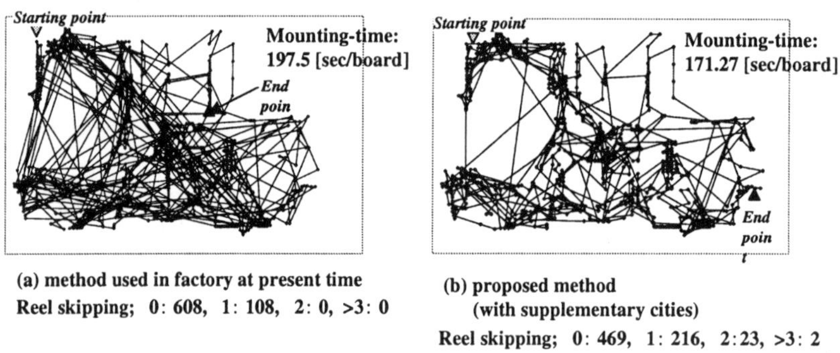


**Fig. 4.** Obtained tracing routes of handler for the problem with 109 kinds and 716 parts; (a) method used in factory at present time and (b) proposed algorithm (five-dimension). The characteristics of the reel skiping are compared.

The printed circuit board with 109 kinds and 716 parts was used for the first evaluation of our proposed method. The mounting-time was 197.5 seconds per board using built-in method on the mounting machine. When our proposed method was used for the same board, the mounting time was reduced to 171.3 seconds (See figure 4). In figure 4, we illustrates the tracing routes for the factory method and our proposed method. It was found that the tracing route obtained by the method had fewer overlaps among lines. From the comparison of these figures (tracing routes), we can estimate at a glance that our method resulted in a better performance.

The relationship between parameter A and mounting time was found through our previous experiments[5]. The mounting time by our method was also found to be optimized with parameter A in the range of 2000 to 3000.

### 4.2 Case of 48 kinds and 376 parts; max. 0.1[sec/parts] machine

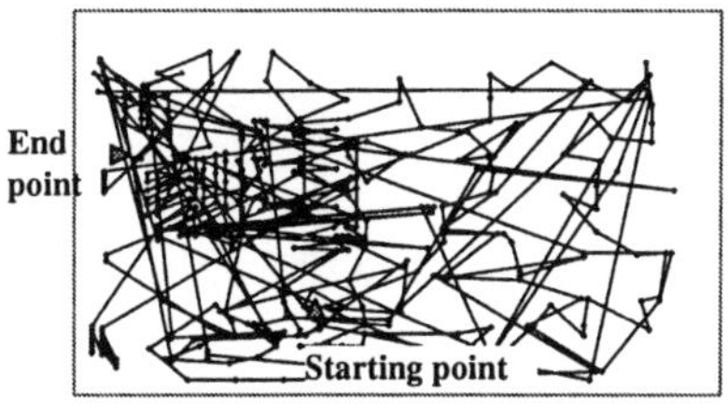


**(a) method used in factory at present time**

| | Mounting-time [sec/board] | Total Tracing Route Length [mm] |
|---|---|---|
| (a) | 98 | 11125 |
| (b) | 108 | 5539 |
| (c) | 84 | 6179 |

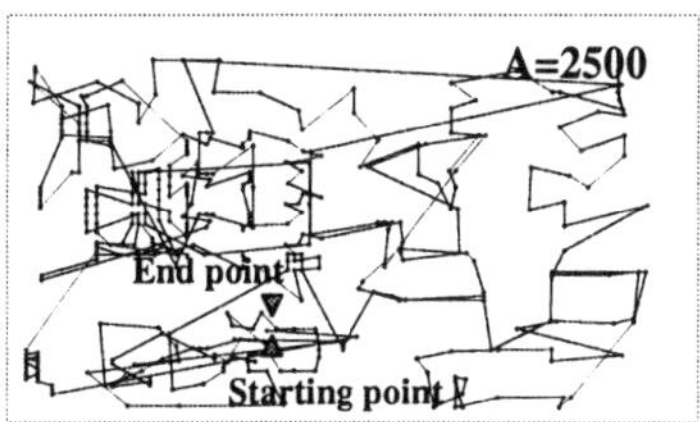


**(b) proposed method (A=2500)**

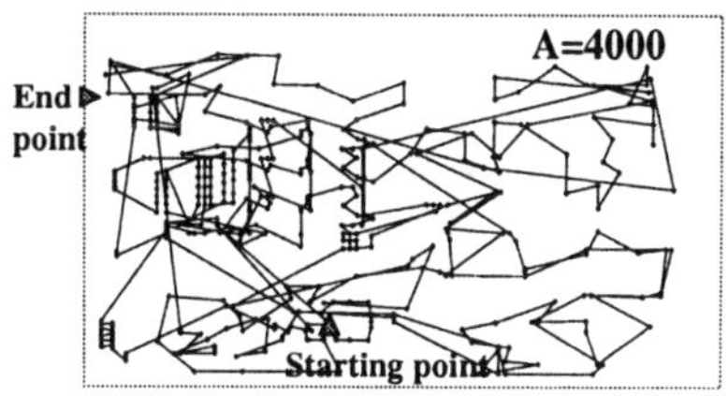


**(c) proposed method (A=4000)**

**Fig. 5.** Obtained tracing routes of handler for the problem with 48 kinds and 376 parts; (a) method used in factory at present time, (b) proposed method with A=2500 and (c) with A=4000. The characteristics of the mounting-time and total tour length are compared.

We experimented with a mounting-machine with a higher X-Y and feeder axis speed. When parameter A was set at 2500, the mounting time recorded was 108 seconds as compared to 98 seconds, which was obtained using the factory method. Figure 5 illustrates the results obtained for this board. It was observed that the high speed at the XY stage caused an increase in the reel waiting time. The value for parameter A was the increased to 4000 (route length increased and reel skipping decreased). This reduced the mounting time from 108 seconds to 84 seconds. It can be said that the value of parameter A affects the performance of the mounting-machine hence for optimum performance, an appropriate value for the parameter A must be selected.

## 5 Conclusion

The SOM-TSP method was applied to the electronic parts mounting problem of a printed circuit board. The algorithm which consists of 2 processes was applied to the problem. By applying this method to actual boards, a reduction in mounting time was achieved. If this technique is applied to the mounting process of the printed circuit board at an actual factory, it is certain that the productive efficiency of the factory will improve appreciably.

## References

1. Kohonen T. Self-Organizing Maps. Springer-Verlag, 1995.
2. Angeniol B, de C. La Vaubois G, Y. Le Texier J. Self-Organizing Feature Maps and the Travelling Salesman Problem. Neural Networks 1988; 1:289–293.
3. Fujimura K, Tokutaka H, Ohshima Y, Kishida S. The Traveling Salesman Problem Applied to the Self-Organizing Feature Map. Proc. of ICONIP'94-Seoul 1994; 427–432.
4. Padberg M, Rinaldi G. OPTIMIZATION OF A 532-CITY SYMMETRIC TRAVELING SALESMAN PROBLEM BY BRANCH AND CUT. Operation Research Lett. 1987; 6:1–7.
5. Fujimura K, Tokutaka H. : SOM-TSP:An approach to optimize surface component mounting on a printed circuit board. In:KOHONEN MAPS. ELSEVIER, AMSTERDAM, 1999, pp.219–230

# Self-Organising Maps for Condition Assessment of Paper Insulated Cables

José M. Rodríguez Arroyo [*#], Andy J. Beddoes [*], Nigel M. Allinson[#]

[*] EA Technology Ltd, Capenhurst Technology Park, Capenhurst, Chester, CH1 6ES, UK
[#] Dept. of Electrical Engineering and Electronics, UMIST, Manchester, M60 1QD, UK

Abstract

This paper concerns the feasibility of using Self-Organising Maps for the insulation assessment of paper insulated cables. In order to identify novel states in the insulation of the sample, the authors present a variation of the SOM based on the expansion of the trained map. With this modification, SOM can be used for the condition monitoring of the cables and the prediction of incipient faults.

## 1. Introduction

Due to the growth in energy demand and the high cost associated with the installation of new assets, interest is being focused on condition monitoring techniques, which are able to estimate the remaining life of existing systems. In the field of paper insulated cables, a common method for the evaluation of electrical assets is the detection and analysis of partial discharges (PD) throughout their degradation process.

Discharge activity in the cavities of the paper insulation produces a constant bombardment of ions over the walls of the gaps and the creation of gases that increase the pressure within the cavity. The result of this bombardment and pressure is the increase of the size of the voids and the development of carbonaceous paths between the fibres of the paper layers that may lead to a fatal breakdown of the cable.

Although this physical deterioration process is well known, the PD behaviour in the different stages of the degradation and the relevant parameters needed for the assessment of the cables remains uncertain [1]. Therefore, in order to develop an efficient and reliable condition assessment system for paper insulated cables, the analysis and monitoring of a high number of variables over a long period of discharge activity is required.

Self-Organising Feature Maps (SOM) are one mathematical approach to the processing and analysis of high dimensional problems involving large amounts of data, recognising trends and groups of similar readings. They are the basis of the condition monitoring technique presented in this paper.

# 2. Data acquisition system

The results presented in this paper relate to a 20 m length, 11 kV 3 core belted construction paper insulated cable, which was recovered from the field after 30 years of service. During the experiments, only one phase of the sample was energised with 50 Hz ac voltage. The apparent charge of the discharges and the phase at which they appeared were obtained by means of a data acquisition system depicted in Figure 1.

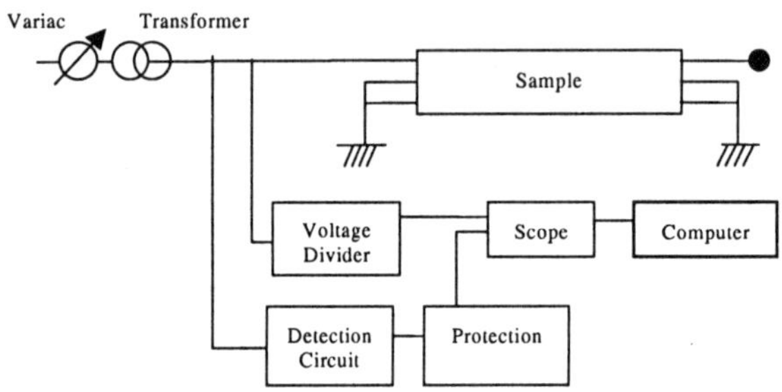


Figure 1. Experimental Set Up and Data Acquisition System

To degrade the cable and obtain the discharge activity at different condition states, the sample was subjected to several over-voltage ageing processes. The cable was energised at 1.5, 1.75, 2.0, 2.25 and 2.5 times their nominal level, following the cycle 1.5→1.75→1.5→2.0→1.5→2.25→1.5→2.5→ 1.5. Each voltage levels was maintained for at least two days. This ensured that a minimum of 288 readings (6 logs per hour x 24 hours x 2 days) were obtained for each voltage period. Due to experimental limitations, the acquisition of data was not continuous in some periods, particularly at the higher voltages. In these cases, the time interval between logs was set to 5 minutes in order to obtain the required data set in the least possible time.

Using the Phase Resolved Partial Discharge Method, the discharges were represented as a 3 dimensional function (ϕ-q-n function) depending on their magnitude, inception phase and number of pulses obtained. This methodology has often been used for the recognition of defects in HV Equipment, including cables [2-5]. However, in our work, this approach is applied to describe the changes in the behaviour of the PD throughout the degradation of the samples.

From these ϕ-q-n functions, a second data processing stage reduced the information to a set of 21 parameters. These parameters formed the input for the SOM.

# 3 Experimental Work

After energising the sample to its nominal voltage for 600 hours in order for the PD to settle, the cable was subjected to two ageing cycles. However, only the discharge activity at each voltage level during the second degradation process was considered. The number of readings and ageing time for each voltage level is shown in Table 1.

Table 1. Number of readings and ageing time for ageing process

| *P.U. Voltage* | *Label* | *Readings* | *Ageing time* |
|---|---|---|---|
| 1.5 | A1 | 295 | 49 h |
| 1.75 | B | 296 | 49h 10 min |
| 1.5 | A2 | 370 | 61h 30 min |
| 2.0 | C | 239 | 56 h 40 min |
| 1.5 | A3 | 346 | 75 h |
| 2.25 | D | 289 | 50h 30 min |
| 1.5 | A4 | 294 | 49 h |
| 2.5 | E | 389 | 110h 40 min |
| 1.5 | A5 | 294 | 24h 40 min |

Figure 2 shows a SOM of size 15x15 nodes, using 2000 epochs, obtained for the data from the second ageing process. The SOM obtained showed a clearer partition of the maps into two areas, which was confirmed when the maps were calibrated. The labels of each voltage level were concentrated in recognisable zones within the map, although this label clustering was not reflected by grey-scale areas.

After relaxing the sample for 20 days, the cable was again energised to its nominal voltage for 3,300 hours. At the end of this period a fan heater was positioned 2 cm from the cable sheath. The heating source was used to increase the temperature in this zone of the sample, causing the dielectric compound to liquefy and migrate. The migration of the insulating oil enhances a localised degradation in the cable and therefore an increase in the breakdown likelihood of the sample.

The voltage was increased to 1.5 p.u, and the discharge activity was again monitored. A third ageing process was initialised after 200 hours, maintaining the heating degradation. The sample finally broke down within 50 minutes of being energised at 2.25 p.u. voltage.

When the PD readings before the final breakdown were presented to the trained SOM, the trajectories obtained were situated at the border of the map (Figure 2). This result was not conclusive in showing how the insulation condition of the cable varied along the degradation process, mainly because of the limitations of basic SOM algorithm.

SOM always associate a node to an input vector, independently of the distance between the new data and the training set. Therefore, if the new reading corresponds to a novel and probably abnormal state of the system, SOM will not recognise it as irregular, but it will activate a node that could have been previously associated with a healthy condition. This case is particularly important for the nodes at the border of the map, which are normally associated with the topological frontier of the training set.

Therefore, from the nodes activated in the maps, no conclusions could be reached in terms of similarity between data obtained just before the final breakdown and others acquired in the ageing cycles. Similarly, the behaviour of the trajectories in the SOM prior to the failure (in terms of sudden changes or areas visited) proved to be worthless for the condition monitoring of the sample, as they were localised at the border of the map.

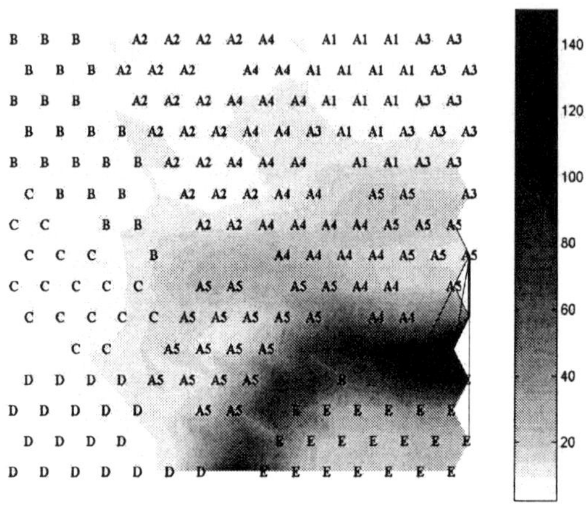


Figure 2. SOM obtained from the second ageing cycle and trajectory associated with the readings taken from the 1.5 p.u. voltage level prior breakdown

# 4 Extension of SOM

A possible solution for the problem of this border localisation would be the expansion of the map with more nodes at each border, defining the new weight vectors in such manner that they covered a broader extension of the input space. When a new input, coming from a novel state, is presented to the network, the node activated would be allocated in this new extended area. Ideally, this node should be as far away from the central core (the original map) as the input is distant from the training set used in the original SOM.

The first approach taken has been based on the definition of the new weight vectors as a linear combination of those associated with the nodes in the border of the map. Special considerations were taken in the selection of the coefficients and in the individual contribution of each original weight, trying to achieve a uniform expansion of the original subspace. Furthermore, an *increment parameter* $\beta$ was used to control the distance in the input space between the parent weights and the resulting vectors associated to the new nodes. Hence, if $w_1$, $w_2$ and $w_3$ are the original vectors, the new weight is defined as

$$w_{new} = \frac{Dir(w_1, w_2, w_3)}{\left\| Dir(w_1, w_2, w_3) \right\|} \left( \eta \left( 1 + \frac{\beta}{100} \right) \right)$$

where $Dir(w_1, w_2, w_3)$ is the linear combination defining the direction of the vector $w_{new}$ and $\eta$ is the average norm of $w_1$, $w_2$ and $w_3$.

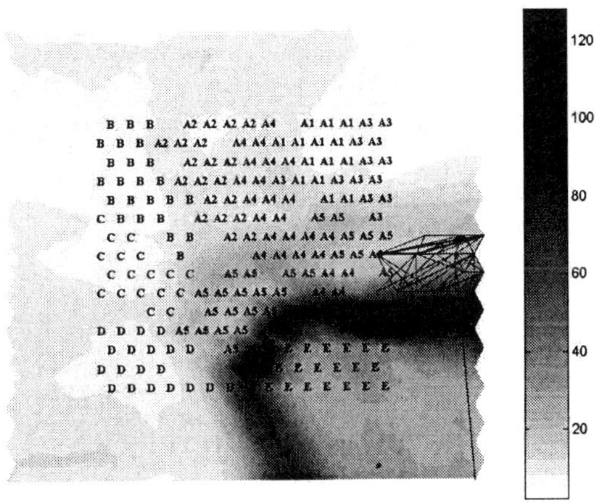


Figure 3. Trajectory of readings prior breakdown in extended SOM

Figure 3 shows the result of applying this technique to the readings taken before the breakdown of the sample. The original SOM is the map represented in Figure 2, which has been extended to a map of size 25x25 using a increase parameter of $\beta = 5$. The trajectory plotted corresponds to the data obtained from the previous voltage level (1.5 p.u) to the final failure of the sample.

The nodes activated by these readings are situated in the extended area of the map, confirming that the data obtained before the breakdown came from a novel and, finally, fatal condition of the insulation.

# 5 Future Work

These promising findings have encouraged further research into the map extension techniques and recognition of novel states using trained SOM. The current scope in the project is focused in the improvement of the expansion algorithm to enhance the coverage properties of the new vectors, in particular in those dimensions which less contribute to the total variability of the training data.

The possible relationship between the number of additional nodes, the increase parameter β and the distance of the new input to the training set is also subject to analysis, since a positive outcome of this field would imply the possibility of defining the confidence level of the map to identify abnormal readings.

# 6 Acknowledgements

The authors would like to thank the UK regional electrical companies associated with the Strategic Technology Programme Module "Cable Networks" for their financial support of this project and the provision of samples.

This project is being developed under the Postgraduate Training Partnership scheme run by the EPSRC, DTI, EA Technology and UMIST.

## References

1. Olsson L, Bengtsson G: PD Mapping of Paper Insulated MV Cables in Gothemburg, 14th Intern.ational Conference and Exhibition on Electricity Distribution CIRED'97, Birmingham, UK, pp. 3.12. 1-5.

2. Gulski E: Computer-aided Measurement of Partial Discharges in HV Equipment, IEEE Trans. Electrical Insulation, Vol 28, No 6, pp. 969-983, 1993

3. Schnettler A, Tryba V: Artificial Self-Organising Neural Network for Partial Discharge Source Recognition, Archir für Elektrotechnik 1993, 76: 149-154

4. Bozzo R, Gemme C, Guastavino F, Sciutto L: Diagnostic by Means of Neural Networks Having Inputs Derived from the Statistical Analysis of PD Patterns, 1996 IEEE International Symposium on Electrical Insulation, Montreal, Canada, 16-19 June, 1996, pp. 389-392

5. Krivda A: Recognition of Discharges. Discrimination and classification, Delft University Press, Delft, 1995

# An Essay in Classifying Self-organizing Maps for Temporal Sequence Processing

Gabriela Guimarães
Department of Computer Science and CENTRIA (Centre of Artificial Intelligence), New University of Lisbon, Portugal
gg@di.fct.unl.pt

Fernando Moura-Pires
Department of Computer Science, University of Évora, Portugal
faje@di.uevora.pt

Abstract

This paper presents a possible classification of modifications and adaptations of Self-organizing Maps (SOMs) for temporal sequence processing. Four main application areas for SOMs and temporal sequences have been identified. These are prediction, control, monitoring and mining. In order to model temporal relations among the data items within these application domains, usually an adaptation of the original learning algorithm, a modification of the network topology, or a combination of SOMs with special visualization techniques is made. Distinct approaches of SOMs for temporal sequence processing are classified into this scheme. Often, and in order to handle more complex domains, several adaptation forms are combined.

## 1 Introduction

Self-organising Maps (SOMs), as proposed by [1], have been successfully applied to a wide-ranging number of applications [2]. Since SOMs are neural networks based on unsupervised learning, they have been widely used for exploratory tasks [2, 3, 4]. The main aim is, here, to discover patterns in high dimensional data and possibly find novel patterns that might be hidden in the data. This often assumes an adequate visualization of the network structures, such as component maps [5], U-matrices [6], or a hierarchical and colored visualization of the SOM contraction process [7].

SOMs are also well suited for processing temporal sequences, often using the capability of SOMs for an exploratory analysis of the data. We identify four main application areas for SOMs and temporal sequences. These are prediction [8, 9, 10], control [11], monitoring [12] and mining [13]. Since temporal relations among the data items have to be considered, usually an adaptation of the original learning algorithm, a modification of the network topology, or a combination of SOMs with special visualization techniques is needed. In order to model the underlying process, data is gathered by measuring values of characterizing variables sequentially in time. If more than one sequence is measured simultaneously, then a multivariate analysis of the temporal sequence is made.

In this paper, a classification of different modifications and adaptations of the original SOM algorithm with regard to temporal sequence processing is given. We mainly identify three adaptation possibilities. These are the visualization of temporal paths on SOMs, the adaptation of the learning algorithm, and the modification of the network structure. In section 2 several approaches are presented, where temporal paths, i.e. trajectories, are displayed on the map. Here, we distinguish between visualizations of trajectories on the map itself, and trajectories on the visualized weights using component maps and U-matrices. Section 3 discusses adaptations of
the learning algorithm that are mainly used for control and prediction tasks. In section 4, modifications of the network topology for temporal sequence processing, such as recurrent and hierarchical SOMs, are discussed. Some conclusions and a classifying of the approaches into a scheme are presented in section 5.

# 2 Visualisation of Trajectories on SOMs

First, no modification of the original Kohonen algorithm is needed. At each time point, the best match representing the input vector is searched and visualized on the map. In order to visualize time-related input vectors, succeeding best matches connected through paths are visualized on the map. These paths are often named as trajectories. This approach was first applied to speech recognition [14]. Here, a decomposition of a.continuous speech signal is performed in order to recognize phonetic units. Before presenting the data to the network, a transformation into the frequency domain is made. A phonotopic map was generated with various neurons of the network sensitized to spectra of different phonemes. This approach has been widely used in medicine, such as the identification of coarticulation variation and voice disorder [15], the detection of fricative-vowel coarticulation [16], and the acoustic recognition of „s“ misarticulation enabling a distinction between normal, acceptable and unacceptable articulations [17]. In [18] trajectories have been used in a speech recognition system with SOMs at different hierarchical levels, where each layer operates on a different time scale and deals with higher units of speech, such as phonemes, syllables, and word parts.

Trajectories on SOMs have also been used for EEG signal monitoring [19]. A multivariate signal was gathered from 22 electrodes, enabling the identification of six typical EEG phenomena, such as well organized alpha frequencies, eye movement artifacts and muscle activity. Within another experiment with 16 subjects, different sleep/awake stages have been identified with trajectories [12]. SOMs have been proposed to model low dimensional non-linear processes, such as non-linear time series obtained from a Mavkey-Glass system, following three steps [20]. These are the reconstruction of the state space from the input signal, the embedding the state space in the neural field, and the estimation of the locally linear predictors. Trajectories are used to present the results of the prediction. Within another application firing activities in monkey´s motor cortex have been measured and presented to a SOM in order to predict the trajectory of the arm movement, especially while the monkey was tracing spirals and doing center-out movements [21]. From the map, three circle-shaped patterns representing the spiral trajectory have been identified through paths on the map. The results showed that the monkey´s arm movement directions are clearly encoded in firirng patterns of the motor cortex.

However, for all these applications the trajectories have been directly displayed on a map without a visualization of the network structures. This is possible, if a prior classification is known, for example, using classified phonemes in speech recognition or distinct sleep stages in EEG signals. If no classes are known beforehand, other visualization techniques for SOMs are needed. For instance, in [22] components maps are used for process state monitoring. Here, the values of one parameter are visualized as gray values. The lighter the unit on the map, the higher the parameter value is. The trajectory of the state changes detects a fault in the system as a transition to a forbidden area on the map marked with a very dark color. This approach was also applied to process control in chemistry, monitoring a distillation process [23]. Visualization of trajectories on U-matrices have been used for monitoring chemical processes [24], complex processes, such as the dynamic behaviour of a computer systems with regard to utilization rates and traffic volume [25], industrial processes, such as a continuous pulp digester, steel production and pulp and paper mills [26], and different subjects with distinct sleep apnea deseases [13]. The main advantage using U-matrices lies in the identification of inherent patterns within the time series, and in observing the temporal development of a complex system on a single map.

# 3 Adaptation of the Learning Algorithm

Another possibility for processing temporal data with SOMs lies in the adaptation of the original Kohonen learning rule. Therefore, the input vector is decomposed into two distinct parts, a past or context vector and a future or pattern vector. This approach was introduced by [27] and named as hypermap architecture. It was applied to the recognition of phonemes in the context of cepstral features. Each phoneme is then formed as a concatenation of three parts of adjacent cepstral feature vectors. The idea was to recognize a pattern that occurs in the context of other patterns, where $x(n) = [x_{patt}(n), x_{cont}(n)]$. Here, a two phase recognition algorithm was proposed. First, the best match was searched using only the pattern vector, but in a context domain, where a subset of units on the map were selected through the context vector. Second, a two step learning of the weights was performed. At first an unsupervised learning of the context weights was made, succeeded by an unsupervised learning of the pattern weights. Next, in order to provide the different units with phonemic labels, a supervised learning algorithm was used. This architecture was generalized by [28] to realize hierarchical relationships having n-1 levels defining the context for the classification of EEG signals from acoustical and optically evoked potentials. In [29] this approach is modified in the sense that temporal knowledge is coded by a pre-activation of the units. The pre-active area is defined to be an environment of the previous best match. The main advantage lies here in shifting the attention of the map from some input signal set to another, such that different sources are associated to different parts of the map enabling an independent processing of different signal sources. This type of model was also studied for simulating a sensory-motor task [30], for robot control [31, 32, 33], as well as for phoneme recognition using the LASSO model [34].

Hypermaps have also been used for prediction tasks, for instance, using SOMs for local models in the prediction of chaotic time series [9]. The time series is embedded in a state space using delay *coordinates* $\mathbf{x}(n) = [x(n), x(n-\tau), \ldots, x(n-(N-1)\tau)]$, where $\tau$ is a suitable time delay and $N$ the order of the embedding. The embedded vector is then used to predict the next value of the series $x(n+\tau)$. The following vector $y(n) = [\gamma x(n+\tau), x(n)]$ is presented to the map during the learning phase. However, in searching the best match the target value is left out. This means that only the first part (past) of the whole vector is used during the training phase for the determination of the best match. The weights are adapted using the whole input vector. During the classifying of a new vector, only the first part of the vector is used for the best-match search. The missing values then correspond to the predicted future part. They are obtained through an associative mapping. In [8] a two step implementation this approach was used for the prediction of hailstorm. First, it was used to identify distinct types of hailstorm developments. After the classification part, prediction was made using the completed vector.

# 4 Modification of the Network Topology

The third possibility in handling temporal data lies in modifying the network topology, introducing recurrent connections into the network or several hierarchical layers, each with one or more SOMs. The latter is used when a segmentation of complex and structured problems is needed in application domains, such as image recognition, speech recognition, time series analysis, process control and protein sequence determination.

Recurrent SOMs have been studied in temporal sequence processing, and have been successfully applied to EEG based epileptic activity detection and to time series prediction with local models [35, 9]. Recurrent SOMs are a modification of the temporal Kohonen Map [36], where a recursive difference equation defines the actual unit activity as a function of previous activations and the actual input vector. This model overcomes the known drawbacks of traditional temporal sequence approaches with time delay embedding, namely the difficulty to determine a suitable length for the delay. Recurrent SOMs, however, model the context of the temporal sequence without the need of specifying external delay mechanisms. In addition, SOMs could also reveal useful information inherent to the data without the need of a pre-classification. Briefly, recurrent SOMs allow learning temporal context, where the information is stored within a difference vector for each unit of the map. It is used for the best match search and the weights adaptation. In recurrent SOMs the leaky integrators are moved from the output to the input with $y_i(n) = (1 - \alpha) \cdot y_i(n-1) + \alpha \cdot (x(n) - w_i(n))$, $0 < \alpha < 1$. Large $\alpha$ corresponds to short memory, while small values of $\alpha$ correspond to long memory and slow decay activation. The best match unit $b$ is searched by $y_b = mini\ \{\|\ y_i(n)\ \|\}$, and the map is trained with a modified Hebbian learning rule, where the difference vector is replaced by $y_i$. The advantages of recurrent SOMs lie in their ability to find statistical temporal dependencies from the underlying process, as well as in the visualization of state changes of the process.

Hierarchical SOMs are often used in highly complex application fields, where one or more than one SOMs are located at each layer, usually operating on a different time scales. The main difference between approaches that use hierarchical SOMs lies in the codification of the results of the SOMs of a lower level. Hierarchical SOMs in temporal sequence processing have been successfully applied to speech recognition, complex process monitoring and temporal data mining. In speech recognition, each layer deals with higher units of speech, such as phonemes, syllables, and word parts. The methods differ mainly in their codification of the map output. In [18] a segmentation of the signal is made at different time scales using the longitude of the "jumps" between succeeding best matches on the map, as well as the occurrence frequency. At each level, a fixed number of classifiers exists. The activities of all classifiers are fed up to the input of the next layer. However, not only the best match is used as input, but instead the four best matching cells are taken into account. [37] propose a speaker recognition system based on the auditory model. Therefore, they introduce several hierarchical levels, where the output of the map is $y_i(n) = \alpha\cdot(k\ /\ (1 + d_i^2(n)^{1/2}\ ) + (1 - \alpha\ )y_i(n - 1),\ 0 < \alpha < 1$ and $d_i(n) = |w_i\text{-}x(n)|$. Both, units of the second and the third layer have recurrent connections with units at the lower level. In [38] hierachical SOMs are also used at different time scales. The codification for the input of the next layer is based, however, on a classification of the signal at different levels. A class is associated to each unit on the map. In order to calculate the output for a given input vector, the mean of the weight for all units belonging to the class is calculated, and used as input to the map of the next higher level map.

Hierarchical SOMs have also been applied to monitoring and modeling the dynamic behaviour of complex industrial processes, such as the dynamic behaviour of a computer system [25]. The main problem in process analysis is to find characteristic states or clusters of states, that determine the general behaviour of the system. Here, a hierarchical SOM with SOMs on two levels was constructed: a state map, that is used to track the operating point of the process with trajectories; and a dynamics map, that is used to predict the next state on the state map. Each unit on the dynamics map represents a "path" leading into the corresponding state. The training set of the state map for the dynamics map is formed by concatenating subsequent vectors into a single vector, representing the history of state transitions. In prediction, the state map vector having the best matching trajectory in its dynamics map is the predicted state.

An application of hierarchical SOMs in medicine, namely in sleep apnea research, is given in [13]. Here, SOMs are used at different hierarchical levels, in order to handle the complexity given by the large number of signal channels. At the lowest level primitive patterns in multivariate time series are discovered for distinct time series selections, while more complex patterns are identified at the higher level SOM. This approach, however, considers the structures obtained by the low-level maps, using the U-matrix for the calculation of the input of the next-higher level map. Therefore, all weights are approximated towards the cluster center $c_k$ according to the following adaptation rule: $w_{i,new} = w_i + \alpha \,||w_i - c_k||$, if $c_k > w_i$ and $w_{i,new} = w_i - \alpha \,||w_i - c_k||$, if $c_k \leq w_i$.

# 5 Conclusions

In this paper, an essay was made in classifying distinct approaches with Self-organizing Maps (SOMs) and their modifications for temporal sequence processing. We mainly identified three possibilities for temporal sequence processing with SOMs. These are: the visualization of temporal paths, the adaptation of the learning algorithm; and the modification of the network structure. Table 1 gives an overview to the different kind of adaptations and modifications that have been used by the different approaches. Often, one of the adaptation form of SOMs has been used isolated, for instance, just using trajectories or recurrent networks. However, there are also approaches that combine distinct adaptation forms of SOMs, such as trajectories on U-matrices at different hierarchical levels. At last, we would like to remark that no an exhaustive reference could be made to all existing approaches for temporal sequence processing, since there are other approaches, not mentioned here, that fit into this scheme.

| | Trajectories | Adaptation of the Learning Rule | Modification of the Network Topology |
|---|---|---|---|
| Trajectories | 12, 14, 15, 16, 17, 19, 21, 22, 23, 24, 26 | 20 | 13, 18, 25 |
| Adaptation of the Learning Rule | | 8, 27, 28, 29, 30, 31, 32, 33, 34 | 11 |
| Modification of the Network Topology | | | 9, 35, 36, 37 |

Table 1: Classification of distinct approaches of SOMs for temporal sequence processing. The numbers within the table refer to the respective reference.

# References

1 Kohonen, T. (1982), "Self-organized formation of topologically correct feature maps". Biological Cybernetics, 1982; 43: 141-152
2 Kohonen, T.: Self-Organizing Maps, Springer Verlag, New York, 1995
3 Deboeck, G., Kohonen, T. (Eds.) . Visual Explorations in Finance with Sef-organizing Maps, Springer, 1998

4 Kaski, S. Data exploration Using Self-organizing Maps. PhD Thesis, Helsinki University of Tecnology, 1997

5 Olli Simula and Esa Alhoniemi. SOM Based Analysis of Pulping Process Data, Proceedings of International Work-Conference on Artificial and Natural Neural Networks (IWANN '99), Springer, 1999; Vol. II: 567-577

6 Ultsch, A., Siemon, H.P.: Kohonen's Self-Organizing Neural Networks for Exploratory Data Analysis. In: Proc. Intl. Neural Network Conf. INNC90, Paris, Kluwer Academic, 1990: 305-308

7 Himberg, J. A SOM based cluster visualization and its application for false coloring. Proceedings of IEEE-INNS-ENNS Intl. Joint Conf. on Neural Networks (IJCNN'2000), Como, 24-27 July, Italy , 2000; Vol III: 587-592

8 Ultsch, A. Guimarães, G., Schmidt, W. : Classification and Prediction of Hail using Self-Organizing Neural Networks, Intl. Conference on Neural Networks, ICNN'96, Washington, 3-6 June, 1996: 1622-1627

9 T. Koskela, M. Varsta, J. Heikkonen, and K. Kaski. Recurrent SOM with Local Linear Models in Time Series Prediction. Proc. of ESANN'98, 6th European Symposium on Artificial Neural Networks, D-Facto, Brussels, Belgium, 1998: 167-172

10 Vesanto, J. Using the SOM and Local Models in Time-Series Prediction. Proc. ofWorkshop on Self-Organizing Maps (WSOM'97), Espoo, Finland, June 4-6, 1997: 209-214.

11 Jörg Walter, Helge Ritter. Investment Learning with Hierarchical PSOM. In: D. Touretzky and M. Mozer and M. Hasselmo (ed.): Advances in Neural Information Processing Systems 8 (NIPS*95), 1996: 570--576.

12 Joutsiniemi, S.L., Kaski, S., Larsen, T.A.: Self-Organizing Map in Recognition of Topographic Patterns of EEG Spectra. IEEE Transactions on Biomedical Engineering, Vol. 42, No. 11, November, 1995; 1062-1068

13 Guimarães, G.: Temporal knowledge discovery with self-organizing neural networks. In: Part I of the Special issue (Guest Editor: A. Engelbrecht): Knowledge Discovery from Structured and Unstructured Data, *The International Journal of Computers, Systems and Signals (IJCSS)*, 2000: 5-16

14 Kohonen, T.: The „Neural" Phonetic Typewriter, In: Computer, 1988: 11-22

15 Utela, P.,Kangas, J., Leinonen, L.: Self-Organizing Map in Acoustic Analysis and On-Line Visual Imaging of Voice and Articulation. In: I. Aleksander, J. Taylor (Eds.): Artificial Neural Networks, Elsevier Science Publisher, 1992; 2: 791-794

16 Leinonen, L., Hiltunen, T., Torkkola, K., Kangas, J.: Self-organized Acoustic Feature Map in Detection of Forcative-Vowel Coarticulation. In: J. Acoust. Soc. Am. , 1993; 6: 3468-3474

17 Mujunen, R.,Leinonen, L, Kangas, J., Torkkola, K.: Acoustic Pattern Recognition of /s/ Misarticulation by the Self-Organizing Map. In: Folia Phoniatr., 1993; 45: 135 - 144

18 Behme, H., Brandt, W.D., Strube, H.W.: Speech Recognition by Hierarchical Segment Classification. In: S. Gielen, B. Kappen (Eds.): Proc. Intl. Conf. on Aritificial Neural Networks (ICANN 93), Amsterdam, Springer Verlag, London, 1993: 416-419

19 Kaski, S., Joutsiniemi, S.L.: Monitoring EEG Signal with the Self-Organizing Map. In: S. Gielen, B. Kappen (Eds.): Intl. Conf. on Artificial Neural Networks (ICANN 93), Amsterdam, Springer Verlag, London, 1993: 974-977.

20 Principe, J.C., Wang, L.: Non-Linear Time Series Modeling with Self-Organizing Feature Maps. In: Proc. IEEE Workshop on Neural Networks for Signal Processing, Piscataway, NJ, IEEE Service Center, 1995: 11 -20.

21 Lin, S, Si, J. Self-organization of Firing Activities in Monkeys Motor Cortex: Trajectory Computation from Spike Signals. Neural Computation , 1998; 9: 607-621

22 Kasslin, M.,Kangas, J., Simula, O.: Process State Monitoring usinf Self-Organizing Maps. In: I. Aleksander, J. Taylor (Eds.): Artificial Neural Networks, Elsevier Science Publisher, 1992; 2: 1531-1534

23 Tryba, V.,Goser, K.: Self-Organizing Feature Maps for Process Control in Chemistry. In: T. Kohonen, K. Mäkisara, O. Simula, J. Kangas (Eds.): Artificial Neural Networks, Elsevier Science Publisher, North Holland, 1991: 847-852

24 Ultsch, A.: Self-Organized Feature Maps for Monitoring and Knowledge Acquisition of a Chemical Process, Gielen, S., Kappen, B. (Eds.): Proc. International Conference on Artificial Neural Networks (ICANN 93), Amsterdam, September, Springer-Verlag, 1993: 864-867

25 Olli Simula, Esa Alhoniemi, Jaakko Hollmén, Juha Vesanto. Monitoring and modeling of complex processes using hierarchical self-organizing maps, Proceedings of the {IEEE} International Symposium on Circuits and Systems (ISCAS'96), Supplement, May, 1996: 73-76

26 Esa Alhoniemi and Jaakko Hollmén and Olli Simula, Juha Vesanto. Process Monitoring and Modeling using the Self-Organizing Map, Integrated Computer Aided Engineering, John Wiley & Sons, 1999; 6, No.1: 3-14

27 Kohonen, T.: The Hypermap Architecture. In: T. Kohonen, K. Mäkisara, O. Simula, J. Kangas (Eds.): Artificial Neural Networks, Elsevier Science Publishers, North Holland, 1991: 1357-1360

28 Brückner, B., Franz, M., Richter, A.: A Modified Hypermap Architecture for Classification of Biological Signals. In: I. Aleksander, J. Taylor (Eds.): Artificial Neural Networks, Elsevier Science Publisher, 1992; 2: 1167-1170

29 Kangas, J.: Temporal Knowledge in Locations of Activations in a Self-Organizing Map. In: I. Aleksander, J. Taylor (Eds.): Artificial Neural Networks, Elsevier Science Publisher, 1992; 2: 117-120

30 Ritter, H., Martinetz, T., Schulten, K. Topology-conserving maps for learning visuomotor coordination, Neural Networks 1989; 2: 159-168

31 Ritter, H.: Parametrized Self-Organizing Maps for Vision Learning Tasks. In: M. Marinaro, P.G. Morasso (Eds.): Proc. Intl. Conference on Artificial Neural Networks (ICANN 94), Italy, Springer Verlag, 1994: .803-810.

32 Walter, J. A. PSOM Network: Learning with Few Examples. Procs Intl. Conf. On Robotics and Automation (ICRA), IEEE, 1998

33 Walter, J.A., Schulten, K.J.: Implementation of Self-Organizing Neural Networks for Visual-Motor Control of an Industrial Robot. In: IEEE Transactions on Neural Networks, January, 1993; 4, No. 1: 86-95.

34 Midenet, S., Grumbach, A.: Learning Associations by Self-Organization: The LASSO Model. In: Neurocomputing, Elsevier Science Publisher, 1994; 6: 343-361..

35 T. Koskela, M. Varsta, J. Heikkonen, and K. Kaski, "Temporal Sequence Processing using Recurrent SOM", KES'98, 2nd Int. Conf. on Knowledge-Based Intelligent Engineering Systems, Adelaide, Australia, April 1998, 1: 290-297

36 Chappel, G.J.,Taylor, J.G.: The Temporal Kohonen Map. Neural Networks, 1993; 6: 441-445

37 Jiang, X., Gong, Z., Sun, F., Chi, H.: A Speaker Recognition System Based on Auditory Model. In: World Congress on Neural Networks (WCNN 94), Hillsdale, NJ., Lawrence Erlbaum, 1994; 4: 595-600

38 Kemke, C.,Wichert, A.: Hierarchical Self-Organizing Feature Maps for Speech Recognition. In: Proc. of the World Congress on Neural Networks (WCNN 93), Hillsdale, 1993; 3: 45-47

# Evaluating SOM-based models in Text Classification Tasks for the Greek Language

G. Tambouratzis, N. Hairetakis, S. Markantonatou, G. Carayannis
Institute for Language and Speech Processing
Epidavrou & Artemidos 6, 151 25 Maroussi, Greece

Abstract

In the present paper, the Self-Organising Map (SOM) is applied to the problem of categorising a corpus of Modern Greek texts according to the style of their authors. A number of variants of the SOM model are used in a series of experiments, in order to compare and contrast their behaviour in the specific task. The experimental results indicate that the SOM possesses the ability to analyse such data, successfully uncovering the differences among authors.

## 1. Introduction

When attempting to retrieve relevant texts from a collection of textual data, any information about the topic of each text or the style employed by its author may be useful in specifying the search more accurately and thus improving its precision. The task of manual categorisation of documents and author discrimination is a labour-intensive one. A number of approaches have been employed to perform these tasks in an automated manner, ranging from statistical techniques [1][2] to neural network models, such as multi-layer perceptrons and radial basis functions [3][4]. Here, the SOM is applied to the task of author attribution for texts written in Modern Greek. This model has been selected because it operates in an unsupervised manner and can successfully process the amount of information available through the extensive digital libraries that exist nowadays. SOM applications range from analysing texture in images to planning the path of autonomous robots [5]. In this article the data employed is textual in nature, resulting in an analysis more similar to that of the WEBSOM system [6].

The experiments reported here form part of a wider set of studies on stylistics. These have focused on (i) separating between different registers of the Greek language (the term "register" being used as a cover term for speech varieties defined by their situational characteristics [7]) and (ii) separating a set of documents according to the style of their author. In these analyses, statistical methods have shown that the separation of registers is easier than the classification of documents belonging to the same register according to their author's style [8][9]. In this article we focus on the harder task, aiming to separate sets of documents on the basis of their author style by employing the SOM model and comparing its results to these of statistical techniques.

# 2. The SOM Neural Network

The origins of the SOM [5] are traceable to the study of biological neural networks, where neurons are organised so that neighbouring neurons become sensitive to similar stimuli. The SOM model consists of a set of nodes (neurons) organised into a usually one- or two-dimensional regular structure, most frequently forming a rectangular or hexagonal neighbourhood. During the training phase, neighbouring nodes are similarly adapted towards the input patterns. Thus, following the completion of the training phase, these nodes respond similarly to each input. The SOM is trained by an unsupervised adaptation law that generates a global ordering of the input patterns through competition between the nodes. The node weight vectors are initialised either randomly or with pre-defined values, introducing a partial order in the map. Each input vector is then broadcast throughout the network and the node whose weight best matches the input vector is updated (together with its neighbours) by the following adaptation rule:

$$\begin{aligned} m_i(t+1) &= m_i(t) + a(t)[x(t) - m_i(t)], \quad \forall i \in N_c(t) \\ m_i(t+1) &= m_i(t), \quad \forall i \notin N_c(t) \end{aligned} \tag{1}$$

where $m_i(t)$ is the weight of node $i$ at time instance $t$, $x(t)$ is the input pattern, $a(t)$ is the gain factor and $N_c(t)$ is a function denoting the nodes which belong to the neighbourhood of the winning node. Both the neighbourhood function $N_c(t)$ and the adaptation gain $a(t)$ decrease as training progresses. The SOM training is divided into (i) a rough training phase, to achieve the ordering of the node weights, and (ii) a fine-tuning phase, allowing each node weight to reach its optimal value.

Based on the SOM, hierarchical feature maps (HFM) have been proposed [10], to provide an increased level of accuracy in data representation. The key idea of HFM is the use of a hierarchical structure of layers. At the first layer a single SOM is used. Each subsequent layer is formed by a number of independent SOM maps, one for each node in the previous layer. This approach has been used [11] to generate a self-organised classification of documents and results in an improved visibility of the patterns space clusters in comparison to the standard SOM.

Vesanto et al. [12] have proposed using the SOM to create a set of prototypes, which are then clustered using clustering techniques. This approach is particularly useful during the exploratory phase of data mining, in applications where the number of SOM units is large. Then, rather than processing each node separately, it is advantageous to group similar units and process them in the same way.

# 3. Classification Task Description

Here, the SOM model is employed to distinguish among Modern Greek texts on the basis of their author's style. Five members of parliament were selected, one from each of the five political parties represented in the Greek Parliament over the period 1996-2000. The speeches were retrieved from the Parliament Minutes. For each speech, a set of 85 features were extracted:

- 14 variables depicting the diglossia phenomena on Modern Greek verbal morphology [13], which have been shown to distinguish between formal Greek (Katharevousa) and casual Greek (Demotiki) [8];
- 11 variables expressing Part-of-Speech counts in percentages;
- 7 variables capturing macroscopic linguistic features such as average number of letters per word, average number of words per sentence and counts of punctuation marks;
- 8 variables concerning the use of negative words and one variable reflecting diglossia phenomena (the contrast between Katharevousa and Demotiki) in the use of anaphoric pronouns;
- 2 variables quantifying the use of nominal expressions in the genitive case, indirectly counting the tendency to use nominal rather than verbal structures [7];
- 8 verbal tags reflecting discourse tendencies;
- 5 variables describing the distribution of word length in letters and 12 variables describing the distribution of sentence length in words and
- 17 variables counting the occurrences of specific words/lemmata in the texts. These words were selected because each one of them seemed to differentiate one speaker from another.

# 4. Experimental Results

## 4.1 Application of the Standard SOM Model

To simulate the Self-Organising Map, the SOM Toolbox for thc MATLAB environment has been used [14]. As the features selected for our experiments were diverse in both their nature and their range of values, they were pre-processed using the "normalisation of variance" technique. Two-dimensional SOM maps were used, which were initialised in each run via a linear normalisation technique along the two largest eigenvectors. Then, training was performed, using a Gaussian-type neighbourhood function. In the first phase the size was decreased linearly, from a maximum equal to the smallest network dimension towards 1. In the second phase, the neighbourhood size was fixed to 1. The rough training phase consisted of $4 \times mpd$ epochs, while the fine-tuning phase consisted of $16 \times mpd$ iterations, all performed using the batch-training mode. The value of $mpd$ is defined as the ratio of the number of map units over the length of the training data. Maps of different sizes were used, ranging from 2x3 up to 6x12 nodes, in order to evaluate the SOM performance more extensively.

| | Corpus I | | Corpus II | |
|---|---|---|---|---|
| | *Texts* | *Size (words)* | *Texts* | *Size (words)* |
| *Speaker A* | 36 | 90,090 | 255 | 343,215 |
| *Speaker B* | 16 | 73,988 | 48 | 136,344 |
| *Speaker C* | 30 | 64,905 | 91 | 156,945 |
| *Speaker D* | 21 | 60,071 | 92 | 141,219 |
| *Speaker E* | 24 | 75,550 | 55 | 149,644 |
| **Total** | **127** | **364,604** | **541** | **927,367** |

*Table 1 – Composition of the two datasets.*

To evaluate the SOM effectiveness, two different corpora were used. The first one (Corpus I) consisted of the opening speeches of each speaker for any given session when at least two speakers presented their views, for the period 1999-2000. The second one (Corpus II) consisted of all opening speeches given by each of the 5 speakers (irrespective of the number of speakers presenting their views in each session) over the entire life span of the specific Parliament, 1996-2000. Thus, Corpus II can be expected to be more diverse that Corpus I. The details of the two datasets are shown in Table 1.

| | Clustering Accuracy | | | | | | Cells | |
|---|---|---|---|---|---|---|---|---|
| *SOM size* | *A* | *B* | *C* | *D* | *E* | *Total* | *Total* | *Unused* |
| *4x2* | 83.3 | 56.3 | 90.0 | 81.0 | 50.0 | 74.8 | 8 | 0 |
| *3x6* | 69.4 | 87.5 | 90.0 | 76.2 | 75.0 | 78.7 | 18 | 0 |
| *4x8* | 86.1 | 87.5 | 83.3 | 85.7 | 70.8 | 83.5 | 32 | 2 |
| *5x10* | 86.1 | 93.8 | 86.7 | 81.0 | 79.2 | 85.0 | 50 | 9 |
| *6x12* | 86.1 | 100.0 | 86.7 | 85.7 | 87.5 | 88.2 | 72 | 16 |

*Table 2 – Comparative results for different sizes of standard SOMs with Corpus I.*

Initially, a standard SOM network was trained using the textual data. Each node was then labelled according to a majority-voting scheme and the patterns were classified by the labelled SOM. The experimental results indicate that the discrimination accuracy - both for the entire dataset and for every individual speaker - is reduced as the map size gets smaller. This is expected because in a smaller map the samples need to be distributed into fewer cells. Inadvertently, some samples are assigned to cells that already contain patterns from other categories. It is important to note that Speaker E is found to have systematically lower recognition rates both with the statistical method [9] and the SOM analysis. Conversely, speakers B and C are recognised easier, using both statistical and neural network methods.

## 4.2. Introducing Hierarchical Feature Maps

As the SOM attempts to map a high-dimensional data space into a two-dimensional display, some salient information may be lost in the process. As an alternative, a two-layer HFM was used to classify Corpus I. The decision not to use HFMs with more than 3 layers was due to the resulting cost of substantially increased processing time as well as memory requirements. The HFM consisted of a 4x2 SOM as its first layer and either 3x2 or 4x2 SOMs in the second layer. In terms of nodes used, the 4x2/3x2 topology is broadly comparable to a 5x10 SOM, while the 4x2/4x2 topology can be compared with a 6x12 SOM. The results of this analysis are summarised in Table 3. It can be seen that the HFMs give similar results to the SOMs with the same number of nodes.

| | *Clustering Accuracy* | | | | | | *Cells* | |
|---|---|---|---|---|---|---|---|---|
| *Configuration* | *A* | *B* | *C* | *D* | *E* | *Total* | *Total* | *Unused* |
| *4x2/3x2* | 86.1 | 93.8 | 96.7 | 81.0 | 75.0 | 86.6 | 56 | 9 |
| *4x2/4x2* | 86.1 | 100.0 | 100.0 | 76.2 | 79.2 | 88.2 | 72 | 21 |

*Table 3 – Results obtained using a two-layer hierarchical feature map model for corpus I.*

The advantage of the shorter training times for the higher layer is counterbalanced by the extra time needed to simulate the second layer topology. More importantly, the first layer was found not to provide a higher level of abstraction, since the separation of authors was only achieved at the second layer. Therefore, for the given data, the more elaborate HFM scheme does not provide an improvement in the clustering performance over the SOM.

## 4.3. Clustering the SOM-generated Codebooks

The final SOM variant evaluated was that proposed in [12] (hereafter denoted as CSOM), which involves clustering the SOM node codebooks to reveal the existing classes in the dataset. To simulate the CSOM, initially a standard SOM was trained. Then, the node codebooks were grouped using Ward's hierarchical cluster analysis technique. The resulting clusters were processed using a majority-voting technique to determine the speaker each cluster corresponded to. To simplify the description as well as the comparison to other techniques, in table 4 the results for the texts of Corpus I are illustrated using a fixed target of ten clusters. For comparison purposes, the results of the standard SOM are also included. Finally, the clustering results obtained by Ward's method on the original feature vectors are also shown. According to table 4, the standard SOM gives the best results (though this is achieved by considering each SOM node as one distinct cluster). However, the CSOM hybrid method has a similar performance. In contrast, the pure clustering approach gives clusters with patterns from more classes and thus a lower accuracy.

| *Method* | | | *Clustering Accuracy* | | | | | |
|---|---|---|---|---|---|---|---|---|
| | *SOM size* | Clusters | *A* | *B* | *C* | *D* | *E* | *Total* |
| *Standard SOM* | *3x6* | (18) | 69.4 | 87.5 | 90.0 | 76.2 | 75.0 | *78.7* |
| | *4x8* | (32) | 86.1 | 87.5 | 83.3 | 85.7 | 70.8 | *83.5* |
| *CSOM* | *3x6* | *10* | 69.4 | 87.5 | 90.0 | 66.7 | 79.1 | *78.0* |
| | *4x8* | *10* | 86.1 | 81.3 | 86.7 | 71.4 | 66.7 | *79.5* |
| *Cluster Analysis* | -- | *10* | 77.8 | 75.0 | 96.7 | 0.0 | 45.8 | *63.0* |

*Table 4 – Comparative results obtained for Corpus I with (i) the standard SOM, (ii) the CSOM and (iii) Ward's hierarchical clustering method.*

| *Method* | | | *Clustering Accuracy* | | | | | |
|---|---|---|---|---|---|---|---|---|
| | *SOM size* | Clusters | *A* | *B* | *C* | *D* | *E* | *Total* |
| *Standard SOM* | *4x8* | (32) | 84.3 | 64.6 | 84.6 | 60.9 | 40.0 | *74.1* |
| | *6x12* | (72) | 77.3 | 79.2 | 82.4 | 70.7 | 60.0 | *75.4* |
| | *8x16* | (128) | 74.1 | 83.3 | 82.4 | 77.2 | 78.2 | *77.3* |
| *CSOM* | *4x8* | *10* | 76.5 | 66.7 | 68.1 | 58.7 | 34.6 | *66.9* |
| | *6x12* | *10* | 58.8 | 85.4 | 70.3 | 65.2 | 16.4 | *59.9* |
| | *6x12* | *20* | 88.2 | 60.4 | 70.3 | 72.8 | 27.3 | *73.9* |
| | *8x16* | *10* | 75.7 | 62.5 | 44.0 | 79.4 | 5.5 | *62.7* |
| | *8x16* | *20* | 83.9 | 52.1 | 73.6 | 53.3 | 49.1 | *70.6* |
| *Cluster Analysis* | -- | *10* | 47.1 | 64.6 | 72.5 | 19.6 | 34.6 | *47.0* |
| | -- | *20* | 23.5 | 60.4 | 68.1 | 56.5 | 69.1 | *44.6* |

*Table 5 – Comparative results obtained for Corpus II with (i) the standard SOM, (ii) the CSOM and (iii) Ward's hierarchical clustering method.*

To further study the behaviour of the different models, the same experiments were repeated with the more extensive dataset of Corpus II. The results obtained are shown in table 5. It can be seen that the accuracy of the CSOM is now substantially reduced. Therefore, larger map sizes were also studied and the SOM/clustering scheme was allowed to settle in 20 clusters, generating the results of table 5. As expected, the results obtained using 20 clusters are considerably improved compared with the results of the 10 clusters and at the same time not considerably lower than the standard SOM results. Of course, one should take into account the fact that each node of the standard SOM is treated as a distinct cluster and thus the standard SOM accuracy would unavoidably be higher. It was found that throughout the experiments the CSOM clusters consist of localised nodes indicating the successful topology preservation on the SOM map (a typical example of the cluster formation on a 4x8 SOM lattice is shown in figure 1).

Finally, it should be noted that the processing time required for the cluster analysis on the feature vectors is much larger than that for the SOM methods, due to the high-dimensionality of these vectors.

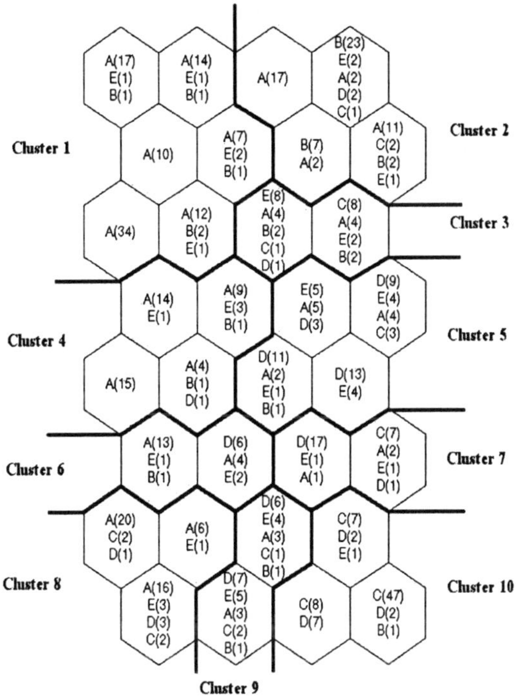


*Figure 1 – 4x8 CSOM for Corpus II indicating the assignment of speeches on the network nodes and the corresponding clusters.*

# 5. Conclusions

In this paper a number of SOM-based clustering methods have been compared when applied to the author classification problem for texts in Modern Greek. Both the standard SOM and the hierarchical feature map have generated broadly equivalent results, thus indicating an equivalence of the two methods. Furthermore the SOM/clustering hybrid method has been shown to produce a successful clustering for the two datasets, while reducing the requirements for the labeling task. However, it should be noted that when using the larger dataset the clustering results become comparable to the SOM results only when the number of allowed clusters is raised, reflecting the higher diversity of patterns. The results obtained indicate that the SOM (as well as its HFM and CSOM variants) is able to successfully cluster and subsequently classify the documents according to their author, making use of stylistic features. Thus, it is believed that further experimentation is warranted on the application of these techniques in the author identification task. At the same time, it is worth studying in more detail the role of the features used in our experiments, in an effort to improve the network model accuracy. Finally, it would be worth investigating the efficiency of alternative SOM variants, whose structure may be modified during learning, allowing them to adapt to the pattern space characteristics.

### Acknowledgements

The authors wish to acknowledge the assistance of the Secretariat of the Hellenic Parliament in obtaining the session transcripts studied in this piece of research.

### References

1. Mosteller, F. & Wallace, D.L. Applied Bayesian and Classical Inference: The Case of the Federalist Papers, Springer, New York, 1964.
2. Karlgren, J. Stylistic Experiments in Information Retrieval. In T. Strzalkowski (ed.), Natural Language Information Retrieval, Dordrecht: Kluwer, 1999: 147-166.
3. Matthews, R. & Merriam, T. Neural Computation in Stylometry I : An Application to the Works of Shakespeare and Fletcher. Literary and Linguistic Computing, 1993; 8(4):203-209.
4. Tweedie, F. J., Singh, S. & Holmes, D.I.. Neural Networks in Stylometry: The Federalist Papers. Computers and the Humanities, 1996; 30:1-10
5. Kohonen, T. Self-Organising Maps (2nd ed.) Springer-Verlag, Berlin, 1997
6. Kaski, S., Lagus, K., Honkela, T. & Kohonen, T. Statistical Aspects of the WEBSOM system in Organising Document Collections. *Computing Science and Statistics*, (Scott, D.W., ed.), Interface Foundation of North America, Inc.: Fairfax Station, VA, 1998:281-290
7. Biber, D., Conrad, S. & Reppen, R. Corpus Linguistics: Investigating Language Structure and Use. Cambridge University Press, 1998
8. Tambouratzis, G., Markantonatou, S., Hairetakis, N. & Carayannis, G. Automatic Style Categorization of Corpora in the Greek Language. Proceedings of the LREC-2000 Conference, Athens, Greece, 31 May – 2 June, 2000;1:135-140.
9. Tambouratzis, G., Markantonatou, S., Hairetakis, N., Vassiliou, M., Tambouratzis, D., & Carayannis, G. Discriminating the registers and styles in the Modern Greek Language. Proceedings of the Workshop on Comparing Corpora, held in conjunction with the 38th ACL Conference, 7 October, Hong Kong, 2000;35-42
10. Miikkulainen, R. Script Recognition with Hierarchical Feature Maps. Connection Science, 1990;2(1-2):83-101
11. Merkl, D. Document Classification with Self-Organising Maps. In Kohonen Maps, Oja, E. & Kaski, S., (eds.), Elsevier, Amsterdam, 1999
12. Vesanto, J. & Alhoniemi, E. Clustering of the Self-Organising Map. IEEE Transactions on Neural Networks, 2000; 11(3):586-600
13. Clairis, C. & Babiniotis, G. Grammar of Modern Greek – II Verbs. Ellinika Grammata, Athens (in Greek), 1999
14. Vesanto, J., Himberg, J., Alhoniemi, E. & Parhankangas, J.. SOM Toolbox for Matlab 5. Report A57. SOM Toolbox Team, Helsinki University of Technology, 2000

# Nonlinear Blind Source Separation Using SOMs and Applications to Image Denoising

Michel Haritopoulos, Hujun Yin and Nigel Allinson
Department of Electrical Engineering and Electronics
UMIST, Manchester, M60 1QD, England

Abstract

A SOM-based approach is applied to blind source separation for nonlinearly mixed signals corrupted with multiplicative noise. The simulation results, and a practical illustration of this approach applied to the image denoising problem, show that despite its high computational cost, this method can provide improved results in terms of separation quality. Furthermore, faster implementation and optimisation of this method have been shown possible and could lead to a practical tool for the image processing.

## 1 Introduction

Processing of real world data sets using neural network techniques based on unsupervised learning can provide useful information, such as principal components, clusters, prototypes or feature maps, about the data (input) space. Self-Organizing Maps (SOM) also known as Kohonen or topographic maps [1] are examples of neural networks employing this principle. Despite their wide range of applications and implementations, there are areas where the use of SOMs can be shown to be useful, for example Blind Source Separation (BSS) or the closely related Independent Component Analysis (ICA).

BSS consists of recovering the unknown statistically independent signals (sources) given only a set of their observations (sensor response) while the ICA technique [2] is the underlying mathematical tool providing a solution to the BSS problem. The case of linear mixtures of the unknown signals has been widely treated in the literature, but the nonlinear case is still a topic of active discussion. Early work on Nonlinear ICA (NLICA) can be found in [3]. It can be shown that when the distribution of the sources is uniform, SOM networks can provide an approximate solution to the Nonlinear BSS (NLBSS) problem. Some work on the application of SOM to NLBSS with sub-Gaussian sources has been recently addressed in [4].

The use of SOM-based separating structures can be justified as SOMs perform a nonlinear mapping from an input space to an output space usually represented by a low dimensional lattice. Using some suitable interpolation method (topological or geometrical), the map can be made continuous providing estimates of the unknown signals. However, there are difficulties associated with the NLBSS, such as the intrinsic indeterminacy and the unknown distribution of the sources, as well as the condition of the mixing, which depends on the strength of the unknown nonlinear function involved in the mixing process and on the presence of noise (correlated or not) with the observations. All these make it difficult for a complete analytical study of SOM behaviour when applied to this particular context.

The purpose of this paper is to investigate the performance of SOMs when applied to source separation with noisy observations. Hence, we present some results obtained by applying SOMs to the NLBSS in the case of data corrupted with multiplicative noise. A comparison between the standard Kohonen algorithm and its modified version proposed in [5] is validated by simulations and applied to an image denoising problem.

# 2 ICA Formulation

Denote by $S$ the unknown source space and by $\mathbf{X}$ the observations space. Let $\mathbf{x}(t) = [x_1(t),\ldots, x_m(t)]^T$ be the $m$-dimensional observation (mixture) vector of the unknown source signals vector $\mathbf{s}(t) = [s_1(t),\ldots, s_n(t)]^T$ at discrete time $t$. The most common assumptions for ICA are the statistical independence of the source vector components, a number of observations $m$ at least equal to the number of sources $n$ and at most one Gaussian source is allowed.

The generic BSS problem for instantaneous mixtures can then be formulated as:

$$\mathbf{x}(t) = F(\mathbf{s}(t)) \tag{1}$$

where $F$ is the unknown and generally nonlinear real-valued transformation of the source vector. The aim of BSS is to estimate the inverse of the transformation function $F$ without any prior knowledge concerning the sources or the mixtures, so as to recover an estimate of the unobservable source signals $\mathbf{y}(t) = \hat{\mathbf{s}}(t)$:

$$\mathbf{y}(t) = F^{-1}(\mathbf{x}(t)). \tag{2}$$

A very common pre-processing step is "whitening", also known as "sphering", of the mixtures $\mathbf{x}(t)$, providing uncorrelated components: $\mathbf{v}(t) = \mathbf{V}(t)\mathbf{x}(t)$, where the whitening matrix $\mathbf{V}$ is usually based on single- or eigen-value decomposition of the covariance matrix of $\mathbf{x}$. Note that, Principal Component Analysis (PCA) can be used for sphering the data. In many cases, such a pre-processing accelerates convergence of the source separation algorithms.

# 3 Neural Approach to ICA Using SOM

Various architectures of artificial neural networks have been applied to NLBSS problems. In [4] and [6], SOMs have been used to extract independent components from nonlinearly mixed discrete or continuous sources, but the network complexity increases with the number of neurons and the quantization error (interpolation error, in the continuous case). The use of multilayer perceptrons (MLP) motivated by biomedical applications has been studied in [7], while earlier Burel [8] employed a two-layer perceptron. Recently, a radial basis function (RBF) neural network proposed in [9] has been used to recover the unknown sources from their nonlinear mixtures in the presence of cross nonlinearities. This method appears robust against additive white uniformly distributed noise but, if this is the case, a noise suppression technique is necessary to denoise the separated source signals.

## 3.1 Kohonen SOM Algorithm and NLBSS

Kohonen SOM algorithm is based on competitive learning and transforms an input space of arbitrary dimension using a topology preserving nonlinear mapping. In the NLBSS context, the input patterns are randomly selected from the observation space $\mathrm{X}$. Each neuron $j$, $1 \leq j \leq l$ is connected to the input with a synaptic weight vector $\mathbf{w}_j = [w_{j1},\ldots, w_{jm}]^T$. At each iteration, SOM finds the best-matching (winner) neuron $\mathbf{w}^*(\mathbf{x})$ by minimizing the following cost function:

$$\mathbf{w}^*(\mathbf{x}) = \mathrm{argmin}_j \|\mathbf{x}(t) - \mathbf{w}_j\|, j = 1,\ldots, l \tag{3}$$

where $\|.\|$ denotes the Euclidean distance, while the update of the synaptic weight vectors is carried out according to:

$$\mathbf{w}_j(t + 1) = \mathbf{w}_j(t) + \eta(t)h_{j,\,\mathbf{w}^*(\mathbf{x})}(t)[\mathbf{x}(t) - \mathbf{w}_j(t)], j = 1,\ldots, l \tag{4}$$

where $\eta(t)$ is the learning-rate parameter and $h_{j,\,\mathbf{w}^*(\mathbf{x})}(t)$ is the neighbourhood function centred around the winner. For our simulations, we updated dynamically both the $\eta$ and $h_{j,\,\mathbf{w}^*(\mathbf{x})}$ applying the general rules proposed in [10].

## 3.2 SOM in Presence of Noise

A modified version of the original Kohonen algorithm aimed at building nonlinear data models in the presence of noise is given in [5]. In this work, the authors propose an algorithm which provides a good approximation of the principal manifolds modelling the data distributions, especially in the presence of multiplicative noise.

The novel aspect of this algorithmic scheme is the use of an individual neighbourhood for each neuron. The individual widths of the Gaussian neighbourhood function are adjusted in order to keep them near their critical values. Finally, by averaging over the deviations of the input samples from the winning neurons, an estimate of the manifold in the data space is obtained.

### 3.3 A SOM-Based Signal Denoising Method

Many applications in digital processing of data sets deal with the removal of the noise contained in the signals. Especially, images corrupted by noise may be fitted by different models, like additive, multiplicative, and Poisson. Moreover, image corruptions may occur in a nonlinear way and correlations within the image appear.

As most of the research done in the BSS field concerns additive noise models, this work considers the more difficult problem of multiplicative noise using data-driven nonlinear approximation methods, such as SOMs. Furthermore, when SOMs are used in a NLBSS scheme applied to the image denoising problem, the absence of prior knowledge about the noise-free data and the noise distribution can be compensated in the following manner.

Let $\mathbf{v_e}$ denote the Independent Component (IC) corresponding to the noise vector as estimated by the SOM network. If it is correlated in a multiplicative way with the data, then the noisy ICA model of eqn. (1) can be rewritten as:

$$\mathbf{x}(t) = \mathbf{s}(t)(1 + \mathbf{v_e}(t)). \tag{5}$$

By talking the logarithms of both sides in this equation, one trivially obtains:

$$\log(\mathbf{s}(t)) = \log(\mathbf{x}(t)) - \log(1 + \mathbf{v_e}(t)) \tag{6}$$

That is, by exponentiation, it is possible to retrieve a denoised version of the noisy image frames vector **x**. The experimental results presented in section §4.2 confirm this approach.

## 4 Experiments

### 4.1 Simulation on 2 Continuous Sources

In order to compare the performances of the original Kohonen SOM algorithm and its modified version described in the previous section, simulations using the same signals as in [4] were carried out. The first source is a sinusoid and the second uniformly distributed white noise with both of them being sub-Gaussian signals. They were first mixed linearly with matrix **A**:

$$\mathbf{A} = \begin{pmatrix} 0.7 & 0.3 \\ 0.3 & 0.7 \end{pmatrix} \tag{7}$$

The observation vector **x** ($m = 2$) was obtained after the following nonlinear transformation:

$$F(\mathbf{x}) = \mathbf{x}^3 + \mathbf{x}. \tag{8}$$

Finally, the observations were corrupted with multiplicative noise **v** of arbitrary variance $\sigma^2$, generated randomly with a uniform distribution.

Table 1 summarises the results of this simulation, where $s_1$, $\hat{s}_1$ and $s_2$, $\hat{s}_2$ denote the sinusoid and the white noise original sources and their estimates, respectively, while SOM denotes the original Kohonen SOM algorithm and SOM-m the modified version [5]. In some experiments we also employed a PCA pre-processing before applying a NLBSS using the SOM-based algorithms and used the signal-to-noise ratio (SNR) to quantify the separation results.

Table 1. Comparative results with different multiplicative noise variance

| | | SOM | SOM–m | SOM (with PCA) | SOM–m (with PCA) |
|---|---|---|---|---|---|
| $\sigma^2 = 0.01$ | $\hat{s}_1$ | 5.72 db | 3.36 db | 3.36 db | 4.72 db |
| | $\hat{s}_2$ | 7.7 db | 7.53 db | 8.16 db | 8.36 db |
| $\sigma^2 = 0.1$ | $\hat{s}_1$ | 1.94 db | 4.11 db | 2.85 db | 4.31 db |
| | $\hat{s}_2$ | 6.09 db | 6.61 db | 7.21 db | 6.67 db |

For the source of interest being $s_1$, it can be observed that in the absence of a pre-whitening step and excepting the case of small noise variance ($\sigma^2 = 0.01$), SOM-m enhances remarkably the SNR. While, when applying a PCA pre-treatment to the mixtures, the locally varying neighbourhood width of the SOM-m algorithm in conjunction with the presence of noise helps the map to escape from local minima. Nevertheless, due to the correlation of the noise with each of the source signals, the estimated sources remain fairly noisy, as illustrated in Figure 1.

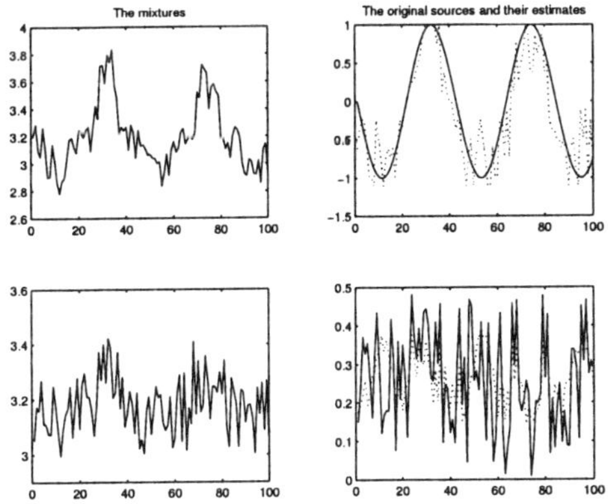


Figure 1. Estimated sources (dotted) using SOM-m in presence of multiplicative noise

## 4.2 Image Denoising Using SOM

The lack of research in the image denoising using SOM-based ICA algorithms may be due to the high computational cost of such approaches. Alternative methods for image denoising related to ICA, have been proposed, e.g., sparse code shrinkage applied to images [11], has been proved to give satisfactory results for additive Gaussian noise. Nevertheless, we propose a NLBSS scheme for image denoising using SOM neural networks, which will enable us to compare the performances of the standard SOM and the modified SOM-m algorithms and provide some evidence in the difficulties encountered by the use of such separating structures.

### *4.2.1 Pre-processing of the image data*

For all our experiments the first source, $s_1$, is a 50×100 pixel region of the Lena image (Figure 2(a)) containing representative features with high contrast, while the second one, $s_2$, is a uniformly distributed random noise of zero mean and arbitrary variance. These two sources, supposed unknown, were mixed in a multiplicative manner, using noise variances of 0.05 and 0.01 to form the observations consisting of two noisy versions of the original Lena image.

A $p \times p$ pixels size windowing is used to decompose each noisy image into a $k = (50 \times 100)/p^2$ dimensional vector containing $p^2$ samples each. The distinct windowing option is justified as a means to minimize the computational cost at the expense of the separation quality. The observation vector of dimension $m = 2$ is formed by taking pairs of the corresponding components of the previously constructed vectors. The local mean is finally subtracted from the data vector whose normalised version is pre-processed with a standard PCA whitening algorithm, before applying the SOM based separation scheme to each one of the $1 \leq j \leq k$ mixture vectors $\mathbf{x}_j(t) = [x_{j1}(t), x_{j2}(t)]$. We obtain two estimated sources $\hat{s}_{j1}$ and $\hat{s}_{j2}$, $1 \leq j \leq k$, corresponding to the denoised image and the noise source.

(a) (b) (c)

Figure 2. Original image (a), estimated independent source (b), and estimated noise (c)

### *4.2.2 Separation results and post-processing*

The choice of the window size is critical in such applications, as it has also been noted in [12]. Experiments have shown that a good choice for this case is around $p = 10$ pixels. For smaller sizes, the reconstructed separated sources are still very correlated. After separation of $n = m$ sources, the estimated denoised images are reconstructed by putting together and rearranging the $k$ pairs of the estimated sources given by the SOM map after linear interpolation for each one of them.

Using a 18×18 neuron map for the SOM-m algorithm, we obtain the estimated sources illustrated in Figure 2 (b) and (c). It is obvious that we have to deal with the indeterminacies associated with the BSS problem and, in particular, with the sign and the amplitude of the separated sources. By adding (or subtracting) an optimal quantity $\alpha$ of the last term of eqn. (6) where $\mathbf{v_e}$ has zero mean, we obtain two denoised versions (b) and (d) of the mixtures (a) and (c), respectively. In this case both the sign and the coefficient $\alpha$ was determined empirically. The improvement in terms of peak signal-to-noise ratio is 1.6db for mixture (a) and 0.5db for mixture (c). Further work will focus on the automatic selection and determination of these parameters. Note, that the classical SOM algorithm could not deal as well as this modified version with the presence of this kind of noise.

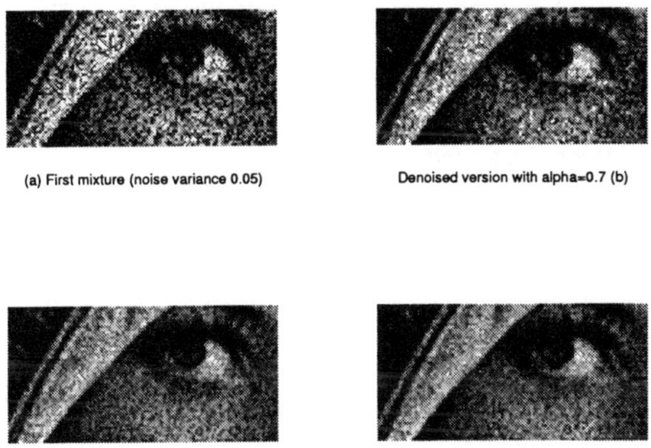

(a) First mixture (noise variance 0.05)

Denoised version with alpha=0.7 (b)

(c) Second mixture (noise variance 0.01)

Denoised version with alpha=0.2 (d)

Figure 3. The noisy images and their denoised versions using SOM based NLBSS

# 5 Discussion

In this work, an approach, based on SOM networks, to the image denoising problem has been proposed. When used with individual neighbourhoods in separating algorithms for the NLBSS problem, SOM can provide satisfactory results in denoising of images corrupted with multiplicative noise. However, the proposed method is computationally expensive and it still has indeterminacy problems to be solved for an automated separation. Subsequently, this will constitute a promising application in denoising of still images and video frames.

**Acknowledgment**

This work is supported by a UK EPSRC grant GR/R 01460.

## References


1. Kohonen T. Self-Organizing Maps. Springer-Verlag, Berlin Heidelberg, 1995 (Springer Series in Information Science no. 30, second edition 1997)
2. Comon P. Independent component analysis – a new concept? Signal Processing 1994; 36(3):287-314
3. Hyvärinen A, Pajunen P. Nonlinear independent component analysis: existence and uniqueness results. Neural Networks 1999; 12(3):429-439
4. Pajunen P, Hyvärinen A, Karhunen J. Nonlinear blind source separation by self-organizing maps. Progress in Neural Information Processing: Proc. ICONIP'96, Hong Kong, 1996, pp 1207-1210
5. Der R, Balzuweit G, Herrmann M. Building nonlinear data models with self-organizing maps. In: van der Malsburg C, von Seelen W, Vorbrüggen JC, Sendhoff B (eds) Artificial Neural Networks. Springer-Verlag, Berlin, 1996, pp 821-826 (Proceedings of ICCAN'96)
6. Hermmann M, Yang HH. Perspectives and limitations of self-organizing maps in blind separation of source signals. Progress in Neural Information Processing: Proc. ICONIP'96, Hong Kong, 1996, pp 1211-1216
7. Lee TW, Koehler BU, Orglmeister R. Blind source separation of nonlinear mixing models. Proc. of IEEE Workshop, Neural Networks for Signal Processing, 1997, pp 405-415
8. Burel G. Blind separation of sources: a nonlinear neural algorithm. Neural Networks 1992; 5:937-947
9. Tan Y, Wang J, Zurada JM. Nonlinear blind source separation using a radial basis function network. IEEE Trans. Neural Networks, 12(1):124-134, 2001
10. Haykin S. Neural Networks – A comprehensive foundation. Prentice-Hall, New Jersey, 1999 (second edition)
11. Hoyer P. Independent Component Analysis in Image Denoising. Master's Thesis, Helsinki University of Technology, Espoo, 1999
12. Hurri J. Independent component analysis of image data. Master's Thesis, Helsinki University of Technology, Espoo, 1997

# Signal-Based Feature Extraction and SOM based Dimension Reduction in a Vibration Monitoring Microsystem

Ingo Jossa, Uwe Marschner, and Wolf-Joachim Fischer

Dresden University of Technology,
Semiconductor and Microsystems Technology Laboratory,
01062 Dresden, Germany
Phone: +49 351 463 4375
Fax: +49 351 463 7021

E-mail: jossa@ihm.et.tu-dresden.de

**Summary.** We describe different signal based feature extraction methods which we are using in a intelligent microsystem for machine vibration monitoring. Each of these methods enables a special view to the machine vibration signal. The statistical methods like the quadratic mean value and the percentiles are useful for a fast global view. Signal processing methods as envelope detection, fourier transforms, wavelet transforms and time frequency methods are special views to the spectral composition of the vibration signal. The dimension reduction is realized by two self-organizing maps, one for the statistical features and one for the time frequency distribution. With a "Fuzzy Neural-Gas" network, which is a integration of the "Neural-Gas" Network and the Fuzzy c-Means algorithm, we are classifying the output of the self-organizing maps. The described hierarchical neural network integrates the dimension reduction and the classification for the machine vibration monitoring.

## 1 Introduction

Traditional systems for the detection of machine failures are mostly working in two stages. The first stage is the Condition Monitoring which is directly applied to the observed machine and delivers a decision between a faulty and a healthy machine. On this first stage only fast and low level feature extraction and classification algorithms are used. As second stage follows the Diagnosis which performes some tasks of more powerful but also computing power consuming feature extraction and multivariate classification. In contrast to the condition monitoring the diagnosis stage in traditional systems is not located directly in the sensor on the machine.

The recent developement of microelectronic and microsystem technology delivers the basis for intelligent microsystems with high computing power and large memories. Such technologies open the door to the integration of the Condition Monitoring and the Diagnosis stage in a single online machine diagnosis microsystem. We have used some of the actually most advanced semionductor and microsystem technologies for the developement of a intelligent microsystem [JMF99] [MJE+99] which is directly mounted to the observed machine.

## 2 Feature Extraction

In fact of the large computing power of our microsystem we are using different methods from statistics and signal processing for the extraction of features. In the time domain the envelope and it's spectra has a long tradition in machine vibration monitoring. The most methods from statistics which we use are traditional methods from the field of machine diagnosis. The last years have shown that with the application of advanced signal processing methods to machine vibration signal analysis a much more better feature extraction is possible. With time frequency distributions and wavelets the search for typical failure "fingerprints" in the machine vibration signal becomes a new quality.

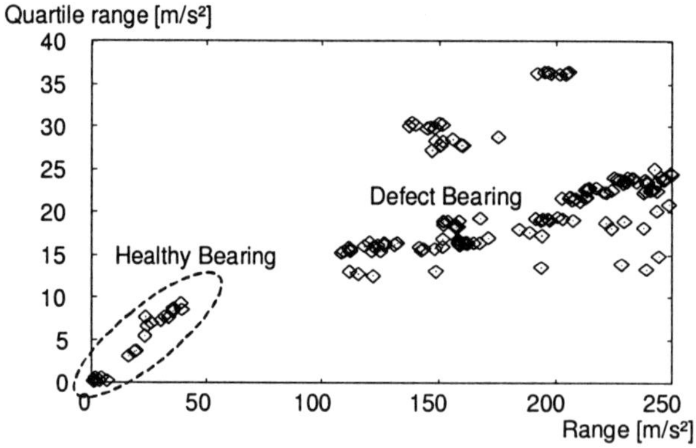


**Fig. 1.** Quartile range and range of the vibration signal as features

### 2.1 Time Domain

We use several traditional statistical features in the time domain as median, range, standard deviation, root mean square value (rms), quadratic mean value, kurtosis, skewness and percentiles (lower quartile, upper quartile, quartile range). With the rms value a good description of the global machine state is possible. The quadratic mean value represents the energy in the vibration signal. When a failure in a machine are growing up the energy in the vibration signal becomes also higher values. A very flexible way for the description of positive and negative peak values in a signal are given by the so called percentiles. The percentile of a distribution of values is a number x such that a percentage p of the population values are less than or equal to x. For example, the 25th percentile (also referred as the 0.25 quantile or lower quartile) of a variable is a value x such that 25% of all the values of the variable fall below that value. Similarly, the 75th percentile (also referred as the 0.75 quantile or upper quartile) is a value such that 75% of the values of the variable fall below that value. Fig. 1 shows the good separation of the faulty and the healthy motor stage by the range and the quartile range of the observed vibration signal magnitudes.

## 2.2 Frequency Domain

One of the most popular methods for the detection of failures in machine vibration signals is the envelope and it's spectrum. We using different meth-

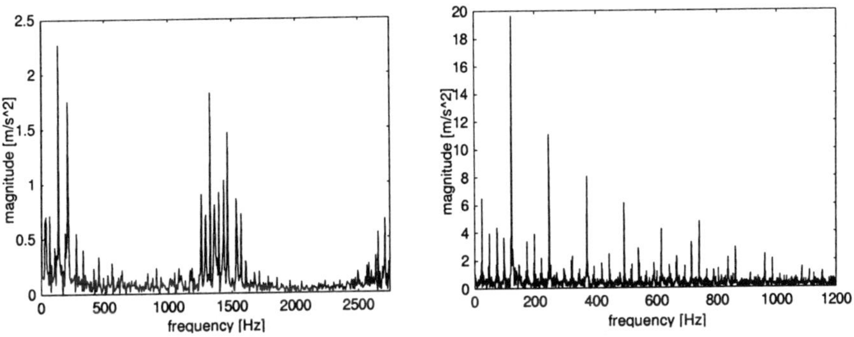


**Fig. 2.** Envelope spectra of a healthy (left) and a faulty (right) bearing signal

ods for the envelope computing, the hilbert transform, the local maximum method and the rectifier with low pass. Our preferred method is the local maximum method because they needs not so much computing power and realizes a compression. With a window size of 16 we become a compression factor of 16 that means the envelope data are 16 times smaller as the original data set. As next step we computing the spectrum of the envelope. Fig. 2 shows the envelope spectra of a healthy (left figure) and a faulty (right figure) bearing. The spectrum of the faulty bearing shows typical spectral lines at the failure frequency and its higher harmonics. The magnitude of the spectral lines are much higher as in the spectrum of the healthy bearing.

## 2.3 Time Frequency Domain

Actually we are also testing the possibility to use a discrete approach of the Wigner-Ville distribution (WVD) (Fig. 3,4) for feature extraction. A typical fault detection procedure are based on visual observation of the WVD contour plots. Dark zones and curved bands in the contour plot are the main features of an impulse caused by a bearing fault. The progression of a fault can be monitored by observing the changes in the WVD contour plot. We are only

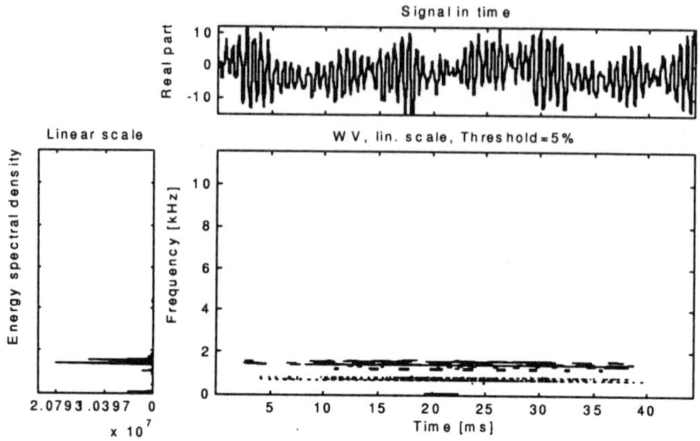


**Fig. 3.** Wigner Ville Distribution of the vibration signal from a healthy bearing

using a part of the time frequency plane and to this part we are applying a dimension reduction method. For this we are subdividing the time frequency

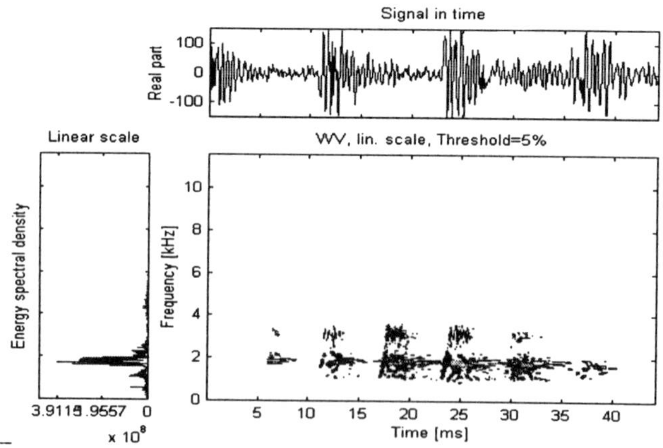


**Fig. 4.** Wigner Ville Distribution of the vibration signal from a faulty bearing

plane (Fig. 3,4) between the frequencies of 0...4KHz in 32 areas. The energy represented by every of this areas is used as input to a Self-Organizing Map for dimension reduction. The SOM output is part of the feature vector applied to the "Fuzzy Neural-Gas" Network for classification.

## 3 Classification

Martinetz and Schulten [MS91] have developed the "**Neural-Gas**" Network which is inspired by Kohonens Self-Organizing Map [Koh82] [Koh95] algorithm. The "Neural-Gas" Network produces a connectivity structure that defines a very good topology preserving map.

Tsao et. al [TBP94] introduced the **Fuzzy Kohonen Clustering Network** (**FKCN** ) which integrates the **Fuzzy c-Means** algorithm (**FCM**) [Bez81] with the SOM and has some advantages over the both separate algorithms. The sequential presentation of input pattern and also the sequential adaption of the weight vectors like the SOM are replaced by the parallel method of the FCM. This algorithm was used as basis for the learning algorithm of the "**Fuzzy Neural-Gas**" Network [Jos98].

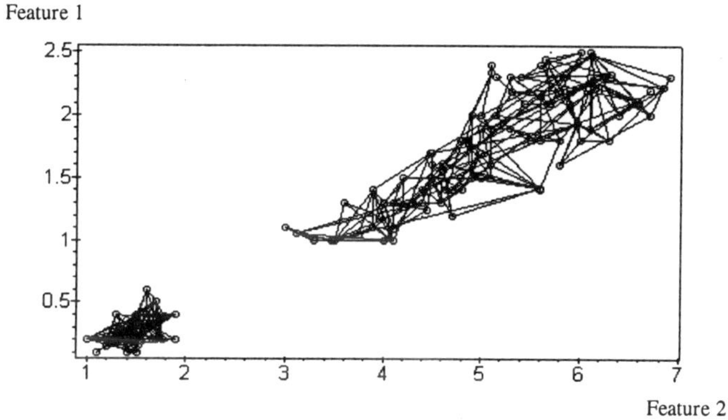


**Fig. 5.** 2D projection of the trained "Fuzzy Neural-Gas" Network structure.

The "Neural-Gas" Network can be used for the observation of the state space of industrial processes and machines. To do this the "Neural-Gas" Network realizes a vector quantization of the machine state space. All the subspaces that are representing faulty or healthy states are concluded to one larger common faulty or healthy subspace. At first the "Neural-Gas" Network state space classificator is trained with the observations of healthy and faulty motors. After this training phase the "Neural-Gas" Network classifies the states of a online observed machine. When the state of a machine is floating from the healty state to the faulty state it passes the border line which is hard defined by the nodes of the "Neural-Gas" Network. This jump from the healthy state to the faulty state is in contrast to the floating state changes in real machines. In fact of this we are using the "Fuzzy Neural-Gas" Network to become a better modelling of the nature of machine state changes.

# 4 Practical Experiences

For practical measurements we have used the large electrical motor test field of our industrial partner. We have taken a large number (nearly 1000) of 64Kbyte samples from different working situations of different motors. Such working situations are the motor startup and warmup, different rotation speeds and different motor loads Typical healthy and faulty motors are observed and analysed. Some of the faults are also realized by special preperartions of bearings. All the sampled data are used for the training of our SOM based machine state classificator.

# 5 Conclusions

In this paper the integration of Self-Organizing Neural Networks and some feature extraction methods in a intelligent microsystem for online machine vibration monitoring is reported. With some of the most advanced semiconductor, micro and intelligent technologies we have developed a microsystem which has enough computing power and memory space to use statistical and advanced signal processing methods for feature extraction and Neural Network classifiers. The prototype of our system is now realized and the first real world application to motors in a chemical plant and a wind farm of our industrial partners is planned for the end of this year.

# References


[Bez81] J. C. Bezdek. *Pattern Recognition with Fuzzy Objective Function Algorithms.* Plenum, New York, 1st edition, 1981.

[JMF99] I. Jossa, M. Marschner, and W.-J. Fischer. A intelligent microsystem for machine vibration monitoring. In L. C. Jain, editor, *Proc. KES'99, Int. Conf. on Knowledge based Engeneering Systems*, pages 389–392, New York, NJ, USA, 1999. IEEE.

[Jos98] I. Jossa. A self-organizing fuzzy neural network for clustering. In *Proc. 6th Zittau Fuzzy Colloquium*, 1998.

[Koh82] Teuvo Kohonen. Self-organizing formation of topologically correct feature maps. *Biol. Cyb.*, 43(1):59–69, 1982.

[Koh95] Teuvo Kohonen. *Self-Organizing Maps.* Springer, Berlin, Heidelberg, 1995.

[MJE+99] U. Marschner, I. Jossa, G. Elender, T. Hoffmann, T. Herrmann, M. Heinrich, and W.-J. Fischer. Decentralized microsystem-based diagnostics of bearings of an electric motor. In *Proc. ECC'99, European Control Conference*, 1999.

[MS91] Thomas Martinetz and Klaus Schulten. A "Neural-Gas" network learns topologies. In T. Kohonen, K. Mäkisara, O. Simula, and J. Kangas, editors, *Artificial Neural Networks*, volume I, pages 397–402, Amsterdam, Netherlands, 1991. North-Holland.

[TBP94] E. C.-K. Tsao, J. C. Bezdek, and N. R. Pal. Fuzzy kohonen clustering networks. *Pattern Recognition*, 27:757–764, 1994.

# AUTHOR INDEX